MARINE MINERAL EXPLORATION

FURTHER TITLES IN THIS SERIES

1 J.L. MERO
THE MINERAL RESOURCES OF THE SEA
2 L.M. FOMIN
THE DYNAMIC METHOD IN OCEANOGRAPHY
3 E.J.F. WOOD
MICROBIOLOGY OF OCEANS AND ESTUARIES
4 G. NEUMANN
OCEAN CURRENTS
5 N.G. JERLOV
OPTICAL OCEANOGRAPHY
6 V. VACQUIER
GEOMAGNETISM IN MARINE GEOLOGY
7 W.J. WALLACE
THE DEVELOPMENTS OF THE CHLORINITY/SALINITY CONCEPT IN OCEANOGRAPHY
8 E. LISITZIN
SEA-LEVEL CHANGES
9 R.H. PARKER
THE STUDY OF BENTHIC COMMUNITIES
10 J.C.J. NIHOUL (Editor)
MODELLING OF MARINE SYSTEMS
11 O.I. MAMAYEV
TEMPERATURE-SALINITY ANALYSIS OF WORLD OCEAN WATERS
12 E.J. FERGUSON WOOD and R.E. JOHANNES
TROPICAL MARINE POLLUTION
13 E. STEEMANN NIELSEN
MARINE PHOTOSYNTHESIS
14 N.G. JERLOV
MARINE OPTICS
15 G.P. GLASBY
MARINE MANGANESE DEPOSITS
16 V.M. KAMENKOVICH
FUNDAMENTALS OF OCEAN DYNAMICS
17 R.A. GEYER
SUBMERSIBLES AND THEIR USE IN OCEANOGRAPHY AND OCEAN ENGINEERING
18 J.W. CARUTHERS
FUNDAMENTALS OF MARINE ACOUSTICS
19 J.C.J. NIHOUL (Editor)
BOTTOM TURBULENCE
20 P.H. LEBLOND and L.A. MYSAK
WAVES IN THE OCEAN
21 C.C. VON DER BORCH (Editor)
SYNTHESIS OF DEEP-SEA DRILLING RESULTS IN THE INDIAN OCEAN
22 P. DEHLINGER
MARINE GRAVITY
23 J.C.J. NIHOUL (Editor)
HYDRODYNAMICS OF ESTUARIES AND FJORDS
24 F.T. BANNER, M.B. COLLINS and K.S. MASSIE (Editors)
THE NORTH-WEST EUROPEAN SHELF SEAS: THE SEA BED AND THE SEA IN MOTION
25 J.C.J. NIHOUL (Editor)
MARINE FORECASTING
26 H.G. RAMMING and Z. KOWALIK
NUMERICAL MODELLING MARINE HYDRODYNAMICS
27 R.A. GEYER (Editor)
MARINE ENVIRONMENTAL POLLUTION
28 J.C.J. NIHOUL (Editor)
MARINE TURBULENCE
29 M. WALDICHUK, G.B. KULLENBERG and M.J. ORREN (Editors)
MARINE POLLUTANT TRANSFER PROCESSES
30 A. VOIPIO (Editor)
THE BALTIC SEA
31 E.K. DUURSMA and R. DAWSON (Editors)
MARINE ORGANIC CHEMISTRY
32 J.C.J. NIHOUL (Editor)
ECOHYDRODYNAMICS
33 R. HEKINIAN
PETROLOGY OF THE OCEAN FLOOR
34 J.C.J. NIHOUL (Editor)
HYDRODYNAMICS OF SEMI-ENCLOSED SEAS
35 B. JOHNS (Editor)
PHYSICAL OCEANOGRAPHY OF COASTAL AND SHELF SEAS
36 J.C.J. NIHOUL (Editor)
HYDRODYNAMICS OF THE EQUATORIAL OCEAN
37 W. LANGERAAR
SURVEYING AND CHARTING OF THE SEAS
38 J.C.J. NIHOUL (Editor)
REMOTE SENSING OF SHELF SEA HYDRODYNAMICS
39 T. ICHIYE (Editor)
OCEAN HYDRODYNAMICS OF THE JAPAN AND EAST CHINA SEAS
40 J.C.J. NIHOUL (Editor)
COUPLED OCEAN–ATMOSPHERE MODELS

Elsevier Oceanography Series, 41

MARINE MINERAL EXPLORATION

Edited by

H. KUNZENDORF

Risø National Laboratory, DK-4000 Roskilde, Denmark

ELSEVIER

Amsterdam – Oxford – New York – Tokyo 1986

ELSEVIER SCIENCE PUBLISHERS B.V.
Sara Burgerhartstraat 25
P.O. Box 211, 1000 AE Amsterdam, The Netherlands

Distributors for the United States and Canada:

ELSEVIER SCIENCE PUBLISHING COMPANY INC.
52, Vanderbilt Avenue
New York, NY 10017, U.S.A.

Library of Congress Cataloging-in-Publication Data

Marine mineral exploration.

(Elsevier oceanography series ; 41)
Bibliography: p.
Includes index.
1. Prospecting--Geophysical methods. 2. Geochemical prospecting. 3. Marine mineral resources. 4. Ore-deposits. I. Kunzendorf, H. II. Series.
TN270.M335 1986 622'.17 86-4610
ISBN 0-444-42627-2

ISBN 0-444-42627-2 (Vol. 41)
ISBN 0-444-41623-4 (Series)

Printed in The Netherlands

CONTRIBUTORS:

C. Aage
Department of Ocean Engineering
The Technical University
of Denmark
Building 101E
DK-2800 Lyngby, Denmark

H. Bäcker
Preussag AG
Erdöl und Erdgas
P.O. Box 4829
Arndtstr. 1
D-3000 Hannover
F.R. Germany

K. Boström
University of Stockholm
Kungstensgatan 45
S-10691 Stockholm
Sweden

E.D. Brown
Department of Law
University of Wales
Colum Drive
Cardiff CF1 3EU
Wales

R. Fellerer
Preussag AG
Erdöl und Erdgas
P.O. Box 4829
Arndstr. 1
D-3000 Hannover
F.R. Germany

R. Hansen
Preussag AG
Erdöl und Gas
P.O. Box 4829
Arndstr. 1
D-3000 Hannover
F.R. Germany

H. Kunzendorf
Risø National Laboratory
P.O. Box 49
DK-4000 Roskilde
Denmark

J. P. Lenoble
Institut francais de recherche
pour l'exploitation de la mer
IFREMER
66, avenue d'Iena
75116 Paris
France

W.L. Plüger
Institut für Mineralogie
und Lagerstättenlehre
RWTH
Süsterfeldstr. 22
D-5100 Aachen
F.R. Germany

W. Pohl
Preussag AG
Erdöl und Erdgas
P.O. Box 4829
Arndtstr. 1
D-3000 Hannover
F.R. Germany

H. Richter
Preussag AG
Erdöl und Erdgas
P.O. Box 4829
Arndtstr. 1
D-3000 Hannover
F.R. Germany

W. Ries
Prakla Seismos GmbH
P.O. Box 510530
Buchholzerstr. 100
D-3000 Hannover
F.R. Germany

H. Roeser
Bundesanstalt für Geowissen-
schaften und Rohstoffe
BGR
P.O. Box 510153
Stilleweg 2
D-3000 Hannover 51
F.R. Germany

PREFACE

The outcome of this book is influenced largely by an active engagement in both terrestrial and marine mineral exploration for nearly twenty years. Work in marine mineral exploration was initiated in 1973 when the Editor was invited to participate in large-scale German manganese nodule exploration efforts, conducted at that time in the equatorial North Pacific nodule belt. Later, other areas of the Pacific Ocean were included in the exploration programs. Detailed work was carried out using mainly geochemical techniques. In recent years, work was also carried out in the exploration for marine placers and massive sulfides.

Since the Editor had been involved in geochemical exploration projects in the arctic Greenland for some years and continued this activity in parallel with marine work, the change to manganese nodule exploration surprisingly was not as dramatic as expected. Although operations were convincingly different in that metal accumulations had to be assessed through several kilometers of seawater, general exploration strategies seemed to be related. While field work on land was conducted from exploration camps served by helicopter, at sea the research vessel was a most comfortable operational base. Analogies and methods in terrestrial and marine mineral exploration were found to be numerous. Many geophysical methods are essentially the same, the general difference being that surveys in the former mode had to be conducted through a layer of air which, in marine work is compensated by a layer of seawater; other normal mineral exploration steps, on the other hand are rendered in many different ways in marine research.

For instance, accurate navigation of ships was, and still is, much of a problem in marine investigations. New development of a global positioning system soon to be implemented may however increase marine navigation performance. Also, bathymetric mapping of the ocean floor was very time-consuming in the seventies but is now greatly advanced through the use of multiple-beam echosounder systems. New computer technology certainly will further refine techniques and have an impact on oceanographic research.

Over the years it became evident that much of the marine research conducted by the major industrialised countries could not be regarded as marine mineral exploration. Many cruises that also covered the marine minerals, both in the shallow waters and in the deep ocean, in fact were normal oceanographic research cruises. This can be deduced from the staff lists of these cruises which mostly comprised physical oceanographers, marine geologists, marine biologists, etc., on often crowded research ships. Very many cruises sponsored by governments or research foundations quite often under the heading of marine minerals research had very little to do with systematic marine mineral exploration. A long story could be told here.

It is not surprising that there is an increasing literature on marine hard minerals since about 1960. However, very few were concerned with the exploration and discovery of marine mineral deposits, although there are surely a number of mining company reports on shelves not accessible to the public. Oceanographic literature in the form of textbooks is also increasing. Some of them in detail cover marine minerals but only a few of them are concerned with the methods and techniques to outline them on the ocean floor, and make proper economic evaluations.

To transfer the methods of marine mineral exploration that were used in the past 10 to 20 years in condensed form to a broader audience, the Editor, in the early 1980s proposed a collaborative effort on this subject to Elsevier Science Publishers. Negotiations succeeded in 1984 and a manuscript proposal was delivered to the publisher at the end of 1985. Because marine mineral exploration represents a considerable amount of practical work, authors were invited to contribute who had considerable experience which may have included participation in many exploration cruises. Though most of the authors were asked to opinions freely about their topic, thereby being responsible for their own contribution, the Editor also is obliged to take a fair share of the responsibility.

The book is divided into a preface, 8 chapters and 3 appendices. An Introduction combines history with fundamental principles of marine mineral exploration. Chapter 2 is devoted to a brief overview of known marine hard mineral occurrences. It is followed by a short description of research vessels and submersibles. Marine geophysical exploration methods are outlined in Chapter 4, followed by the geochemical methods described in Chapter 5. Data evaluation procedures are found in Chapter 6, while Chapter 7 describes marine mineral exploration case histories in some detail. Finally, the status of the laws concerning marine minerals is dealt with in Chapter 8.

There is perhaps some overlapping of contents of some chapters, which is difficult to prevent in an undertaking involving multi-discipline multi-national contributions. Limited space does not allow us to go into much detail; therefore, an additional list of selected references, preferentially from after 1970, is included in Appendix I. In Appendix II we attempt to supply the reader with an understanding of the outcome of the United Nations Conference on the Law of the Sea (UNCLOS III) by tabulating hypothetical Exclusive Economic Zones (EEZ) areas of different countries. These data have to be corrected in the near future when actual EEZ "claiming" has been concluded.

They are therefore only an indication of the order of magnitude of the sovereignity gains of the participating countries. Finally, Appendix III outlines a proposed exploration strategy for metalliferous sediment-type marine deposits that according to their size and grade could be of economic interest in the near future.

Environmental topics that are important in any mineral resources work are not included in this volume. They have to be considered in detail at mineral exploitation steps.

During the editing and writing of this volume, I was backed-up by the management of Risø National Laboratory, supplying me with necessary time, especially in 1985, which is greatly acknowledged. Discussions with many colleagues in and outside Risø greatly advanced the production of this book. I especially thank A.M. Eichen, Risø, for a most effective effort during the final typing of the manuscript. The efforts of A Berman (language consulting), B. Pedersen (literature search efforts), P. Nielsen (photographic service), and B. Wallin (computer data management), all Risø staff, provided very valuable help that is much appreciated. Finally, I could not have produced this volume were it not for the patience of my family.

It is hoped that this book enlarges knowledge about the seafloor. The more technical methods and instruments of marine mineral exploration need simply to be known to a broader audience. The text elaborates therefore more on the practical aspects than on theoretical considerations. I wish that this work can be used also as a kind of overview textbook for students who are not too specialised in their graduate work, and that it will stimulate further research.

Risø, December 20, 1985

H. Kunzendorf

CONTENTS

CHAPTER 1

INTRODUCTION

H. KUNZENDORF

Mineral exploration has played a major role in the economy of mining companies and by that of many countries, both industrialised or developing. Exploration includes a number of steps to search for and evaluate a mineral deposit prior to its exploitation. Definition and purpose of mineral exploration are well-defined (e.g. Bailly and Still, 1973). Often, exploration is confused with exploitation and with prospecting, the latter of which is an older expression dating back to the gold rushes. However, even the newly established Law of the Sea still differentiates between prospecting and exploration (see chapter 8 of this volume).

Marine mineral exploration could be regarded as a branch of economic geology, comparable to terrestrial mineral exploration. In contrast to the latter, however, marine minerals, over the past years, paid very little attention to the public and the earth science community, for several reasons.

Firstly, marine resources in a modern society almost exclusively in the public mind were and still are connected with the fisheries industry. Very few decision-making bodies are aware of geologic marine resources. Secondly, terrestrial mineral wealth favoured the development of techniques to explore and exploit relatively easily accessible mineral deposits on land. A limited search for offshore minerals has been devoted to a few private companies only and little interest has been shown by governmental institutions. For these institutions, marine research has always had an image of purely academic efforts with no economic interest whatsoever. It is probably true that we know more about the moon's surface and near space than we do about the ocean floor or the earth's interior. Over a long period of time, therefore, after World War II, no special efforts were made to concentrate governmental research on more economic aspects of the non-living resources of the oceans. As a matter of fact, major national and international oceanographic research up to now was very much devoted to fundamental studies, leaving marine minerals to few oceanographic groups. It could also be argued that the political surroundings of shelf areas and the deep ocean were not in a state to promote exploration for marine minerals.

The situation has changed dramatically some years before, and after the first oil crisis in 1973. Offshore energy resource potentials had then to be re-valuated, and as a result of the world-wide unstable political situation, national offshore areas and even the deep ocean floor attended new consideration. Offshore oil production at present is moving into the outer continental shelf and much of the generated new technology has an impact on exploration and exploitation in deep waters.

Probably because of the oil crisis ocean floor divisions and their probable mineral commodities had their share to greatly delay the United Nations Conference on the Law of the Seas (UNCLOS III). This conference was initiated in 1973, developed to a series of sharp international disputes and finally terminated into the Law of the Sea (LOS) in 1982. The status of LOS is discussed in chapter 8 of this book. The indirect recognition of the importance of presumed marine mineral resources in vast areas of unexplored ocean floor by an international audience of lawyers, economists and politicians at UNCLOS III is remarkable and has no historic counterpart. It is also amazing that much of the mineral wealth of the sea discussed at the conference even at present (1985) is based on sporadic investigations only and its proof needs large periods of time with extensive and systematic marine mineral exploration.

For the future, it is expected that marine mineral exploration will be stimulated considerably. Not so much, perhaps, in the deep oceans where the United Nations efforts have to be settled properly before any international exploration work can be initiated, but more in the Exclusive Economic Zones (EEZs) of many of the coastal States.

The LOS could also be interpreted as the largest instance of colonisation on earth, because the area of sovereignty for many countries was increased considerably through the UN treaty (see Fig. 1.1). A tabulation of countries EEZ areas is given in Appendix II. It shows large increases of sovereignty for especially island-based small States, but also for large coastal States. If not for the marine mineral resources, many countries will therefore be interested in knowing more about these areas, not at least for environmental reasons.

To elaborate on marine mineral exploration, this introductory chapter tries to place the subject into the proper framework, i.e. into the applied sciences of mineral exploration and oceanography. In this context it is necessary to limit marine exploration to hard minerals only and leave the marine hydrocarbon accumulations of the shelf areas to special publications (e.g. Hobson and Tiratsoo, 1975; Tissot and Welte, 1978). Also, dissolved minerals are excluded from consideration because they would, among others, require quite different

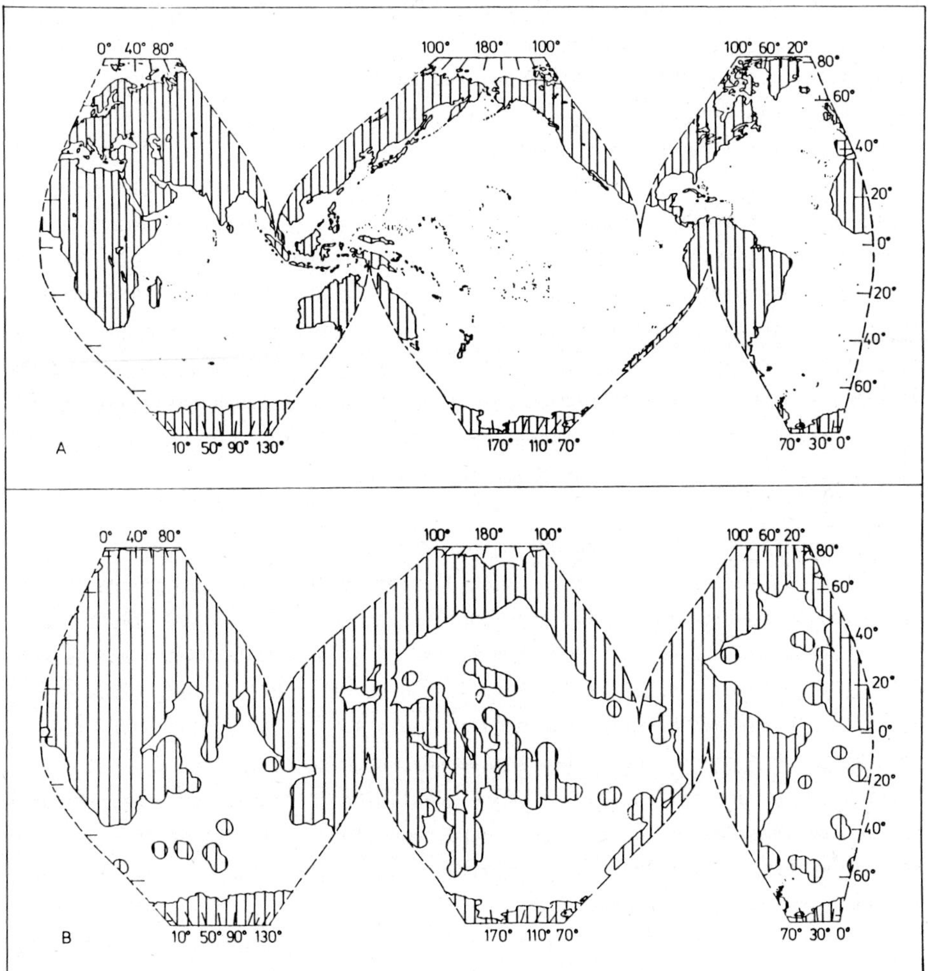

Fig. 1.1. Areas of State sovereignty before (A) and after UNCLOS III (B). Note that the status of the Antarctic shelf is not decided on yet. Situation (B) shows no subdivision into living and non-living resources claims but gives simply the hypothetical EEZ added to a State's sovereignty area. Alterations may be expected in the near future.

exploration techniques. The separation of marine minerals into hard minerals and hydrocarbons is justified by the deep burial of the latter which needs different exploration techniques compared to hard minerals occurring mainly at surfaces and upper substrata.

The following subjects are discussed in this introductory chapter:
- Historical events in oceanography
- National oceanographic institutions and agencies
- International organisations
- Marine mineral exploration strategies.

A historical review is found necessary to briefly mark historical changes in oceanography towards research into non-living resources. A few large national oceanographic institutions are then taken into consideration and their efforts to investigate marine minerals are discussed. The development of international organisations in sponsoring expensive ocean exploration is mentioned and new development trends are given. Finally, exploration strategies for marine minerals are described with main emphasis on the use of techniques known for a long time in terrestrial mineral exploration. These techniques are then discussed in more detail in the following chapters.

Historical events in oceanography

Historical oceanography is a well-documented branch of oceanography and detailed information is available in the literature (e.g. Riley, 1965; Tait, 1968; Wolff, 1968; Burstyn, 1968; Moore, 1971; Deacon, 1971; Deacon and Deacon, 1974; Fye, 1980; Tomczak, 1980; Seibold and Berger, 1982). Some of the events in oceanography are briefly mentioned in Table 1. It is obvious that such a tabulation cannot be complete because the weighing of efforts with time is difficult to accomplish and therefore tabulation is biased by the author's choice and imagination. On the other hand, additional data can be obtained from the literature mentioned. The following short historical review is based mainly on Deacon and Deacon (1974).

Up to the beginning of the first century, ancient philosophers from Greece, Rome and Arabia and others speculated with generally little success about the salinity of seawater. Theoretical considerations quite often ended in unsolvable philosophical disputes. However, Arabian navigators made systematic observations during their voyages and by that increased knowledge about the oceans.

Knowledge about the sea advanced little in the Middle Ages. The salinity problem was discussed in detail when Robert Boyle published an article about the saltness of the sea in 1670. A first textbook on the science of the sea was published by count Marsigli in 1725, although no science of the sea actually existed at that time. The founder of marine geology is probably James Hutton with his publication on "The theory of the earth". Important in this respect is also the voyage H.M.S. "Beagle" with Charles Darwin as naturalist, in 1831. Much of seafloor contouring was accomplished between 1850 and 1860 as a step in

TABLE 1.1

Brief account of historical events in oceanography.

Time	Event(s)
Up to 1000	Greek and Roman philosophers discuss the salinity of the sea; Arabian navigators make systematic investigations.
Middle Ages	Little advancement in knowledge about the sea.
1670	Robert Boyle on the saltness of the sea.
1725	Count Marsigli's textbook on the science of the sea.
1750	Oyster dredge by Marsigli and Donati.
1795	James Hutton, book on "The theory of the earth".
1818	John Murray: water analysis by gravimetry.
1825	v. Lenz: self-registering thermometer.
1831	Charles Darwin, H.M.S. "Beagle".
1850-1860	North Atlantic seafloor contouring for laying telegraph cables.
1873-1876	"Challenger" expedition; return of the ship barely noticed by the public.
1891	J. Murray and A.F. Renard: "Deep sea deposits" as last volume of the Challenger results.
1901	International Council for the Exploration of the Sea (ICES).
1925-1940	Period of national and dynamic ocean surveys; oceanographic expeditions by Norway, Denmark, The Netherlands, U.S.A., Germany and others.
1947-1956	Development of many marine geological, geophysical and other methods; major national oceanographic expeditions; new international organisations: SCOR, IOC.
1959-1965	International Indian Ocean Expedition (IIOE).
1965-1979	Geophysical and commercial marine exploration; manganese nodules and placers.
1973-1982	United Nations Conference on the Law of the Sea (UNCLOS).
1979-	Black smokers and ridge deposits research; EEZ area mapping of mainly manganese/phosphorite crusts and placers.

the laying down of intercontinental telegraph cables.

In regard to marine minerals, the classical oceanographic expedition by

H.M.S. "Challenger" (1873-76) must be mentioned as the most important historical marker (Burstyn, 1968). It is astonishing to realize that the return of the ship in 1876 barely was noticed by the public; it was overshadowed by a British expedition to the Antarctic. Publication of the vast material gathered by the "Challenger" expedition took nearly 20 years and it was only in 1891 that Murray and Renard completed the last expedition volume ("Deep sea deposits"). It was therefore not surprising that when this volume arrived three of the 7 scientists of the expedition had already died. However, its results must be regarded as important for deep-sea biology, marine geology and marine sedimentology. Results in marine chemistry and physical oceanography were, however, not sensational. At that time, Edinburgh had been the world centre for oceanographic research for nearly two decades.

The first part of the twentieth century was later defined as the time of national systematic and dynamic ocean surveys with major oceanographic expeditions conducted by Norway, Denmark, The Netherlands, the United States and Germany. Research ships that should be mentioned in this context were "Enterprise", "Medusa", "Hirondelle", "Princess Alice", "Fram", "Michael Sears", "Ingolf" and "Valdivia". Between the two world wars, few oceanographic research activities were conducted.

Of activities with impact on marine minerals after World War II, could be mentioned many world cruises, but the impact on oceanography is due more to the development and refinement of many marine geophysical and other marine methods. A number of international organisations were initiated during that time, in particular the Special (now: Scientific) Committee on Oceanic Research (SCOR) of the International Council of Scientific Unions (ICSU), and the International Oceanographic Commission (IOC), sponsored by United Nations. Of importance during this time is also the International Indian Ocean Expedition (1959-1965).

The years between 1965 and 1979 are characterised by geophysical and commercial marine exploration. Activities moved from the shelf areas (marine placer deposits) to the deep ocean (manganese nodules). Pioneering work on marine mineral deposits was published by Mero (1965). Other major publications dealt with marine manganese deposits (Glasby, 1977; Bischoff and Piper, 1979).

Much of oceanographic debate took place during the United Nations Conference of the Law of the Sea (UNCLOS III) between 1973 and 1982. In fact, marine minerals were the main reason for the delay of the conference, although at that time there was little known about shelf deposits; polymetallic massive sulfides and Co-rich manganese crusts were practically left out of the discussions, because they were then undiscovered.

After the discovery of hot springs on the Galapagos Spreading Centre in 1977 and of active black smokers in the central parts of the East Pacific Rise in

1979 (e.g. Edmond et al., 1979; Francheteau et al., 1979; Edmond et al., 1982) most marine mineral research in deep waters is now devoted to the global system of spreading centres. However, as a result of LOS and because the U.S. adopted the concept of an Exclusive Economic Zone (EEZ) although it did not sign the treaty, much of economic-related oceanographic research is now devoted to the EEZ areas. In the time to come, both non-signing and signatory countries with small and large EEZ areas will be interested in the mapping and evaluation of probable marine minerals. Of particular importance here will be sand and gravel, heavy minerals (placers), manganese/ phosphorite nodules and encrustations and massive sulfides.

National oceanographic institutions and agencies

Up to quite recently, marine resources research was thought to be a natural part of national oceanographic institutions, the geological surveys and geological departments of the universities. However, because marine mineral exploration by the politicians was regarded to be of interest to private companies, marine resources in the form of hard minerals very seldom achieved an appropriate role within the public and academic institutions. In this section we attempt to describe the national situation in some selected countries which, hopefully, can illuminate the situation of marine mineral research efforts. Although it is quite difficult to collect information on selected institutions, some pertinent data are given in Table 1.2.

Major nongovernmental oceanographic institutions of the United States are Woods Hole Oceanographic Institution (WHOI) and Scripps Institution of Oceanography (Scripps). While WHOI was established in 1930, Scripps is claimed to originate somewhat earlier because a small laboratory was already connected with the University of California in 1912. The annual report notes that in 1983 Woods Hole research can be divided into the areas of marine biology, marine chemistry, marine geology and geophysics, ocean engineering, and ocean policy & management. Of about 900 employees, very few, mainly from the ocean policy and management unit worked in marine hard minerals research, although part of the general mapping efforts with Seabeam (R/V "Atlantis II") could be classified into non-living resources research, as could some of the leading research with DSRV "Alvin". As is with other U.S. oceanographic institutions, however, none of these activities can be termed systematic marine mineral exploration.

Similar conditions are valid for the oceanographic research at Scripps. Of 13 research divisions, one is, at least by name, entirely devoted to marine resources, namely the Institute of Marine Resources. However, on comparing the research activities of this institute (1983), most work is with the living resources of the oceans. Nevertheless, of reported major cruises, 9 out of 60 could be called mineral-related in that year, because they included massive

TABLE 1.2
Selected national oceanographic institutions and comparable research units

Country	Institution	Remarks
U.S.A.	WHOI Woods Hole Oceanographic Institution Woods Hole MA 02543	Mainly academic oceanographic research by many branches; marine minerals covered by small unit: Marine Policy and Ocean Management
	Scripps Institute of Oceanography La Jolla CA 92093	Large-scale oceanographic research comparable to WHOI; Instute of Marine Resources sponsors mainly bio-related resource studies
Canada	BIO Bedford Institute of Oceanography Dartmouth Nova Scotia, B2Y4A2	Central oceanographic unit; non-living resources research mainly devoted to hydrocarbons
France	IFREMER+) Institut francais de recherche pour l'exploitation de la mer 75116 Paris	Central oceanographic institution; clear research efforts into marine hard minerals; significant ocean technology development
F.R. Germany	BGR Bundesanstalt für Geowissenschaften und Rohstoffe Hannover	Geological survey; marine geophysics and marine geology involved in marine minerals studies and exploration
New Zealand	NZOI New Zealand Oceanographic Institute Wellington	Oceanographic central unit; marine minerals research within marine geology unit; marine mineral exploration efforts

+) several institutions, including CNEXO

sulfide deposits in pioneering work at the East Pacific Rise.

Major marine mineral investigations are also conducted and sponsored by other governmental agencies of relatively recent origin. A number of Federal Departments are involved in oceanographic research. To mention are here the National Oceanic and Atmospheric Administration (NOAA) of the Department of Commerce, and the Minerals Management Service (MMS) of the Department of Interior. A critical comparison of both agency activities has recently been given by Broadus and Hoagland (1984). These agencies are broadly concerned with hard minerals, excluding oil, gas and sulphur. While NOAA's Office of Ocean Minerals and Energy is supposed to sponsor work in marine mineral exploration and mineral recovery, MMS is involved merely in marine mineral exploitation steps in the form of bid mineral leases. NOAA is aware that very little is known about marine minerals within U.S. waters and therefore co-operation is envisaged with marine mineral activities at the U.S. Geological Survey (USGS, Department of the Interior) with necessary formal agreements. The USGS on the other hand has been involved in ocean minerals and resources assessment for some years.

With much of the marine activities focussed on the EEZ, leaving less research for the international deep waters, many of the U.S. oceanographic institutions are now forced to co-operate with NOAA and MMS. It is then, however, necessary to alter research efforts to be more economically related than was the case in previous times where funding often was obtained by the National Science Foundation.

In Canada, the Bedford Institute of Oceanography (BIO) acts as one of the two central oceanographic units conducting marine research. BIO is composed of several laboratories/units which are sponsored by the Department of Fisheries and Oceans, the Department of Environment, and the Department of Energy, Mines and Resources. The latter's Atlantic Geoscience Centre is involved in hydrocarbon basin analysis and in resource appraisal, but there is apparently very little marine hard minerals research.

The situation in France is somewhat different. While marine resources research up to 1984 was with the Centre National pour L'Exploration des Oceans (CNEXO), a major concentration effort led to the recent establishment of Institute Francais de Recherche pour l'Exploitation de la Mer (IFREMER). A large part of the merged CNEXO was devoted to non-living resources in 1982. There is a very clear aim to investigate marine resources (living and non-living), ocean energy resources, ocean environmental problems, and to conduct fundamental ocean technology development. The latter research efforts have put IFREMER, besides the U.S.A., into a leading position as regards deepsea diving operations. Ocean technology development terminated in the famous DSRV "Cyana", actively engaged in pioneering ridge deposits research. A new manned diving submersible for operating depths of up to 6 km is under construction (DSRV "Nautile"). In general, French efforts seem to concentrate in a single agency for marine minerals research which probably is better than loosely coordinated marine research.

The same cannot be said for oceanic research in the Federal Republic of Germany. Though oceanography is somewhat concentrated at the university of Kiel, governmental marine minerals research is conducted mainly by the Bundesanstalt für Geowissenschaften und Rohstoffe (BGR), and by some few geology departments of a few universities. At BGR, considerable efforts were made on marine geophysics and marine geology, including marine hard mineral exploration during the past two decades. This effort which still is limited, is by far larger than that of the universities.

In this country, a major fraction of marine mineral investigations is also conducted by private companies. To mention should be the mining company Preussag, Hannover, with a minor government participation, often co-operating with governmental institutions. Considering the economics of deep water marine mineral exploration, this type of co-operation is probably very effective. Over the years, partial governmental funding for marine minerals research was attained and extensive exploration efforts were made since 1970. However, with such a diversified structure, involving many institutions and companies, marine hard mineral exploration is probably difficult to coordinate. Research activities, nevertheless, have been considerable during the past 20 to 30 years (Glasby, 1984).

A good and well-balanced example of oceanographic efforts in the field of marine minerals is that of New Zealand. Non-living resources are covered here by the New Zealand Oceanographic Institute (NZOI). The country's EEZ is very large but marine mineral-related work at NZOI has already been conducted over several decades, either in the form of national efforts or by international co-operation.

By searching the activities of some selected countries, a clear picture evolves. It could be made even clearer by the efforts of other leading oceanographic institutions of, e.g. Japan, United Kingdom, India, and others. Discussions and disputes during UNCLOS III have forced several countries to concentrate oceanographic research into more economic issues, namely on hydrocarbons and marine hard minerals in the EEZ. It seems as if these investigations are well-controlled in some countries, but some States still have no concentrated action in exploring their EEZ areas and others have even placed the marine minerals into the wrong departments. Because many countries, especially the less developed ones (LDCs), will be interested in the mapping and exploration of their EEZ, help will be sought from those countries with a long marine mineral exploration record. It is likely that such efforts will be bilateral involving technology transfer, and therefore industrialised countries with weak organisations for marine mineral commodities will eventually not be considered. It is also likely that those countries with no concentrated efforts in the marine minerals sector will become dependent on marine technology developed elsewhere. It is difficult, however, at this time to predict the speed by which EEZ mineral exploration will proceed in the future as is also the case of mineral discoveries by marine mineral exploration.

International organisations

Taking into consideration that vast areas of the earth (about 70%) are covered by water there is a need for international cooperation to study the oceans. This need has not been reduced by LOS, which transferred about 35% of international waters to national jurisdiction. The marine hard minerals are a relatively new oceanographic topic and therefore, only since the 1970s have there been some significant international efforts. Although it is not easy to unravel the structure of international organisations, especially in the case of the United Nations, a number of relevant international oceanographic agencies are given in Table 1.3. This short review is based mainly on an excellent comprehension given by Wooster (1980).

International oceanographic cooperation was already institutionalised in 1902, when the International Council of Ocean Exploration (ICES) was founded in Copenhagen, Denmark. At present, some 20 States are members, including eastern bloc countries.

In 1978 ICES had eleven committees covering the topics: fisheries, hydrography, statistics and marine environmental quality. Decisions are usually taken more on scientific than political grounds. ICES has no connection to marine geology and geophysics.

International cooperation in science was stimulated by the International Council of Scientific Unions (ICSU), founded in 1931. The council is composed of both national members and of members from international scientific unions. It was ICSU who in 1957 proposed a Scientific Committee on Oceanic Research (SCOR). SCOR has identified 3 major international oceanographic research topics; deepsea use for waste, the ocean as resource for protein, and the ocean's role in climatic changes. SCOR is also acting as an advisory committee to UN-sponsored agencies mainly.

The major United Nations body in oceanographic affairs is the Intergovernmental Oceanographic Commission (IOC), founded in 1960. It is an agency directly under UNESCO, which at present has 100 member States dominated by the LDCs. Because of ongoing political disputes between the industrialised countries and the LDCs, IOC has some difficulty in sponsoring and stimulating effective oceanographic research programs. From the beginning, IOC

TABLE 1.3
International oceanographic agencies.

Year	Agency
1902	ICES International Council of Ocean Exploration. About 20 member states; little marine geology and geophysics; mainly fisheries research.
1931	ICSU International Council of Scientific Unions. Members are national academies and international unions.
1957	SCOR Scientific Committee on Oceanic Research. Branch of ICSU; 34 member states.
1960	IOC Intergovernmental Oceanographic Commission. UNESCO agency; 100 member states, majority by LDCs
1970s	OETB United Nations program on Ocean Economics and Technology. ECOR Engineering Committee on Oceanic Resources. International, non-governmental body.

was thought to initiate large-scale research programs and a number of international projects were conducted over the years since its establishment. One could mention here NORPAC, a study by 19 research vessels from Canada, Japan and the U.S.A. in the North Pacific, between 20 and 60° N, or the International Indian Ocean Expedition, involving 40 vessels from 14 countries in 180 cruises. Other rather successful research programs could be mentioned. Presently, IOC also sponsors international studies, e.g. the project ERFEN, a study of the impact of El Niño, and a number of Training, Education and Mutual Assistance (TEMA) programs. However, nowadays there is limited interest in conducting further large-scale surveys. Merely, for better coordination purposes and to gain the standard of international research efforts, interest is now on integrated oceanographic programs with a smaller number of participants. After LOS, there is a need for permission to work in national

EEZs and therefore, IOC-sponsored research is at a turning point. Meanwhile, UNESCO agencies FAO and IOC try to arrange for training courses for the LDCs with large living and non-living resources in their EEZs, the latter of which certainly has not been investigated by IOC bodies for a satisfactory time span.

Direct United Nations efforts in marine minerals research are of quite recent origin. In the 1970s, a program on Ocean Economics and Technology (OETB) was established. OETB has marine minerals as program activities, which include activities on seabed minerals (manganese nodule studies, OETB, 1979), on near-shore hard minerals and on land-based minerals. It is now recognised that a future marine mineral potential is within the EEZs and that resources beyond national jurisdiction require long-term developmental planning. OETB is aware of this and seems at present to be the only international agency that properly considers marine non-living resources.

Also established in the 1970s, the Engineering Committee on Oceanic Resources (ECOR), is concerned with engineering in the marine field. The committee is international, but non-governmental. ECOR is presented in international working groups and meetings. In general, this committee has little involvement in marine resources research, but is important in the engineering part of ocean technology development.

This short review of international oceanographic agencies cannot be of full value without a comment on the recent report "Ocean science for the year 2000" (OY, 1983), requested by the Executive Council of IOC. The report was written by a group of experts representing SCOR, the Advisory Committee on Marine Resources Research (ACMRR) of the Food and Agriculture Organisation (FAO), and ECOR. Though the group had only one and a half year to work, a considerable and detailed report evolved covering major recent oceanographic knowledge and research trends ahead. Four topics were considered in detail: physical oceanography, chemistry of the oceans, life in the ocean, and the ocean floor. As regards marine minerals, the ocean floor section was particularly concerned with marine geology and marine geophysics as important branches in hydrocarbon and metallic minerals exploration. Marine geophysical tools of the past and the future were elaborated upon and future trends were indicated. The application of marine geosciences is expected to grow as there is an increasing use of the seabed. Especially mentioned were here the increase in manganese nodule surveys, in the investigations on Red Sea metalliferous sediments, and in studies for seabed waste disposal. Marine geoscience applications are also expected to grow in exploration for shelf placers and phosphorites.

A final meeting of the group with extended participation of further experts established the cooperative aspects of future oceanic research and stated that LOS has increased the need of coastal States for scientific research in their EEZ areas. Such research is regarded important regardless of the country's development status. The need for marine institutions on a national, regional and global scale was put forward. It could also be concluded from these discussions that IOC, which has stimulated oceanographic research by both extremely- and less-successful projects, could play a major role in future international oceanographic collaboration.

Finally, perhaps the most important review of oceanographic affairs is the series of "Ocean Yearbook", edited by E. Mann Borgese and N. Ginsburg, and published at appropriate intervals by the University of Chicago Press. Ocean Yearbook 1 appeared in 1978 and at the time of writing (1985) 4 volumes have been distributed in all covering many aspects of oceanographic research. It is expected that this series also will report on new trends and developments in marine mineral exploration.

International journals mostly concerned with marine minerals are those of the Marine Technology Society and Marine Mining.

Marine mineral exploration and strategies

In terrestrial mineral exploration, the working strategies are well-founded (e.g. Peters, 1978). A steady refinement of techniques occurred over the years of extensive use of mineral resources. The science of mineral exploration is complex and utilizes such scientific branches as economic geology, petrology, geophysics, geochemistry, and mathematics. Working schemes of mining companies spending money on mineral exploration include steps like mapping, regional geophysical and geochemical exploration, data interpretation and follow-up surveys on a finer scale terminating in the discovery of a mineralisation, or in special cases, in the evaluation of a mineral prospect. At the end of exploration, exploitation of a deposit starts eventually. Any exploration program consists of two principal steps, namely reconnaissance investigations (searching for targets) and target investigations (detailed studies).

It would be natural to adopt the techniques used in terrestrial mineral exploration to the marine sector. Contact should then be made to literature on this subject (e.g. Bailly and Still, 1973; McQuillin and Ardus, 1977). A marine mineral exploration strategy is given in Fig. 1.2.

A clearer picture is probably obtained by describing the principal steps in marine mineral exploration:

- Mapping of the ocean floor
- Geophysical exploration (sensing of the ocean floor and its substrata)
- Geochemical surveying (sampling and characterisation)
- Deposit evaluation

At the beginning of a marine mineral exploration program, basis information on the prospect area is needed in the form of appropriate-scale maps, known mineral occurrences in the area or close to the area (both on- and

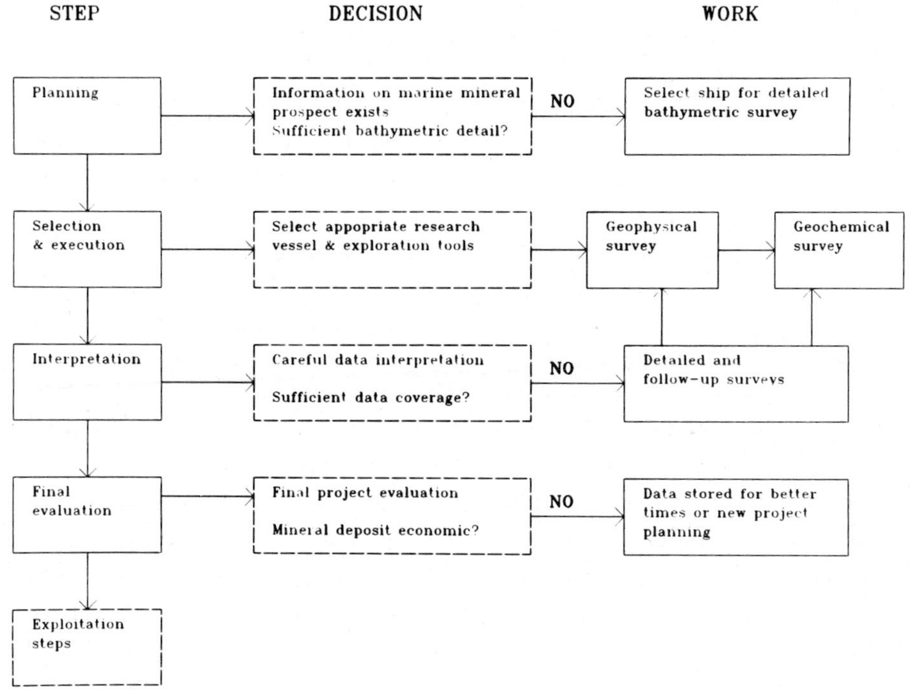

Fig. 1.2. Marine mineral exploration steps.

offshore occurrences) and general oceanographic data are required. Genetic models for the chosen mineral commodity are important here.

Known marine hard mineral resources are discussed in some detail in Chapter 2 of this volume. As regards ocean floor maps, they are supplied by several oceanographic and fisheries agencies, often on a 5- to 10-nautical mile basis. A marine mineral deposit, except for manganese nodules, usually does not exceed about 30 nautical miles in size and therefore, the scales of commercially available bathymetric maps mentioned often are not appropriate. In this case, a bathymetric survey using multiple-beam echo sounder systems (see Chapter 4) should be envisaged. It is often argued that the swath bathymetric systems are a very expensive oceanographic tool and therefore not very useful in marine mineral exploration. Certainly, present-day installation costs of such systems are considerable and only few research vessels are equipped with them. As of 1984, ten research vessels have installed swath bathymetric systems. However, system prices may drop in the future with the application of new computer technology. The limitation of use lies more in political restrictions caused by national defence departments (see Science, 1984). Taken into account that marine mineral occurrences often are connected with morphological features, tectonics and surface relief, the swath bathymetric equipment (Seabeam) is a must in marine exploration. Topographical maps, up to now, simply have not the scale to support careful exploration in deep waters or in the EEZ. While Seabeam certainly cannot be excluded in ridge deposit exploration, it is recommended in delineating manganese nodule fields, seamout encrustations and even for the shelf, a special swath bathymetric system can be used. The bathymetric survey terminates in appropriate scale maps (e.g. 10-m isobaths) and the planning of the exploration survey can then continue.

Vital for the exploration campaign is the choice of an appropriate exploration vessel (see also Chapter 3 of this volume). Size of the vessel is often adjusted to the type of survey carried out, i.e. according to shallow or deep water work. Shelf research can be conducted with smaller and cheaper vessels. Chartering of a research vessel is generally the most expensive part of the survey. Often, ship-time costs are of the order of 10,000 USD per day and very careful planning is therefore necessary. Because of the costs, it is also important which type of routine geophysical, sampling and other instrumentation is available on board ship. Generally, an integrated navigation system is required, mostly based on satellite navigation but also hyperbolic and other ranging systems (e.g. Brown, 1973) are in use. Many electronic systems were developed for the purpose, but a thorough discussion of these systems is not within the scope of this book. The navigation system is, however, important for proper selection of the exploration vessel. A global positioning system (GPS) is at present being implemented, and this is thought to give a better choice for ocean navigation.

As already mentioned, the ship is often equipped with basic geophysical instrumentation. Geophysical exploration determines geological structures of the seafloor or detects mineralisations by the measurement of physical parameters of seafloor and substrata. Most likely, sonar equipment and photographic devices are supplied. The general geophysical survey of an exploration prospect area is usually not carried out together with other ship operations, i.e. it is conducted separately. This is because geophysical surveying often requires different ship speeds and geophysical exploration generally is composed of continuous work along preselected ship tracks, leaving little spare time to other operations. Geophysical equipment used depends greatly on water depth. Many of the standard methods are described in more detail in Chapter 4. In short, techniques include sonar methods (echosounders, side-scan sonar, subbottom profiler) with instruments often mounted on deep-tow instrumentation, seismic techniques (boomer- and sparker seismics, reflection seismics), magnetic methods, gravimetry and others.

Preliminary data evaluation of the geophysical survey is taken into consideration by the subsequent geochemical survey. The geochemical survey includes systematic sampling and chemical analysis (characterisation) of seawater, seafloor and subbottom samples. It makes also use of routine bathymetric devices and optical instruments to guide the sampling. Typical geochemical techniques are described in Chapter 5 of this volume. Most of the sampling devices can be regarded as classic, because some methods of sampling are still the same that was the case several decades ago, although an instrument refinement can be observed. Principally, most sampling equipment for seafloor sampling is cable-mounted, but some free-fall grab and coring devices are also used. Besides grabs and corers, dredges are used extensively for the sampling of manganese nodules, encrustations and rocks. There is a great need of developing new sampling technology, possibly in the form of remotely-operated vehicles (ROV).

The marine exploration project needs rapid and accurate analytical methods, either directly onboard ship or at the home laboratory. The principal analytical techniques are also described in Chapter 5. It is worth mentioning, however, that shipboard analyses of marine hard minerals, rocks and sediments can be conducted most efficiently by non-destructive X-ray fluorescence analysis (XRF). This method requires usually no sample weighing and relatively simple sample preparation. A good exploration vessel should be equipped with modern XRF instrumentation.

The geophysical and geochemical surveys record large numbers of data that have to be manipulated properly. Some of the data evaluation techniques are described in Chapter 6. Data evaluation leads to the production of substrata maps, mineral maps and survey reports. It is often the case that reports point to insufficiently surveyed subareas or to doubtful results. Then, a follow-up survey, either based on geophysics or on an additional sampling/ analysis is necessary.

The marine mineral exploration program terminates in a final report on the surveyed project area. Often, marine minerals that were discovered within the survey area have too small areal extent or grade to be of any economic significance so that further exploration is cancelled. In many cases however, further exploration is necessary before decisions on exploitation can be made.

Many of the marine mineral deposits discovered in the past have a special exploration record. The strategies and techniques used to eliminate them are different for the different mineral commodities. Exploration histories usually gain our understanding of the formation and emplacement of mineral deposits. Therefore, Chapter 7 is confined to 5 exploration case histories, covering manganese nodules, metalliferous sediments, phosphorites and massive sulfides and Co-rich ferromanganese crusts.

Marine mineral exploration in the future certainly requires special efforts. Although it is difficult to predict the trends in a changing world there is a clear requirement for exploration of vast areas of the ocean floor. It would be more easy to explore the seafloor for mineral deposits if fine-scale maps existed for both the EEZ and international areas. Much of the future work should therefore be placed on bathymetric mapping. The trends for the next few decades are, however, clear. EEZ marine mineral exploration will proceed on a national or bilateral basis, i.e. by the EEZ owner alone or by the liasion of companies or agencies of two or several countries, whether neighbours or not. Marine mineral exploration in the international area certainly has to involve several companies from several countries tied together in large consortia because exploration costs will be significantly higher than in EEZs. It cannot be predicted, at present, in which way LOS will stimulate or prevent such initiatives in the near future.

REFERENCES

Bailly, P.A. and Still, A.R., 1973. Exploration for mineral deposits - purpose, methods, management. In: A.B. Cummins and I.A. Given (Editors), SME Mining Engineering Handbook. The American Institute of Mining, Metallurgical, and Petroleum Engineers, New York, Vol. 1, Sect. 5.1.

Bischoff, J.L. and Piper, D.Z. (Editors), 1979. Marine geology and oceanography of the Pacific manganese nodule province. Plenum, New York, 839 pp.

Broadus, J.M. and Hoagland, P., 1984. Rivalry and coordination in marine hard minerals regulation. Proceed. Oceans'84. Marine Technology Society, pp. 415-420-

Burstyn, H.L., 1968. Science and government in the nineteenth century: the Challenger expedition and its report. Bull. Inst. Oceanogr. Monaco, 2: 603-613.

Deacon, M., 1971. Scientists and the sea. Academic Press, London, 445 pp.

Deacon, G.E.R. and Deacon, M.B., 1974. The history of oceanography. In: R.C. Vetter (Editor), Oceanography, the last frontier. Voice of America, pp. 13-30.

Edmond, J.M., Measures, C., McDuff, R.E., Chan, L.H., Collier, R., Grant, B., Gordon, L.I. and Corliss, J.B., 1979. Ridge crest hydrothermal activity and the balances of the major and minor elements in the ocean: the Galapagos data. Earth Planet. Sci. Lett., 46: 1-18.

Edmond, J.M., Von Damm, K.L., McDuff, R.E. and Measures, C.I., 1982. Chemistry of hot springs on the East Pacific Rise and their effluent dispersal. Nature, 297: 187-191.

Francheteau, J., Needham, H.D., Choukroune, P., Juteau, P., Seguret, M., Ballard, R.D., Fox, P.J., Normark, W., Carranza, A., Cordoba, D., Guerrero, J., Rangin, C., Bougault, H., Cambon, P. and Hekinian, R., 1979. Massive deep-sea sulphide deposits discovered on the East Pacific Rise. Nature, 277: 523-528.

Fye, P.M., 1980. The Woods Hole Oceanographic Institution: a commentary. In: M. Sears and D. Merriman (Editors), Oceanography: the past. Springer, New York, pp. 1-9.

Glasby, G.P. (Editor), 1977. Marine manganese deposits. Elsevier, Amsterdam, 312 pp.

Glasby, G.P., 1984. Manganese nodule research in the Federal Republic of Germany: a review. Marine Mining, 4: 355-402.

Hobson, G.D. and Tiratsoo, E.N., 1975. Introduction to petroleum geology. Scientific Press, Beaconsfield, England, 300 pp.

McQillin, R. and Ardus, D.A., 1977. Exploring the geology of shelf areas. Graham & Trotman, London, 334 pp.

Mero, J.L., 1965. The mineral resources of the sea. Elsevier, Amsterdam, 312 pp.

Moore, J.R., 1971. Introduction. In: Oceanography. Scientif. American, pp. 2-10.

OETB, 1979. OETB: The United Nations programme in ocean economics and technology. In: E. Mann Borgese and N. Ginsburg (Editors), Ocean Yearbook 3. The University of Chicago Press, pp. 436-447.

OY, 1983. Ocean science for the year 2000. In: E. Mann Borgese and N. Ginsburg (Editors), Ocean Yearbook 4. The University of Chicago Press, pp. 176-259.

Peters, W.C., 1978. Exploration and mining geology. Wiley, New York, 696 pp.

Riley, J.P., 1965. Historical introduction. In: J.P. Riley and G. Skirrow (Editors), Chemical oceanography. Academic Press, London, pp. 1-42.

Science, 1984. Classification dispute stall NOAA program. Science, 227: 612-613.

Seibold, E. and Berger, W.H., 1982. The sea floor. Springer, Berlin, 288 pp.

Tait, J.B., 1968. Oceanography in Scotland during the XIXth and early XXth centuries. Bull. Inst. Oceanogr. Monaco, 2: 281-292.

Tissot, B.P. and Welte, D.H., 1978. Petroleum formation and occurrence. Springer, Berlin, 527 pp.

Tomczak, M., 1980. A review of Wüst's classification of the major deep-sea expeditions 1873-1960 and its extension to recent oceanographic research programs. In: M. Sears and D. Merriman (Editors), Oceanography: the past. Springer, New York, pp. 188-194.

Wolff, T., 1968. The Danish expedition to "Arabia felix" (1761-1767). Bull. Inst. Oceanogr. Monaco, 2: 281-292.

Wooster, W.S., 1980. International cooperation in marine science. In: E. Mann Borgese and N. Ginsburg (Editors), Ocean Yearbook 2. The University of Chicago Press, pp. 123-136.

CHAPTER 2

MARINE HARD MINERAL RESOURCES

K.BOSTRÖM and H.KUNZENDORF

INTRODUCTION

Marine hard minerals research is clearly overshadowed by marine exploration and exploitation efforts for fuels, particularly the hydrocarbons. After the first oil crisis in 1973, petroleum exploration has been stimulated in the potential offshore regions, and therefore less money was spent on the exploration of marine minerals. Furthermore, whereas in the 1960s and 70s most exploration efforts were focussed on the huge Pacific deposits of manganese nodules, present interest has shifted to hydrothermal massive sulfides at mid-ocean ridges, and particularly to any mineral deposit within Economic Exclusive Zones (EEZ).

In a changing world it is very difficult to predict the mineral requirements of modern societies. In the past it was often possible to foresee metal consumption for some years ahead. However, present electronic and materials sciences revolutions (plastics, composite materials) have led to slower increases in metal consumption; moreover, many metallurgical processes have proved too energy-demanding. Therefore, at present and very likely also in the near future, predictions cannot be made for more than a few years, although it can be said that there will always be some need for minerals.

Thus, metals that were important for many years (e.g. copper) now have a weakened demand due to new technological trends. There is, in other words, a cycle of steady replacement of one mineral commodity by another one or by a newly developed compound. Also, dramatic price changes for several metals occur at present. Strategic metals like cobalt are now of high value, resulting in e.g. that marine mineral exploration at present focusses more on Co-rich manganese crusts than on manganese nodules. These changes are difficult to predict but should be borne in mind when considering marine mineral exploration programs. In general then, for the marine sector all valuable metals in the minerals should be considered in view of probably changing needs.

This chapter briefly outlines the physiographic provinces of the ocean floor since they are of importance for the occurrence of marine minerals. In the following sections the different mineral commodities are then described, but spatial restrictions prohibit detailed treatment of the deposits and descriptions and discussions are therefore generally kept short; additional information can be obtained in the cited sources and additional references given in Appendix I.

PHYSIOGRAPHIC PROVINCES OF THE OCEAN FLOOR

About 70% of the Earth is covered by oceans and oceanographic investigations up to now have discovered considerable relief of the ocean floor. Water depth varies over wide areas and in some places like the famous Marianna Trench water depth (ca. 11 km) exceeds the highest elevations on Earth. There is, however, a clear division of ocean floor provinces (Fig. 2.1). Physiographic provinces can basically be divided into continental margin and deep sea floor. An excellent review on the subject is presented by Heezen and Menard (1963).

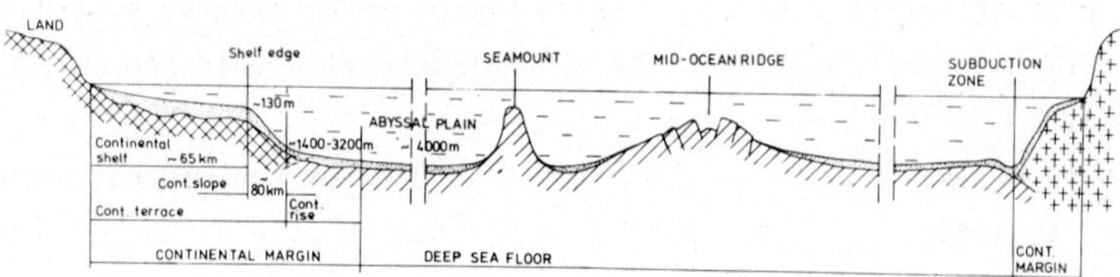

Fig. 2.1. Physiographic provinces of the oceans; not to scale.

The continental margin consists of the continental shelf, separated from the continental slope by the shelf edge or shelf break. The continental slope continues then into the continental rise.

The continental shelves have a relatively flat topography, the average water depths being about 130 m. Very narrow shelves are found near young mountain ranges (subduction zones). The substratum of the shelves in some cases consists of crystalline bedrock, overlain by a thin veneer of sediments. The normal case, however, is that shelves represent several kilometer thick sedimentary deposits from mainly continental weathering caused by the action of wind, waves, currents and ice. These sediment sequences are locally very thick, e.g. in the Gulf of Mexico. The nature of sediments varies considerably even over short distances, and is partly explained by sea level changes. Large parts of shelf areas are occupied by sands and muds, but coral reefs may occur. The main processes acting in shelf areas are transgression, erosion and deposition. A detailed description of marine sediments and marine sedimentary processes can be found in the literature (e.g. Seibold and Berger, 1982). It is often difficult to unravel the history of present shelf areas in view of the differing geological events in the past and because the thickness of the sediments often prohibits complete rock drill penetration.

The shelf edge is usually a very well-developed feature within the continental margins. Principally, this marine marker represents the zone where past transgressions and regressions of the sea diminish their influence on ocean floor topography. Topographical relief at shelf edges usually varies, and may be highly broken where submarine canyons occur.

From the shelf edge the ocean floor continues into the relatively steep continental slope and then into the more gently sloping continental rise, but the border between them is not sharp. Furthermore, the sedimentary cover above the basement rock is generally much thinner on the continental rise. The processes for continental slope formation vary on a world wide basis, but generally continental sedimentary sequences are mixed with marine varieties which are mainly in the form of biogenic shells. Silts, sands and muds are dominant. Sediments are often deposited on gently dipping slopes, but steeper slopes may be the place for redepositional features caused by for instance turbidity current mixing or slumping. Submarine canyons are often observed. Continental slope gradients are between 1:40 to 1:6 whereas at rises, gradients of 1:300 are found. These physiographic patterns are rather well developed in the Atlantic and Indian Ocean basins, whereas the Pacific Ocean continental rises are poorly developed. Continental shelf and slope area cover some 15% of the total ocean floor, continental rises some 5%.

About 42% of the deep sea floor area shows relatively gentle relief. The sediment cover there is considerably thinner (a few hundreds of m) than on the continental margins. It is derived from terrestrial sources and from planktonic shells; hence the sediments are deep sea oozes and deep sea clays. The major part of this area shows a gently rolling relief with occurrences of hills that are approximately 6 km wide at their base and have elevations of about 300 m. In other areas the deep sea floor may be extremely flat with slopes of only 0.001 for hundreds of kilometers. Such areas are generally referred to as abyssal plains, but the total area of the ocean floor covered by such abyssal plains is only a few per cent; mostly, these plains develop near the continents just off the continental rises.

Seamounts are elevations exceeding 1 km in height and being common on the deep sea floor, particularly in certain ocean regions. Flat (eroded) tops and coarse detritus on them show that they were at the sea surface at some time, and later moved with continental drift to their present position in deep waters. It is probable that conical seamounts are extinct or active volcanoes. Seamount chains, especially in the Pacific are regarded as hot spot accumulations on a drifting crust.

At the global system of mid-ocean ridges which is about 60 000 km in length upwelling of mantle material takes place. As a result ocean floor spreading, plate movements and exposure of fresh magmatic rocks to seawater occurs. Spreading rates vary between 1 and 10 cm per year, but at some places of the East Pacific Rise (10-13° S) rates of more than 20 cm/year are reported if the accretion rates at each plate edge are added. The crestal provinces are between 80 to 320 km wide, but the topographical relief at ridges varies considerably. In the Atlantic and Indian Oceans the ridges are rugged and frequently show a central rift which may be a few kilometers wide and 1 km deep. In the Pacific Ocean, on the other hand, the ridges are comparatively smooth and the central rift is generally absent or shallow.

Ridge crests are usually culminating at water depths of about 2.5 km, but deeper crests are found. At mid-ocean ridges flank provinces with intermontane basins occur. Mid-ocean ridge areas have little sedimentary cover. The rocks are principally young basalts (olivine tholeiites) low in elements K, Ti and P. Ocean ridges and rises cover some 33% of the ocean floor.

Another feature of the deep ocean floor are trenches at ocean basin margins, but there are also trenchless continental margins; at least in some cases this is due to sedimentary filling up of former trenches. Trenches occur at the subduction zones where oceanic crustal plates collide with continents or island arcs, but it is still a question of controversy how large a fraction of all sediments that are subducted and how much that are welded on laterally to the older crust as mountain ridges etc. Subduction is inferred among others from earthquake belts along continents, and from the distribution of epicenters along the Benioff zone.

MARINE HARD MINERAL OCCURRENCES

The literature on marine deposits is rapidly increasing and includes studies by Mero (1965), Wilde (1973), Schott, (1976), Emery and Skinner (1977), Glasby (1977), Cruickshank, (1978), Burns (1979), Kent (1980), Earny (1980), Cronan (1980), Crutchfield (1981), Cruickshank, (1982), Charlier (1983), Moore (1983), Holser et al., (1984), Claque et al., (1984) and Galtier (1984).

Various classifications can be adopted for marine hard mineral resources, since they occur in certain physiographic provinces of the oceans or because of their varying origins. Classification according to (a) occurrence in physiographic provinces was made by Crutchfield (1981). These resources may also be divided into (b) dissolved deposits (not covered in this volume), unconsolidated deposits and consolidated deposits (Wilde, 1973), or (c) into commodity, i.e. aggregates, phosphorites, nodules and crusts, massive sulfides (Cronan, 1980). A combination of these possibilities is given in Table 2.1.

TABLE 2.1
Marine mineral deposit grouping according to physiographic province and mineral commodity.

Physiographic province	Unconsolidated sedimentary deposits	Nodular & crust deposits	Hydrothermal massive sulfide deposits
Continental margin	Sand & gravel Calc. shells Placers Bed. Phosphorites	Barites Phosphorites	
Abyssal hills and plains		Mn-nodules & crusts	
Seamount provinces		Co-rich crust Phos. crusts	
Mid-ocean ridges	Metalliferous & brine-rich sediments		Massive sulfides

A number of processes are acting in the marine environment that are important for the accumulation and enrichment of minerals. These processes are:

- Selective sorting of mineral grains
- Selective leaching, and leaching with subsequent deposition of enriched fluids
- Precipitation from seawater
- Diagenetic processes.

These processes are described in some detail by Cronan (1980).

Selective sorting of mineral grains occurs by mechanical winnowing of wind, waves and currents, and by gravity settling and concentration. Sedimentary deposits are mostly exposed to these processes. Selective leaching is not important for unconsolidated sedimentary mineral accumulations but could explain certain secondary enrichments like phosphorites. Large-scale leaching of basalts by penetrating seawater occurs at the mid-ocean ridges; the returning solutions occur as spectacular black smokers on the ocean floor (see section on massive sulfides). Precipitation of metals from seawater by adsorption is an important process in the formation of manganese nodules and crusts. Finally, diagenetic processes in sediments and probably also within some mineral deposits lead to the concentration of certain metals. It should be recalled, however, that much of the matter in deep waters and sediments arrive there in biological debris which stems from productive areas in the surface waters (Boström et al., 1978).

Unconsolidated sedimentary deposits

Following the weathering of terrestrial rocks, fragments are further broken down mechanically and chemically by the action of river waters during the transport to the sea, the most resistant minerals surviving the longest. The unconsolidated sedimentary deposits may be grouped into:

- Aggregates
- Placers
- Bedded phosphorites
- Metalliferous and brine-rich sediments

Aggregates are composed mainly of sand and gravel but they include also shell deposits. Placers are, in general, heavy metallic minerals accumulating in the marine environment. Aggregates and placers are of prime importance onshore. Phosphorite deposits in continental shelf areas are basically connected with major upwelling areas of the oceans. Metalliferous and brine-rich sediments are confined to mid-ocean ridges and their adjacent areas.

Aggregates. A growing construction industry needs large amounts of sand and gravel. Land-based deposits mainly of glacial origin are diminishing and are to some extent exhausted, hence the need to exploit such deposits in the sea. Actually, sand, gravel and shells are the only marine mineral commodities mined on a large scale at present. Being a low-value commodity exploitation is generally dictated by local needs and short transportation distances to the deposits.

Sand and gravel consist of rock fragments, generally representing a complex mineral assemblage, but the predominant mineral is quartz. The erosion products are usually carried by rivers and then transported and graded by waves, tides and currents. The coarsest fragments tend to deposit nearshore while sand, silt and mud as a rule are often transported further offshore.

Some selected recent sand and gravel production figures are given in Table 2.2.

TABLE 2.2

Selected production figures for sand and gravel (Galtier, 1984). All data in 1000 t. Value of sand and gravel is about 2 USD/t.

Country	Marine production 1978	Marine production 1982	Total production 1982
Japan	84700		
U.K.	15000	16500	
Denmark	4600	5040	40800
U.S.A.	4000		595000
France	2800	4200	341000
The Netherlands.	2430		
Belgium	1140		
Iceland	600		
F.R.Germany	20		

As regards marine sand and gravel exploitation, the largest producer Japan has recovered 60 to $80x10^6$ t per year from offshore deposits in the last 10 years (Narumi, 1984), mainly because the exploitation of dry and wet river beds is depleting. Offshore sand and gravel recovery is usually carried out at water depths down to 35 m but Japanese developments have increased exploitation to depths of up to 50 m.

Calcareous shells are exploited at several places in the world. Some shell deposits are recovered for mineral collectors at sea recreational places, but calcareous shell are also used in the construction industry. It is mined as calcareous sands or as coral sands; sand and gravel are non-renewable resources, but calcareous sands are biologically continuously renewed. These deposits are of special importance to countries with small or no limestone deposits, for instance Iceland, where shell production has been carried out for many years. Calcium carbonate deposits are also mined in the Gulf of Mexico region, and at Bahama Banks, fine-grained aragonite being recovered from an artificial island (Kent, 1980).

There is a growing environmental concern with aggregate mining, since removal of large sand and gravel lenses may aggravate coastal erosion. Even offshore bank exploitation may have an impact on beach amenities. Therefore, sand and gravel exploitation is not recommended in tourist areas with attractive beaches.

Placers. Placer deposits are mechanically concentrated detrital minerals, which originate from eroded onshore rocks. Rivers and glaciers discharge large amounts of these minerals and the aggregates mentioned onto continental shelves; due to subsequent reworking in the ocean heavy mineral concentrates may form. Placers also occur in rivers and estuaries. Because detrital minerals are usually relatively heavy they are not transported for long distances from their source; therefore, placers are found mostly in near-shore zones. The areas where placers may occur on the continental shelves are shown schematically in Fig. 2.2.

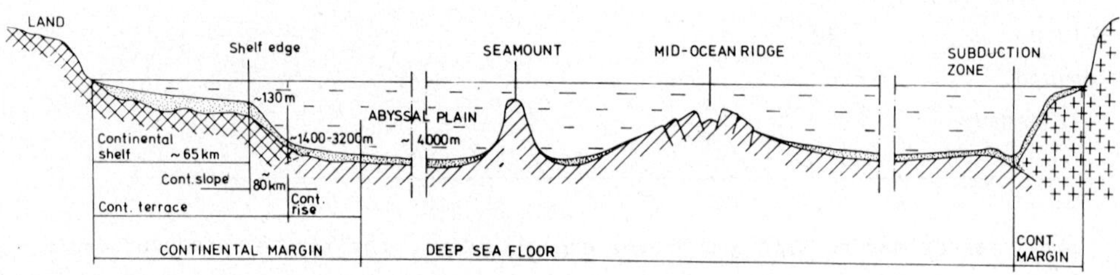

Fig. 2.2. Placer accumulations at shelf areas. I is accumulation at present beaches, II and III are ancient beaches.

Basically, near-shore, heavy mineral accumulations are observed in the form of black or dark stringers and lenses on modern beaches (level I, Fig. 2.2). Past sea level changes, tidal currents and wave action, however, have reworked older placers so that heavy minerals may be found much further offshore (levels II and III, Fig. 2.2). Buried placers are little known but cassiterite exploration in the southeast Asian tin belt undoubtedly will reveal several examples. In general placers are richer on modern beaches than on submerged ones.

Plate tectonics places some constraints on the morphology of continental shelves. For example, collision-type continental shelves like the west coast of the Americas are relatively narrow and are therefore opposed to high-energy transportation and concentration processes. On the other hand, wide shelves that are not very much influenced by subduction (e.g. the east coast of the Americas) are areas where placers may survive at the area of deposition for longer periods. The distinction is important because the former shelf type is then the locus of a greater number of placers which, however, can be reworked at some time.

Corrigendum

Marine Mineral Exploration, by H. Kunzendorf (Editor)

p. 28, Figure 2.2 should be replaced by the following figure:

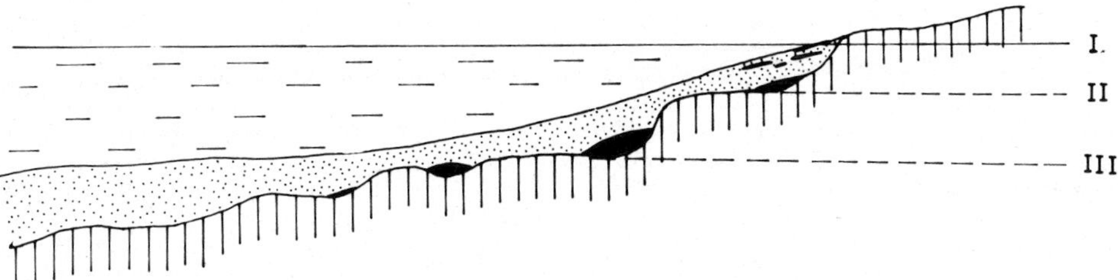

Fig. 2.2. Placer accumulations at shelf areas. I is accumulation at present beaches, II and III are ancient beaches.

data are modified from Cruickshank (1981).

Placers have been divided according to the specific gravity of their mineral components into heavy heavy mineral placers, light heavy mineral placers, and gemstone placers, with specific gravities of 6.8-21, 5.2-5.3, and 2.9-4.1 g/cm^3, respectively. However, such divisions are often difficult to establish because of the multi-mineral behaviour of most placers. Therefore, placers in the following sections are described as:

- Tin placers
- Complex heavy mineral sands
- Precious metal placers
- Diamond accumulations

TABLE 2.4
Selected detrital minerals and their source rocks.

Rock	Heavy minerals present
Granites, granodiorites, rhyolites (acid rocks)	Magnetite, monazite, tourmaline, zircon, cassiterite
Basalts, dolerites, gabbros (basic rocks)	Chromite, ilmenite, leucoxene, magnetite, rutile, zircon
Metamorphic rocks	Garnet, leucoxene, magnetite, tourmaline
Pegmatites	Monazite, tourmaline, zircon
Hydrothermal veins	Cassiterite, pyrite, wolframite, gold, platinum

A comprehensive discussion of United States Exclusive Economic Zone (EEZ) placer deposits was recently given by Beauchamp and Cruickshank (1983). This work also includes economic estimates of the relevant deposits.

Tin placers. Cassiterite has been the most important placer accumulation, exploited for nearly one hundred years. Offshore tin dredging is known to occur since 1907 in Thailand waters (Kulvanevich, 1984). The southeast Asian tin

The placer minerals tend to have a higher specific gravity anc mechanical and chemical attack better than ordinary rockforming r comprehensive work on detrital minerals was given by Baker (1962). shows the characteristic features of selected heavy minerals.

TABLE 2.3
Economically important heavy minerals ordered after decreasing s gravity in g/cm^3.

Mineral	Specif. gravity	Hardness	Remarks
Tourmaline	3.1	7	$NaMg_3Al_6B_3Si_6O_{27}(OH,F)_4$, various
Diamond	3.5	10	C; colorless
Topaz	3.6	8	$Al_2SiO_4(F,OH)_2$; colorl.-brown
Garnet	3.8-4.2	7-7 1/2	$(Fe,Mn,Mg)_3Al_2(SiO_4)_3$, various
Corund	4.0	9	Al_2O_3; multi-colored
Rutile	4.2-4.3	6 1/2	TiO_2; reddish-brown, black
Chromite	4.3-4.6	5 1/2	$FeCr_2O_4$; black
Ilmenite	4.3-5.5	5 1/2	$FeTiO_3$; iron-black
Xenotime	4.4-5.1	4 1/2	YPO_4; yellowish-brownish
Zircon	4.4-4.8	7 1/2	$ZrSiO_4$; colourless, red-brown
Pyrrhotite	4.5-4.6	4	$Fe_{1-x}S$; bronze-brownish
Magnetite	4.9-5.2	6	Fe_3O_4; iron-black
Monazite	4.9-5.3	5 1/2	$(REE,Th)PO_4$; yellow-reddish
Pyrite	5.0	6-6 1/2	FeS_2; brass-yellow
Columbite	5.4-6.4	6	$(Fe,Mn)(Nb,Ta)_2O_6$; brown.-black
Scheelite	5.9-6.2	5	$CaWO_4$; yellow-white, brownish
Thorite	6.7	4 1/2	$ThSiO_4$; brownish-yellow
Cassiterite	6.8-7.0	6 1/2	SnO_2; brown to black
Wolframite	7.1-7.5	4 1/2	$(Fe,Mn)WO_4$; brown, black, red
Uraninite	7.5-10	5-6	UO_2; black
Cinnabar	8.0-8.2	2 1/2	HgS; red-brownish
Platinum	14-19	4	Pt
Gold	15.6-19.3	2 1/2	Au

Heavy mineral concentrations are usually derived from the weathering of normal igneous or sedimentary rocks, whereas ore deposits seldom are the source. Typical mineral paragenesis of igneous rocks is given in Table 2.4. The belt is a north-south trending feature between the Himalaya mountains, Burma, Thailand, Malaysia and Indonesia. This belt of biotite and biotite-muscovite granites is several thousand km long and has been subject to extensive onshore placer tin mining. Tin has probably been concentrated in biotite with the help of volatiles but has possibly also accumulated between muscovite sheets (Burns, 1979). Exploration is now concentrated at offshore regions since onshore deposits are more or less exhausted. Exploration is usually carried out by geophysical methods (seismic, side-scan sonar) down to 45 m into the substrata.

In general, however, quaternary eustatic sea level changes have resulted in rather complex offshore deposits (Sujitno, 1984). Besides cassiterite, the placers typically also contain pyrite, ilmenite, zircon, tourmaline and monazite, in decreasing order of importance. An important by-product of Thailand tin smelters is tantalum, Ta being thought to occur in solid solution with Sn in cassiterite and rutile. According to Galtier (1984), offshore tin production in Indonesia and Thailand amounted to about 8000 t annually in 1978, which is about 20 to 40% of the total tin production for these countries (both on- and offshore).

Of interest are also tin offshore accumulations of Cornwall, United Kingdom. Tin dredging has been carried out off Cornwall in past times so that large amounts of cassiterite from numerous tin lodes and tailings have been transported to the sea; as a result artificially produced placers occur off Cornwall. Highest tin concentrations are found off Red River.

Complex heavy mineral sands. Complex heavy mineral sands contain one or several of the heavy minerals ilmenite, rutile, zircon, monazite, garnet, magnetite, and chromite, but also other ones may be present. Many of these deposits were identified in the past and they are also described in some detail in the literature; their distribution together with sand and gravel, and calcareous shells is shown in Fig. 2.3. In general these deposits are small with mining life times of probably no more than 10 years, and are therefore often of little economic interest.

Principal iron-titaniferous sands were identified in eastern and western Australia, California, Oregon, Alaska, New Zealand, India and Sri Lanka. Commercial exploitation is known at present only off southern Japan, although some near-shore deposits could be exploited by present dredging technology. In Japan, iron sands were mined during the 1960s off south Kyushu Island, at Ariake and Kagoshima Bays. Iron-titaniferous sands were also exploited during the 1970s in the Lingyan Gulf, Philippines. These deposits extended 7 to 8 km offshore (Burns, 1979). Offshore exploration for complex heavy mineral sands has been conducted in many countries during the years 1978-83 (Galtier, 1984), for instance in U.S.A., Mexico, Argentine, Brazil, India, Japan, Australia, Poland, USSR, Sri Lanka, Mozambique, Italy, Bangladesh and Phillipines.

As regards monazite, marine placers were identified especially offshore India and Sri Lanka. Claque et al. (1984) give an estimate of about 2x10^9 m^3 heavy mineral placers for the U.S. Pacific continental shelf.

Chromite-bearing sands are known from the U.S. west coast continuing from Coos Bay, Oregon, to northwestern California. Occurrences are confined to present day and ancient marine terraces. The deposits were of considerable strategic interest during World War II and exploration has also been carried out in the 1970s (Burns, 1979).

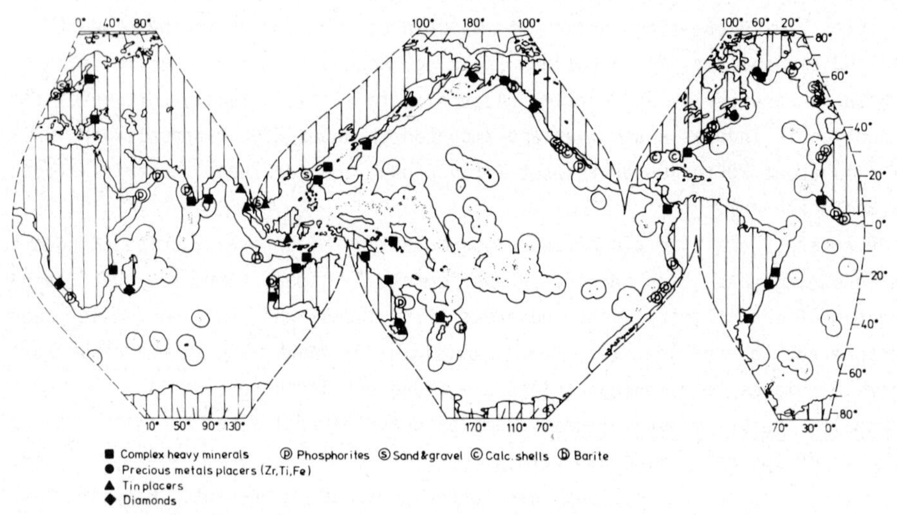

Fig. 2.3. Identified sand and gravel, calcareaous shell, phosphorite and placer deposits on the world's continental shelves. Position is approximate, detailed information can be found in the relevant literature.

Precious metal placers. Precious metal placers mainly consist of economic accumulations of gold, platinum and related metals. Gold and platinum are resistant and very heavy, and they occur therefore in beach sands free from admixtures of other placer minerals. There has been an ancient exploitation of these high-value metals. Famous examples are beach operations at Nome and Bluff, Alaska, where erosion of Anvil Creek basin schist resulted in gold-rich beach accumulations. Exploitation often was hindered by harsh weather conditions. Other U.S. Pacific coast deposits were identified in Californian waters and Pt was discovered in Goodnews Bay and Chagvan, Alaska (Charlier, 1983). Drilling operations for gold were also conducted in the Shelbourn-New Harbour area, Nova Scotia. Gold has been mined in the EEZ of the USSR, e.g. in the Laptev Sea, Arctic Ocean. Deposits have also been identified in southeast Australia and off Oregon, U.S.A; however these deposits are not economic at present. Of importance for Au placers are submerged river channels and gravel bars. Very fine-grained gold was discovered in the Bering Sea, but assessment of such deposits is very difficult. These low-energy placers require quiescent depositional conditions that are found in protected coastal bays and estuaries. Grain size of the native metals may be as low as 5 μm.

Silver and copper accumulations should exist offshore since they have been identified in onshore placers, but such deposits have not been discovered yet.

Diamond accumumulations. The only productive diamond accumulation was that at Hottentot Bay near Orange River, Namibia. During the years 1961 to 1971, diamond-bearing sands were dredged under harsh weather conditions with severe storms resulting in loss of dredging apparatus. The main kimberlite area lies more than 800 km inland, but the offshore diamonds more likely derived from nearby terrigeneous deposits and from the weathering of nearshore volcanic pipes. The diamond sands also contain minerals like epidote, ilmenite, magnetite and pyrite (Burns, 1979).

Bedded phosphorites. Marine phosphate occurrences are widespread and are essentially found on continental margins, isolated islands and on seamounts provinces. Phosphates in the form of guano (excreta of birds) were and are still exploited from deposits on mainly Nauru and Christmas Islands, but the most important source of phosphate are marine bedded phosphorites. Land deposits of this type supply about 70% of the world phosphate market. Famous onshore phosphorite deposits are found in Morocco and Florida, but the worlds largest phosphorite operation is the Western Phosphate Field of the U.S.A. (Idaho, Montana). Phosphorite deposits of the United States were described in detail by Rowland and Cruickshank (1983)

Phosphorite in general is a marine sediment that contains economic amounts of calcium phosphate minerals. The major phosphate mineral is carbonate fluorapatite (Manheim and Gulbrandsen, 1979).

Bedded phosphorite deposits are not identical with nodular or slab-like phosphorite deposits that occur at water depths of 300 to 500 m on certain coastal plateaus. They also differ from P-rich crusts on seamounts (see the respective sections in this chapter).

Both onshore and offshore bedded phosphorites are basically thought to be formed in present and past upwelling areas of the oceans, but there is still an ongoing dispute about the details of their origin. In some areas, usually located on the westerly sides of continents, nutrient rich water returns to the surface from the regions around the thermocline. This upwelling enhances the productivity in these areas and as a result an increased extraction of phosphorus from seawater takes place, mainly by algae (e.g. Seibold and Berger,

1982). After the death of these organisms a rapid flux takes place of decaying organisms to the seafloor, and simultaneously, the dead organisms return phosphorus to the water column. However, due to the timelag much phosphorus is not released until the dead organic matter has reached considerable depth, cften a few hundred, sometimes several thousand meters; as a consequence phosphorus-rich waters in the oceans occur at depths from about 200 m down to abyssal regions.

In the upwelling regions interstitial waters of sediments have higher P content than seawater, and therefore replacement of $CaCO_3$ in the sediments by apatite is initiated. Formation cf pellet-like phosphorite deposits is possible, however, only in areas with little dilution of the calcareous sediments by terrigeneous matter. This means that phosphorites are not observed in areas with nutrient-rich waters and rapid high terrigeneous input such as in the Antarctic ocean.

Phosphorite formation has varied widely in geological time, but was pronounced especially during Tertiary. Tertiary deposits probably underly many of the worlds present continental margins (e.g. Teleki, 1986).

Upwelling areas with well known phosphorite deposits are the Peru-Chile slope, southwest Africa, northwestern Africa and western Australia. Most of these phosporites are relicts, i.e. they are not forming at present, although both ancient and more recent phosphorites have been reported off southwest Africa.

Phosphorite deposits are usually large and confined either to the continental shelf (Congo, Baja California) or continental slope (Peru-Chile), but none of these deposits is exploited at present. Additional exploratory work is required to identify ancient-bedded phosphorite deposits on the contineta1 margin.

Metalliferous and brine-rich sediments. Spreading centers at mid- ocean ridges are clearly loci for metalliferous and brine-rich sediments which have been observed in crestal areas of many mid-ocean ridges and their adjacent environments. However, rich metalliferous deposits are only known at the Red Sea spreading center, these deposits being rich in residual brine. A distinction is made between metalliferous and brine-rich sediments because the latter deposits contain more than 80% brine.

The metalliferous sediments which are a mixture of fine-grained or colloidal precipitates with normal deep-sea sediments are often well-stratified and carry anomalously high amounts of Fe, Mn, Zn, Ba and some other metals. Counterparts to these deposits have been mined onshore, but the metal assemblage makes them economically unattractive. It is generally thought that Fe-rich sediments are genetically connected to ridge hydrothermal activity. Metalliferous sediments may sometimes occur well developed hundreds of kilometers away from the spreading center in areas where the terrigeneous matter input rates are low, e.g. in the Bauer Deep. At present none of the sediments have any economic importance but they are of importance for the genesis of other mineral deposits, both in the sea and on land. The overwhelming fraction of all oceanic volcanogenic iron, manganese, barium, copper and zinc occurs furthermore in these sediments. A detailed description of metalliferous sediments can be found in Boström (1980), Cronan (1980) and Meylan et al. (1981).

Exceptionally rich metalliferous sediments were discovered in the Red Sea during the International Indian Ocean Expedition. For a long time they were the only known marine hydrothermal sulfide mineralisation, but the discovery of black smokers in the Pacific some ten years ago has made these Red Sea deposits less unique. A more detailed discussion of the Red Sea hot brines and the historical aspects in connection with the exploration of the deposits is found in chapter 7 of this volume.

Basically, seawater enters substrata of the Red Sea basin. The water in this region is locally probably in contact with Miocene salt deposits resulting in a particularly aggressive solution. Due to underlying hot volcanic rocks the fluids are heated and extract metals from rocks and sediments. Near the spreading center the hydrothermal solutions discharge their metals upon re-entry into depressions on the ocean floor as metal sulfide particles. In this way stratiform sulfide deposits have formed in some depressions of the Red Sea central area. Because this process of sulfide generation is similar to that at black smokers on the East Pacific Rise, the reader is also referred to the respective sections and to the large number of scientific reports in the literature.

The layered Red Sea deposits consist of several mineral assemblages including clays, hydrous oxides, carbonates and sulfides, the latter being the most important. The hydrothermal fluids discharge normally into sedimentary strata and not above the sea floor because they are effectively sealed by the brines.

The most important brine deposits are found in the Atlantis II Deep where economic estimates suggest about $3x10^6$ t of metals (Iron excluded). The deposits contain much iron (ca. 29%), copper (ca. 1.3%), zinc (ca. 3.4%) and some silver and gold. The deposit is at present under consideration for mining by the Saudi Arabian-Sudanese Red Sea Joint Commission.

Nodular and crust deposits

Marine barite in the form of nodules was already reported by the "Challenger" expedition, by Danish expeditions in the East Indies and by Russian expeditions off southern India; such deposits and associated marine minerals are described by Church (1979). These occurrences are presently uneconomic, but of interest in the study of the marine chemistry of Ba and related elements.

Phosphorite nodules and crusts. Concretionary layers (condensed sequences) of phosphorites are found on several plateaus or rises, such as the Blake Plateau, the Californian Banks, the Agulhas Bank, and on the Chatham Rise. Another type of phosphorites are crusts on seamounts which often coexist with Co-rich manganese crusts. The latter deposits generally occur in water depths of 800 to 2500 m. They are described in the section on manganese crusts.

Phosphorite nodules probably form by accretion and replacement of carbonate on hiatal surfaces, but the process of formation is not well understood in detail. Phosphorite accretion is furthermore favored by low or no sedimentation. Some of the deposits on submerged seamounts may have formed when initial guano deposits interacted with limestone. Nodular and slab-like phosphorite deposits of late Oligocene to early Miocene age formed probably in organic-rich waters and anoxic sediments in water depths of about 200 m (Manheim and Gulbrandsen, 1979) as replacement of carbonate particles by phosphate phases and winnowing away a clay and dolomite matrix. By such diagenetic alteration of limestone and chalk a solution-resistant lag deposit is formed. The lack of abrasion and erosion, lack of mixing of nodules with rock fragments and the absence of nodule coating by glauconite further supports this interpretation. The hard surfaces of such replacement phosphorites frequently show manganese encrustations developed under oxidising conditions.

The phosphorite deposits on the Chatham Rise are particularly abundant in waters of about 400 m depths (Cullen, 1984) and are best characterised as nodular gravels found within certain sands and muds. Most nodules have a size of 10 to 40 mm in diameter. A surface nodule coverage of about 40 kg/m^2 is reported.

The Chatham Rise phosphorites have been thorougly explored by the New Zealand Oceanographic Institute in cooperation with German institutions and companies, but mining has yet to start (see chapter 7). Water depths pose no problems per se since there exists a feasable mining system (adopted from manganese nodule mining), but the topographical relief and the type of surficial sediment may cause severe difficulties. Thus, underlying Oligocene chalk may clog dredging apparatus, seriously hampering dredging operations (Cullen, 1984).

Manganese nodules. Deep sea manganese nodules were discovered already during the "Challenger expedition (1871-73), and since that time numerous additional deposits have been discovered in the deep sea. As a consequence, the literature on manganese nodules, regarding both scientific studies and economical assessments, has grown enormously particularly since 1960. An update of this literature has been given by Meylan et. al. (1981). Deposits with enormous areal extent were in part the reason why the United Nations Conference on the Law of the Sea (UNCLOS III) was greatly delayed.

The ferromanganese concretions are characteristic of many abyssal plain and abyssal hill regions of the oceans characterized by water depths varying between 3000 and 6000 m; the nodules are particularly abundant on the slopes of abyssal hills. Very large deposits are given in Fig. 2.4 but also other areas with nodule occurrences are known, as shown by, e.g. Cronan (1985) on a more specific map.

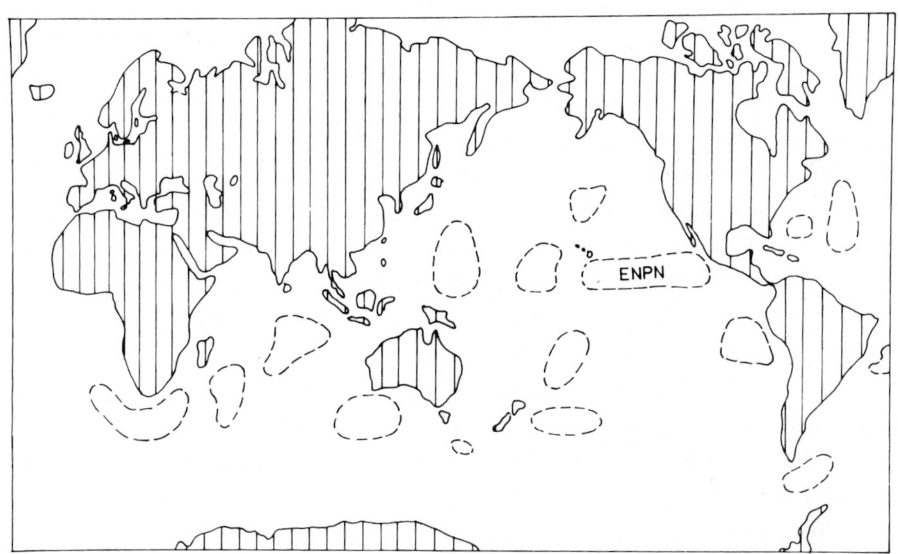

Fig. 2.4. Important manganese nodule areas in the deep oceans. Other areas with generally less nodule coverage are not given in the figure.

Viewed as a deposit, the mineralisation is unique in that it forms one single layer on the present-day ocean bottom surface, the thickness of this layer being about 10 to 50 mm. The deposits show relatively high content of the economically important metals Ni, Cu, Co and Mn, in decreasing order of importance, but other metals may be considered as by-products, such as Mo, V and Ti. According to OETB (1984), the manganese nodule occurrences pose a serious problem to mining because nodule substrata are soft (fine-grained ooze) and they are often affected by bottom water circulation. Comprehensive descriptions and evaluations of these deposits can be found in, e.g., Glasby (1977), Cronan (1980), Fellerer (1980) and McKelvey et al. (1983).

The nodules grow at the sea floor-sediment interface probably by precipitation from seawater. The growth requires low to very low sedimentation rates and oxygen-rich bottom waters, usually below the calcium carbonate compensation depth (CCD).

However, seawater most likely is not the ultimate source of metals in the nodules. Most studies now rather suggest that biological metal uptake in surface waters, subsequent descent of dead biological matter and following diagenetic remobilisation of the deposited elements provide local waters in and on the bottom with the elements in question. As a consequence, some of the trace-metal richest nodules occur at about 10^{0} N, just north of the equatorial high productivity belt in the Pacific. These nodules are usually rich in Ni and Cu, thus the combined content of Ni and Cu in this equatorial North Pacific nodule belt (ENPN) may locally be at or slightly exceed 3%.

However, our general understanding of manganese accretion is much influenced by the deposits in this nodule belt because most detailed studies were conducted in this area. Thus, nodules in the ENPN that are economical, i.e. with a Ni+Cu content larger than 2.5%, occur in areas with significant sedimentary substrata. Those from the slopes and tops of major elevations often have a smaller content of Ni+Cu, making them less economically attractive, but may on the other hand be richer in Co (see further below). This nodule occurrence pattern is the reason why it is difficult to evaluate the economic potential of the deposits from a cursory mapping. Therefore, before any exploitation can be conducted it is necessary to map the ocean floor in detail and determine nodule characteristics both regarding the areal coverage and grade of the nodules.

Nodule growth in general is initiated around nuclei like rock fragments, biogenic remnants (shark tooth, ear boones) or nodule fragments. Erosional surfaces likewise seem to have a large number of seeds needed for nodule growth. Growth rates of manganese nodules are usually reported to be of the order of 1-10 mm/10^6 y.

Nodules were grown artificially at the University of Hawaii, reaching sizes of about 10 mm in diameter in one year. However, these rapidly grown nodules did not adsorb significant amounts of Ni and Cu, which could suggest that a very slow growth is necessary to create normal manganese nodules with a content of Ni+Cu > 0.6% and that special conditions most likely must exist for the growth of economic nodules. As a basic requirement for the growth of the latter deposits it could be said, in a summarised form, that geologically stable areas are necessary with low water temperatures. For further aspects, see Raab and Meylan (1977).

There is a general agreement that mid-ocean ridge hydrothermal activity and related phenomena do not participate directly in nodule formation processes; thus, in ridge areas nodules are remarkably rare although there has been an extensive search for them. Furthermore, the "nodules" that do exist on the ridges are very different from ordinary nodules, both in rate of formation (they apparently form very fast), structure, occurrence, mineralogy and geochemistry. Most probably they are best characterized as hydrothermal slabs or pavements. There are no indications that they will be of any economic value, since they are only known from a few localities.

The paradox that nodules have growth rates that may be 3 orders of magnitude lower than sediment deposition rates is best explained by bottom currents and biologic activity in the sediment surface. However, these explanations rely also on the assumption that growth rate measurements were interpreted correctly. correctly.

TABLE 2.5
Known manganese nodule deposits in the world's oceans (Cronan, 1980).

Ocean	Deposit
Pacific	Equatorial North Pacific nodule belt (ENPN)
	South-western Pacific Basin
	Peru Basin
	Chile Basin
	South Tasman Basin
	Northern Tahiti Basin
Indian	South equatorial basin
	Crozet Basin
	Central Indian Basin
	Western Australia
Atlantic	Western Atlantic
	Drake Passage
	Western South Atlantic
	Blake Plateau

Manganese nodules may be divided into mononucleate and polynucleate species, the latter being formed by coalescence of several nodules. Surface texture of nodules is often mammilated, but there are a great number of nodules that have very smooth surfaces. In this respect, it could be mentioned that many of the economically important nodule occurrences from the ENPN area show smooth surfaces on the sides exposed to seawater while surfaces resting in the sediment often are gritty. An equatorial nobby ring or welt is often observed around the central part of the nodules. A morphological classification of nodules is difficult to make in view of their great diversity in the world oceans (Cronan, 1980) and a generally accepted solution to the problem has not as yet been reached. For practical purposes it must therefore suffice to separate them into mono and polynucleate species with some shape descriptions and division into predefined nodule size classes.

Manganese nodules occur in near-shore, lacustrine, marginal seas and lake environments but the by far largest deposits are those found in the deep ocean. However, lake nodules are the only ones that have been mined on a larger scale and for some centuries, e.g., in Sweden and Finland. An attempt is made in Table 2.5 to comprehend the most important manganese nodule deposits in the world's oceans.

From an economic point of view, areal nodule density (weight of nodules per area unit, see Fig. 2.5) and nodule grade are of paramount importance before any mining step can be carried out. However, the nodule density in the equatorial North Pacific nodule belt is patchy and values of greater than 10

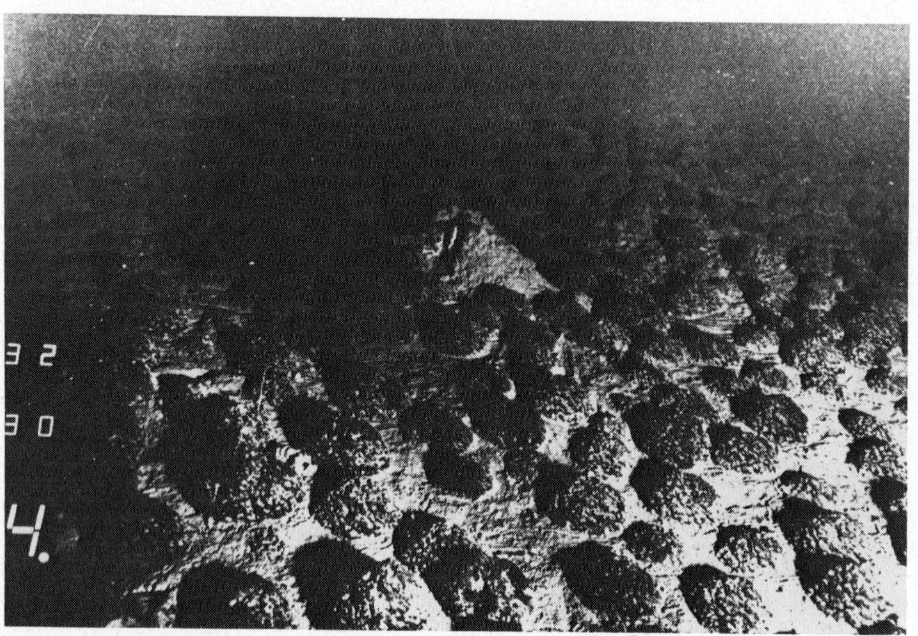

Fig. 2.5. Manganese nodule field in the South Pacific. Nodule size is about 30 mm.

kg/m^2 are common locally, but on a larger scale of several hundreds of km^2, the manganese nodule density is seldom greater than 5 kg/m^2.

The grade of nodules also varies very widely. A geochemical classification of nodule fields (Kunzendorf, 1986) shows that most nodules display a compositional trend for Ni+Cu versus Mn/Fe, see Fig. 2.6, regardless of their depositional environments in the oceans. Economic important nodule deposits have Ni+Cu contents greater than 2% and show Mn/Fe ratios generally exceeding 2. However, no nodules are known with Ni+Cu contents greater than 3.5%, probably due to restricted adsorption capacity of the ferromanganese minerals. Geochemical classifications are especially important during manganese nodule exploration. A discussion about the detailed chemistry of manganese nodules and their mineral phases falls outside the scope of this chapter.

It should be mentioned however that the main manganese phases involved in nodule growth are todorokite, birnessite, and vernadite (Burns and Burns,

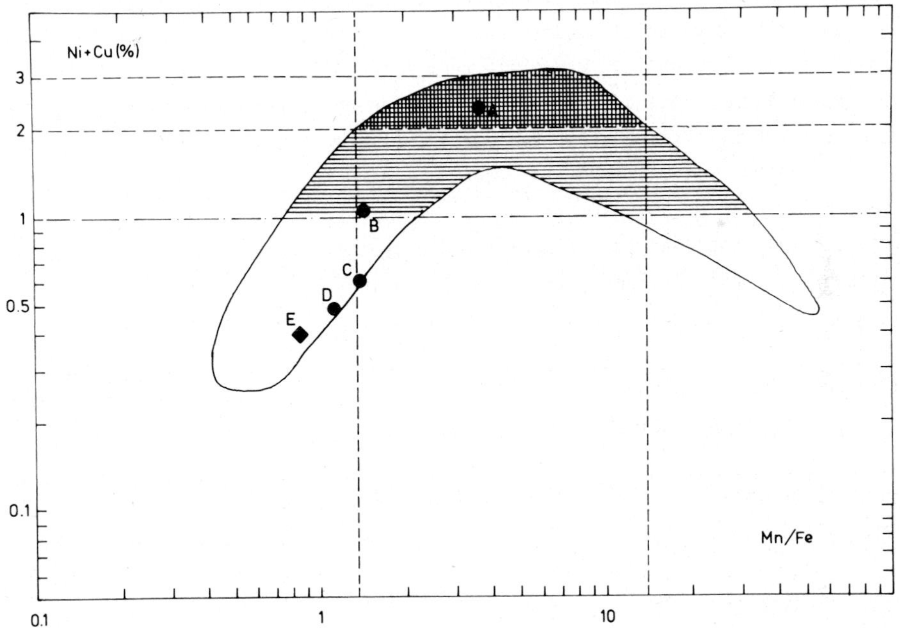

Fig. 2.6. Plot of Mn/Fe vs. Ni+Cu of manganese nodules from the Pacific. The plot is based on over 2000 manganese nodule chemical analyses. The right area of the plot represents nodules from the more near-margin areas of the America. Average data are also plotted: A: ENPN, B: other abyssal plains, C: Mid-Pacific seamounts, D: other seamounts; A to D from Haynes et al.,(1983); E: fossil nodules from Timor (Margolis et al., 1978).

1979). The minerals very often have structural defects. While todorokite (also named buserite and "10Å manganite") is thought to be a main carrier of Ni, Cu, Co and Zn (atomic substitution for Mn^{2+}), birnessite or "7Å manganite" is reported from nodules that occur on topographical elevations. The occurrence of vernadite (δ-MnO_2) in nodules is often invoked by characteristic X-ray diffraction patterns in the region 1.4 to 2.4Å. Because of its known association with feroxyhyte (d-FeOOH) vernadite is the Fe-rich ferromanganese mineral phase observed in many abyssal hill nodules. According to Burns and Burns (1979) this mineral is also characterised by high cation absorption properties leading to enhanced uptake of Co, Ce and Pb. It has also been observed that post-depositional recrystallisation of vernadite occurs within manganese nodules terminating in conversion to todorokite.

The economic deposits of the equatorial North Pacific nodule belt (ENPN, see Fig. 2.4) represent an estimated total of about 10^{10} t of manganese nodules in an area of about $4x10^6$ km^2 (Fellerer, 1980), but only a fraction of these deposits will be mined in the future because of nodule coverage considerations and practical mining aspects. These deposits are nevertheless so large that they have no terrestrial counterpart. Regarding the grade of the nodule deposits most data publicly known are from scientific expeditions with little or no exploration sampling involved. An impression of nodule contents from the nodule belt can be obtained from Table 2.6 (Fellerer, 1980).

TABLE 2.6
Average metal content in selected nodule types from the equatorial North Pacific nodule belt (Fellerer, 1980). All data in %.

Nodule type		Number of samples	Mn	Fe	Co	Ni	Cu	Mn/Fe	Ni+Cu
Mononu-cleate	Discoidal	134	29.3	6.1	0.23	1.38	1.15	4.8	2.53
	Elliptic.	303	28.7	5.8	0.19	1.36	1.15	5.0	2.51
	Späroid.	64	28.8	5.2	0.21	1.32	1.21	5.5	2.53
	Polymorph	78	26.0	6.9	0.21	1.21	1.01	3.8	2.22
	Unspecif.	1447	28.5	6.1	0.19	1.36	1.15	4.7	2.51
Polynu-cleate		19	24.4	8.4	0.23	1.12	0.81	2.9	1.93
Crust		7	24.8	6.1	0.19	1.25	0.99	4.1	2.24

Extensive exploration in the nodule belt has been carried out by several multinational consortia consisting of leading mining companies and metal

TABLE 2.7
Deep sea-mining consortia (after Cruickshank, 1982). Consortia constructions as of 1985, see Chapter 7.

Consortium	Participating company	Status
Kennecott Group	Kennecott, USA Rio Tinto Zinc, UK Consolid. Goldfields, UK British Petroleum, UK Noranda Mines, Canada Mitsubishi, Japan	Nodule belt exploration; mining test
Ocean Mining Associates OMA	US Steel Corp., USA Union Miniere, Belgium Sun Co., USA	Nodule belt exploration; mining test
Ocean Management Inc. OMI	Inco, Canada Sedco, USA Metallgesellsch., FRG Preussag, FRG Salzgitter, FRG 23 Japan. companies	Nodule belt exploration; mining test
Ocean Minerals Co.	Lockheed, USA Billiton, Netherl. Bos Kalis, Netherl. Standard Oil, USA	Nodule belt exploration mining test
AFERNOD	IFREMER, all France CEA BRGM Soc. le Nickel Empain Schneider	Systematic nodule belt exploration
CLB Syndicate	20 companies including above mentioned	Mining ship tests
Eurocean	24 European companies (France, Belgium, UK, Netherl., Italy, Sweden, Norway, Spain, Switzerl.	Non-commercial, scientific purpose

producers. Presently known or recently active consortia are given in Table 2.7, but an up-to date comprehension can be found in chapter 7. Most of these consortia have already conducted mining tests and developed manganese nodule processing schemes, but no decision has been taken (as of 1985) to start commercial managanese nodule mining.

Co-rich manganese crusts. Much interest is presently focussed on occurrences of Co-rich ferromanganese crusts on seamounts and oceanic islands, some of which occur in the EEZ areas of the United States. As cobalt is regarded a strategic metal to be largely imported to the U.S., mining of these deposits would lessen the foreign dependence on Co. Their potential however is at present impossible to evaluate because of lack of sufficient data.

These ferromanganese crust deposits occur in particular on terraces and submerged platforms around the Hawaiian Islands, the western Atlantic and the Carribean (Commeau et al., 1984). They form generally in waters with depths between 800 and 2500 m, but highest Co content was observed in crusts from still larger depths, below 2500 m. Commeau et al. (1984) calculate the central Pacific crust potential as to be about $5x10^8$ t of Co. The largest deposits were found in the Palmyra/Kingman Reef and the Johnston Island regions. The average thickness of the crusts is about 2 cm with average Co contents of in general 0.6-1%, but much exploratory work is left before these deposits can be evaluated properly. Exploration will probably be stimulated considerably, since Co prices are more than twice than those of Ni, and because of probable Pt contents in the crusts.

TABLE 2.8
Phosphorus content of Blake Plateau deposits (Commeau et al., 1984).

Deposit	Thickness (cm)	P (%)
Fe/Mn phosphor. pavement	0.5-3	0.5-0.75
Nodule-slab pavement	1-5	0.5-0.75
Nodules, subregional	2-4	0.02-0.2

While the deposits on the New England Seamounts and in the Carribean need to be investigated more thoroughly, much exploratory work from the 1960s and 1970s was reported from the Blake Plateau (off Florida and South Carolina). The Blake Plateau is a terrace composed of mainly Mesozoic and Tertiary sedimentary strata where erosional surfaces have formed since Middle Tertiary time. As a result "lag" deposits of variable composition and form were created in Oligocene-Miocene, consisting of nodules, slabs and unbroken pavement. In the southwestern part of the plateau nodules were discovered at water depths of about 800 m and later the U.S. Geological Survey observed that the substrata of the nodule fields were ferromanganese-encrusted phosphorite pavements. Such pavements were also reported from the western margin of the plateau. Typical phosphorus contents of Blake Plateau deposits are given in Table 2.8.

The nodules commonly have about 0.3% Co. However, economic calculations are difficult to accomplish due to lack of representative samples of the unbroken pavement which covers large areas of the plateau. Although the Blake Plateau deposits were discovered already in 1885, much is therefore left in the evaluation of the economic potential of this newly recognized marine resource. For many other ocean regions with similar physiographic conditions no data on ferromanganese deposits exist even now.

Massive sulfide deposits at mid-ocean ridges

Hydrothermal activity was suspected already in the early 1960-ies to occur at mid-ocean ridges because of the heat production budgets and the composition of sedimentary deposits on the ridges; only hydrothermal processes and components were thought to explain these phenomena (Boström et al., 1969; Bonatti, 1975; Boström, 1980). However, not until in 1977 were hot springs localised and studied on site on the ocean floor near Galapagos Islands. The discovery and the chemistry, and the activity of these springs are excellently described by Edmond and Von Damm (1983). Many additional discoveries of hot springs, black smokers and associated massive sulfide deposits have taken place since that time. These discoveries have revolutionised the whole branch of economic geology and greatly stimulated economically related marine exploration. Investigation and mapping of these deposits and the mid-ocean ridges in general is a still ongoing research and the near future may well see the discovery of deposits from other ridges than the East Pacific Rise.

Once massive sulfides were discovered on the ridges a number of efforts were directed to compare them with polymetallic sulfide deposits mined on land, since many of these deposits could be ascribed a marine origin. However, almost all of the newly discovered ridge deposits are rather small in size, making them uneconomical at present. Speculations exist that ongoing hydrothermal activity is only one step in marine massive sulfide formation. There are most probably other metal accumulations as well in the deep sea environment, not yet discovered because shielding strata and lack of proper exploration tools; so far less than 1% of the global spreading ridge system has been investigated in some detail. Intensified and internationally coordinated efforts are obviously needed if the exploration pace shall increase significantly.

Numerous investigations and articles were published on hydrothermal massive sulfides connected with seafloor spreading since the first discovery of hot springs at the Galapagos spreading center. Most of the literature is concerned with the deposits at the East Pacific Rise (EPR) or the Mid-Atlantic Ridge (MAR) (e.g. Rona, 1978; Edmond et al., 1979; CYAMEX, 1979; Hekinian et al., 1980; RISE, 1980; Hekinian, 1980; Macdonald, 1982; Malahoff, 1982; Duane, 1982; Bischoff et al., 1983; Rona et al., 1983; Bonatti, 1983; Rona, 1984; Janecky and Seyfried, 1984; Bäcker et al., 1985, Teleki, 1986).

A comprehensive study in describing and classifying the hydrothermal mineralisation at seafloor spreading centers in general can be found in Rona (1984), who distinguished three types of sulfide deposits, see Table 2.9.

TABLE 2.9
Classification of hydrothermal sulfide deposits at mid-ocean ridges (Rona, 1984). All values refer to total sulfide content.

Type	Characteristic	Example
Stratiform sulfide bodies	Deposition in a tight hydroth. system; early opening center reducing depos. condit. metallif. sediments	Atlantis II Deep; ca. $3x10^7$ t (including Fe)
Sulfide mounds surmounted by chimneys	Deposition under oxid. conditions; along linear section of axial zone; intermed.-fast spread.	EPR deposits; Cu-Fe-Zn; ca. 10^3 t
Coalesced mounds surmounted by chimneys	Ibid. along inner marginal zone	Galapagos Spreading Center; relatively large deposits

While stratiform sulfide bodies in the Red Sea (see section on metalliferous sediments) represent deposition of hydrothermal solutions under reducing conditions connected with slow spreading (early stage of opening of a spreading center) the two other types are clearly deposited under oxidising conditions at intermediate to fast spreading centers. Red Sea deposits are nearly three orders of magnitude larger than the two other deposit types. EPR-deposits are found in the form of sulfide mounds surmounted by chimneys along linear sections of the axial zone of volcanic extrusion. Galapagos Spreading Center massive sulfide deposits which generally are larger than EPR-deposits are found along the linear marginal zone of active extrusion in the form of coalesced mounds surmounted by chimneys.

Bonatti (1983) classifies the known deposits according to the discharge stage of hydrothermal solutions into pre-discharge, syn-discharge and post-discharge deposits. Pre-discharge hydrothermal deposits are those that are found mainly as stockwork-disseminated metal sulfides, metal sulfides and disseminated oxides exemplified by massive sulfides occurring in metabasalts and veins, e.g. as observed in DSDP drill cores. Syn-discharge hydrothermal deposits are accordingly black smokers composed of mainly sphalerite, pyrite and chalcopyrite, and oxides and hydroxides in the vicinity of the vents. Post-discharge deposits are formed from those metals that stay in solution after discharge and are precipitated after some residence time. Considering the residence time for Mn, about 50 years, hydrothermal plumes may travel for 1000 km in seawater. The metalliferous deposits of the Nazca Plate and the Bauer Deep are examples of post-discharge deposits according to this classification.

Edmond (1986) also defines three distinct physiographic areas of the mid-ocean ridges for massive sulfide deposit formation:

1) mature, open ocean, sediment-starved ridge axes,
2) rifted margins, intracratonic rifted basins, extensional back arc regimes and areas close to active margins, and
3) young seamounts.

Depositional environment (1) produces ophiolites, black smokers (EPR and Juan de Fuca Ridge) and metalliferous sediments. Examples of depositional environments (2) are Red Sea and Gulf of California deposits. Massive sulfides were recently also found to be connected with young seamounts (e.g. 13° N, EPR).

It should be realised, however, that the quantities of volcanogenic iron, manganese, copper and zinc implied in Table 2.9 are dwarfed by the total quantities in ordinary unconsolidated hydrothermal deposits on the ridges and their flanks in the whole world oceans. The exact figures are not yet possible to calculate, but the order of magnitudes could be about as follows: Fe = 1000, Mn = 500, Cu = 50 and Zn = 5, all in units of 10^{10} tons. It is thus obvious that only a tiny fraction of all volcanic material forms potential deposits of economic value.

While the question of a valid classification has to be settled in the future, there is still a considerable consensus on the processes that generate ocean-ridge-related massive sulfides. These chemical reactions at the sea floor are schematically shown in Fig. 2.7.

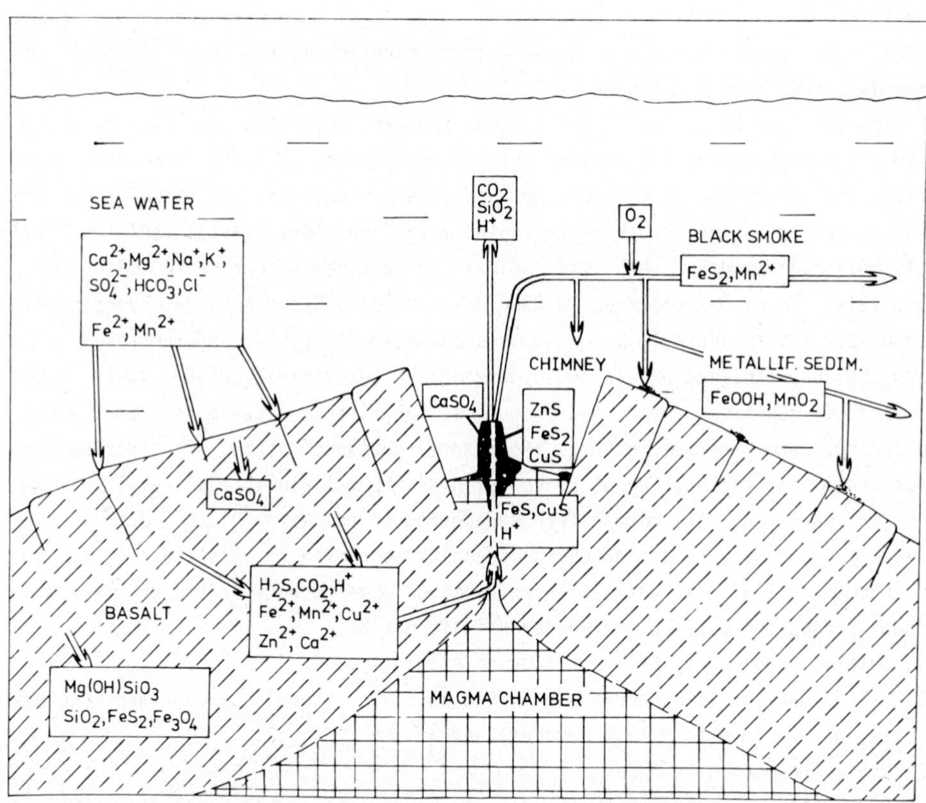

Fig. 2.7. Schematic diagram of processes at the disposition centers for massive hydrothermal sulfides.

Briefly, cold seawater penetrates basaltic substrata via cracks and tectonic structures; penetration depths of several kilometers have been conceived (Lister, 1977). Seawater is progressively warmed up and reacts with basalt leading to the extraction of metals, such as Li, K, Rb, Ca, Ba, Cu, Fe, Mn, Zn and Si, whereas SO_4^{2-} and Cl are added to the basalt. Mg and Na are as a rule added to the basalts in large quantities, but also cases of losses of these metals are known. Chloride greatly facilitates the metal transport because of the solubility of metal-chloride complexes. Additional volatiles are brought to the solutions from the magma and ascension of a metal-enriched hydrothermal solution is the final result. The metals precipitate either within basalts (layer 3) as massive, disseminated and stockwork polymetallic sulfides (Cu, Fe, Zn, Ag and Au), on top of the basalts or in overlying sediments (layer 2). Under oxidizing conditions Fe-rich silicates, oxides and hydroxides, and Mn-rich oxides and hydroxides are precipitated. Amorphous ferric and manganeous hydroxide particles with high adsorption capacities settle as metalliferous sediments. Because black smoker massive sulfide deposits are generated in a predominantly oxidized mid-ocean ridge environment, a decay of sulfide deposits must occur after the cessation of the hydrothermal deposition. It is inferred that hydrothermal activity is of the order of years to tens of years and therefore, if the massive sulfide chimneys not are covered by lava flows, they will simply disintegrate with time. However, a given hydrothermal field is likely to be active for some 100-1000 years as a comparison with Icelandic conditions suggest (Barth, 1950).

According to Rona (1984), the following conditions have a major influence on hydrothermal deposition at spreading centers:

- Thermal gradients
- Permeability
- System geometry
- Timing of events

The heat from a magma chamber creates high thermal gradients which are the driving force of the hydrothermal system. High permeability is created by fractures and faults. The timing of the geological events, i.e. the relation between the deep-seated magma and the extrusive activity probably has an influence on the hydrothermal processes. It also appears likely that system geometry, e.g. sea floor topography will have an effect on these hydrothermal processes. Thus Rona (1984) has presented convincing arguments that suggest that hydrothermal orifices on MAR are structurally well controlled to certain features of the rift valley of the MAR.

However, this problem is more complex than may appear at first. Thus, it is of interest to note that MAR with its very rugged topography has far weaker hydrothermal activity than is the case of the EPR with its comparatively smooth topography. Bischoff et al. (1983) have compared the metal contents of some massive sulfide deposits (Table 2.10).

It is seen that these are Zn deposits with some silver and gold, while the Galapagos spreading center deposits have typically 5% Cu and little Zn. These authors also compare the metal values of the sulfide deposits with manganese nodules and arrive at metal values per ton of 338, 73 and 182 USD for EPR (21° N), Galapagos rift and manganese nodule deposits, respectively. However, such comparisons must also involve other economic factors not known at present.

TABLE 2.10
Selected average metal contents of some selected massive sulfides (Bischoff et al., 1983).

Metal	EPR 21° N basal mound	EPR 21° N CYAMEX	Juan de Fuca
Zn (%)	32.3	40.8	54.0
Cu (%)	0.81	0.61	0.22
Pb (%)	0.32	0.05	0.25
Ag (ppm)	156	380	260
Au (ppb)	170	-	130
As (ppm)	489	-	323
Cd (ppm)	560	500	775

Because there is ongoing exploration both in the Pacific and Indian Oceans, the massive sulfide deposits possibly occurring at the mid-ocean ridges and on young submarine volcanoes will further extend our knowledge on these deposits. It is therefore not within the scope of this chapter to speculate longer than necessary, but only refer to the additional literature in Appendix I.

REFERENCES

Bäcker, H., Lange, J. and Marchig, V., 1985. Hydrothermal activity and sulphide formation in axial valleys of the East Pacific Rise crest between 18 and 22° S. Earth. Planet. Sci. Lett., 72: 9-22.

Baker, G., 1962. Detrital heavy minerals in natural accumulates. The Australian Institute of Mining and Metallurgy, 145 pp.

Barth, T.F.W., 1950. Volcanic geology, hot springs and geysers of Iceland. Carnegie Inst. Washington Publ., 587, 174 pp.

Beauchamp, R.G. and Cruickshank, M.J., 1983. Placer minerals on the U.S. continental shelves - opportunity for development. Proceed. Oceans`83, Marine Technology Society, pp. 698-702.

Bischoff, J.L., Rosenbauer, R.J., Aruscavage, P.J., Baedecker, P.A. and Crock, J.G., 1983. Sea-floor massive sulfide deposits from 21° N; Juan de Fuca Ridge; and Galapagos Rift: bulk chemical composition and economic implications. Econ. Geol., 78: 1711-1720.

Bonatti, E., 1975.Metallogenesis at oceanic spreading centers. Ann. Rev. Earth and Planet. Sci, 3: 401-431.

Bonatti, E., 1983. Hydrothermal metal deposits from the oceanic rifts: a classification. In: P.A.Rona, K.Boström, L.Laubier and K.L.Smith (Editors), Hydrothermal processes at seafloor spreading centers. Plenum, New York, pp.

491-502.
Boström, K., 1980. The origin of ferromanganoan active ridge sediments. In: P.A. Rona and R.P. Lowell (Editors), Seafloor spreading centers: hydrothermal systems. Hutchinson & Ross, Stroudsburg, PA, pp. 288-332.
Boström, K., Peterson, M.N.A., Joensuu, O. and Fisher, D.E., 1969. Aluminium-poor ferromanganoan sediments on active oceanic ridges. J.Geophys.Res., 74: 3261-3270.
Boström, K., Lysen, L. and Morre, C., 1978. Biological matter as a source of authigenic matter in pelagic sediments. Chem. Geol., 23: 11-20.
Burns, R.G., 1979. Marine minerals. Reviews in Mineralogy, Vol. 6. Mineralogical Society of America, 379 pp.
Burns, R.G. and Burns, V.M., 1979. Manganese oxides. Ibid., pp. 1-45.
Burns, V.M., 1979. Marine placer minerals. Ibid., pp. 347-380.
Charlier, R.H., 1983. Water, energy, and nonliving resources. In: E.Mann Borgese and N.Ginsburg (Editors), Ocean Yearbook 4. The University of Chicago Press, pp. 75-120.
Church, T.M., 1979. Marine barite. In: R.G.Burns (Editor), Marine minerals. Reviews in Mineralogy, Vol. 6. Mineralogical Society of America, pp. 175-209.
Claque, D., Bischoff, J, and Howell, D., 1984. Nonfuel mineral resources of the Pacific Exclusive Economic Zone. Proceed. Oceans`84, Marine Technology Society, Vol. 1, pp. 438-443.
Commeau, R.F., Clark, A., Johnson, C., Manheim, F.T., Aruscavage, P.J. and Lane, C.M., 1984. Ferromanganese resources in the Pacific and Atlantic oceans. Ibid., pp. 421-430.
Cronan, D.S., 1980. Underwater minerals. Academic Press, London, 362 pp.
Cronan, D.S., 1985. A wealth of sea-floor minerals. New Scientist, 6 June, pp. 34-38.
Cruickshank, M.J., 1978. Technological and environmental consideration in the exploration and exploitation of marine minerals. Ph.D. dissertation, University of Wisconsin-Madison, 217 pp.
Cruickshank, M.J., 1982. Recent studies on marine mineral resources. Oceanology International, Vol. 1, paper OI 82 1.2.
Crutchfield, J.A., 1981. Mineral resources. In: G.M.Brown and J.A.Crutchfield (Editors), Economics of ocean resources. Washington Sea Grant Publication, pp. 14-26.
Cullen, D.J., 1984. Comments on the economic potential and mining feasibility of the Chatham Rise phosphorite deposits. Proceed. Offshore prospecting and mining problems: current status and future directions. Germinal, Brest, France, pp. 287-299.
CYAMEX, 1979. Massive deep-sea sulphide ore deposits discovvered on the East Pacific Rise. Nature, 277: 523-528.
Duane, D.B., 1982. Elements of a proposed five-year research program on polymetallic sulfides. Mar. Tech. Soc. J.,16,3: 87-90.
Earney, F.C.F., 1980. Petroleum and hard minerals from the sea. Winston & Son, London, 287 pp.
Edmond, J.M., Measures, C., McDuff, R.E., Chain, L.H., Collier, R., Grant, B., Gordon, L.I. and Corliss, J.B., 1979. Ridge crest hydrothermal activity and the balance of the major and minor elements in the ocean: The Galapagos data. Earth Planet.Sci.Lett., 46: 1-18.
Edmond, J.M. and Von Damm, K., 1983. Hot springs on the ocean floor. Scientif.Am., 3: 70-85.
Edmond, J.M., 1986. The chemistry of submarine ore-forming solutions. In: P.Teleki (Editor), Marine minerals: resource assessment strategies. Reidel, in press.
Emery, K.O. and Skinner, B.J., 1977. Mineral deposits of the deep ocean floor. Marine Mining, Vol. 2, No. 1/2: 1-71.
Fellerer, R., 1980. Manganknollen. In: W.Schott (Editor), Die Fahrten des Forschungsschiffes "Valdivia" 1971-1978. Geol. Jb. D38: 35-76.
Galtier, L., 1984. Activites minieres en mer: le point au plan mondial. Proceed. offshore prospecting and mining problems: current status and

future directions. Germinal, Brest, France, pp. 29-45.
Glasby, G.P., (Editor), 1977. Marine manganese deposits. Elesevier, Amsterdam, 312 pp.
Haynes, B.W., Law, S.L. and Barron, D.C., 1983. Mineralogical and elemental description of Pacific manganese nodules. U.S. Dept. of the Int., Bur. Mines, Inform. Circ. 8906, 59 pp.
Heezen, B.C. and Menard, H.W., 1963. Topography of the deep-sea floor. In: M.N. Hill (Editor), The sea. Interscience Publ., Vol. 3, chapt. 12.
Hekinian, R., Fevrier, M., Bischoff, J.L., Picot, P. and Shanks, W.C., 1980. Sulfide deposits from the East Pacific Rise near 21° N. Science, 207: 1433-1444.
Holser, A.F., McGregor, B.A., Rowland, R.W. and Goud, M.R., 1984. United States government initiatives in the assessment and development of mineral resources of the Exclusive Economic Zones of the United States. Proceed. Offshore prospecting and mining problems: current status and future directions. Germinal, Brest, France, pp. 75-91.
Janecky, D.R. and Seyfried, W.E., 1984. Formation of massive sulfide deposits on oceanic ridge crests: incremental reaction models for mixing between hydrothermal solutions and seawater. Geochim.Cosmochim. Acta, 48: 2723-2738.
Kent, P., 1980. Minerals from the marine environment. Edward Arnold, London, 88 pp.
Kulvanevich, S., 1984. Present and future offshore tin prospecting and mining in Thailand. Proceed. Offshore prospecting and mining problems: current status and future directions. Germinal, Brest, France, pp. 247-258.
Kunzendorf, H., 1986. Geochemical methods in manganese nodule exploration. In: P.Teleki (Editor), Marine minerals: resource assessment strategies, Reidel, in press.
Lister, C.R.B., 1977. Qualitative models of spreading-center processes, including hydrothermal penetration. Tectonophysics, 37: 203-218.
Macdonald, K.C., 1982. Mid-ocean ridges: fine-scale tectonic, volcanic and hydrothermal processes within the plate boundary zone. Annu.Rev.Earth Planet.Sci., 10: 155-190.
Malahoff, A., 1982. A comparison of the massive submarine polymetallic sulfides of the Galapagos Rift with some continental deposits. Mar. Tech. Soc. J., 16, 3:39-45.
Manheim, F.T. and Gulbrandsen, R.A., 1979. Marine phosphorites. In: R.G.Burns (Editor), Marine minerals. Reviews in Mineralogy, Vol. 6. Mineralogical Society of America, pp. 151-174.
Margolis, S.V., Ku, T.L., Glasby, G.P., Fein, C.D. and Audley-Charles, M.G., 1978. Fossil manganese nodules from Timor: Geochemical and radiochemical evidence for dee-sea origin. Chem. Geol., 21: 185-198.
McKelvey, V.E., Wright, N.A. and Bowen, R.W., 1983. Analysis of the world distribution of metal-rich subsea manganese nodules. U.S. Geolocial Survey Circular 886, 53 pp.
Meylan, M.A., Glasby, G.P., Knedler, H.E. and Johnston, J.H., 1981. Metalliferous deep-sea sediments. In: K.H.Wolf (Editor), Handbook of stratabound and stratiform ore deposits. Elsevier, Amsterdam, pp. 77-108.
Meylan, M.A., Glasby, G.P. and Fortin, L., 1981. Bibliography and index to literature on manganese nodules (1861-1979). Department of Planning and Economic Development, Hawaii, 530 pp.
Mero, J.L., 1965. The mineral resources of the sea. Elsevier, Amsterdam, 312 pp.
Moore, J.R., 1983. Marine hard mineral resources - progress and problems. Proceed. Oceans`83, Vol. III, pp. 1145-1150.
Narumi, Y., 1984. Latest marine aggregates collection method from dee sea in Japan. Proceed. Offshore prospecting and mining problems: current status and future directions. Brest, France, pp. 119-142.
OETB, 1984. Aanalysis of exploration and mining technology for manganese nodules. United Nations Ocean Economics and Technology Branch, Graham & Trotman, London, 140 pp.

Raab, W.J. and Meylan, M.A., 1977. Morphology. In: G.P. Glasby (Editor), Marine manganese deposits. Elsevier, Amsterdam, pp. 109.146.
RISE, 1980. East Pacific Rise: hot springs and geophysical experiments. Science, 207: 1421-1444.
Rona, P.A., 1978. Criteria for recognition of seafloor hydrothermal mineral deposits. Econ.Geol., 73: 135-160.
Rona, P.A., 1984. Hydrothermal mineralization at seafloor spreading centers. Earth Science Reviews, 20: 1-105.
Rona, P.A., Boström, K., Laubier, L. and Smith, K.L. (Editors), 1983. Hydrothermal processes at seafloor spreading centers. Plenum, New York, 791 pp.
Rowland, T.J. and Cruickshank, J., 1983. Mining for phosphorites on the United States outer continental shelf. Proceed. Oceans`83, Marine Technological Society, pp. 703-707.
Schott, W, 1976. Mineral (inorganic) resources of the oceans and ocean floors: a general review. In: K.H.Wolf (Editor), Handbook of stratatbound and stratiform ore deposits. Elsevier, Amsterdam, 3, pp. 255-294.
Seibold, E. and Berger, W.H., 1982. The sea floor. Springer, Berlin, 288 pp.
Sujitno, S., 1984. Exploration for offshore tin placers in Indonesia. Proceed. Offshore prospecting and mining problems: current satatus and future directions. Germinal, Brest, France, pp. 211-244.
Teleki, P., (Editor), 1986. Marine minerals: resource assessment strategies. Reidel, in press.
Wilde, P., 1973. Physical oceanography. In: A.B.Cummins and I.A.Given (Editors), SME Mining Engineering Handbook. Society of Mining Engineers of the American Institute of Mining, Metallurgy and Petroleum Engineers, New York, Vol. 2, chapter 20.1.1-20.1.3.

CHAPTER 3

RESEARCH VESSELS AND SUBMERSIBLES

C. AAGE

INTRODUCTION

The world's fleet of ocean research vessels comprises about 700 ships from about 80 countries, and the number is rapidly growing. The most important ocean research fields covered by ships are: oceanography, meteorology, hydrology, biology, fishery, geology, and seismology. Ocean research takes place also from aeroplanes, satellites, buoys, and land-based stations.

Fig. 3.1 Geotechnical Research Vessel.

STATISTICS OF RESEARCH VESSELS AND SUBMERSIBLES

This chapter is restricted to the description of marine mineral research ships and submersibles and their navigational systems. But as most research vessels are indeed multipurpose vessels, or can easily be converted from one function to another, some statistics covering all types of research vessels are given in the following tables, based on data from Ocean Yearbook 2 (1980).

Research vessels

Table 3.1 leads to the following conclusions: Three nations, U.S.A., U.S.S.R., and U.K. own nearly half of the world's total ocean research fleet, and ten nations own two-thirds. A typical research vessel is about 40 m long, it has a displacement of 1800 t, and a crew of 20.

The average numbers in this and the following tables can be used only to provide a general picture, because the statistical spreading is very large. Data on specific vessels can often be more informative than average numbers. For this purpose, reference is made to Ocean Yearbook, Jane's Ocean Technology, and registers of the classification societies.

TABLE 3.1

Research vessels - numbers and average dimensions. Note that the statistical material is incomplete. The average dimensions are based on less than half of the vessels. Usually, the most complete data have been supplied for the larger ones. Most data are from 1977.

Country	No.	Displ.(t)	Length(m)	HP	Crew	Berths
U.S.A.	140	1272	42	1994	27	36
U.S.S.R.	118	2816	73	2336	54	88
U.K.	48	1061	49	1189	18	32
India	31		18	210	11	11
Brazil	28		26	63		
France	24	800	29	792	13	12
Italy	18	572	21	463	7	7
Canada	16	1645	45	1993	26	
Chile	16		17			
Argentina	14		38			
Others	241					
World	694	1789	39	1110	19	25

TABLE 3.2

Research vessels - usage in different research fields. Note that the multipurpose group also includes ships with data missing regarding their specific research fields. Most data are from about 1977.

Country	No.	A	B	C	F	G	H	M	O	S	X
U.S.A.	140		2	26	9		9		36	1	57
U.S.S.R.	118	2		1	10	3	20	4	19		59
U.K.	48		2	3	4	8	5		6		20
India	31			7	10						14
Brazil	28				3				1		24
France	24		1		2		1		10		10
Italy	18		1	2	3				1		11
Canada	16				9		3		3		1
Chile	16										16
Argentina	14								1		13
Others	241		2	18	17	2	12	1	11	3	175
World	694	2	8	57	67	13	50	5	88	4	386

Legend: A: satellite/space B: biology C: coastal work F: fishery
G: geology H: hydrology M: meteorology O: oceanography
S: seismology X: multipurpose

For each of the countries in the table and for the world as a whole the largest group of research vessels is by far the multipurpose one. Even though many vessels have fallen into this group simply because of lack of information regarding their specific use, it is still true that the majority of ocean research vessels are indeed multipurpose. In many cases several different research tasks will have to be carried out during the same expedition, or the ship will have to be converted between the voyages for new jobs. A vessel may often be required to carry out research tasks that were unforeseen at the time of construction.

Other large groups of research vessels are used for oceanography, fishery research, coastal work, and hydrology. Research vessels used in marine mineral exploration work fall mainly in the geology and seismology groups with a total of 17 vessels (1977). But many of the vessels actually working in this field will be found in the multipurpose group.

Submersibles

There are several types of research submersibles in use for different purposes. The main two groups are the manned and unmanned submersibles.

The manned can be of the free-swimming or the tethered type, and they can be either dry or wet.

In a one-atmosphere dry submersible the pilot and scientists are encapsulated in a pressure hull which allows their bodies to remain at surface pressure during the diving operation. The pressure hulls, which may have to resist very large external pressures depending on the diving depth, are for structural reasons usually cylindrical for relatively shallow depths and spherical for deeper depths. The pressure hull will most often be only a small part of the whole submersible. Machinery, ballast tanks and the outer hull can be exposed to the ambient pressure without problems.

In an ambient pressure dry submersible the crew will be subjected to the ambient pressure at the work site. This allows them to leave the submersible if they carry proper diving equipment. The ordinary diving bell may be called an ambient pressure tethered submersible.

A wet submersible requires that the crew wear diving equipment. Very small

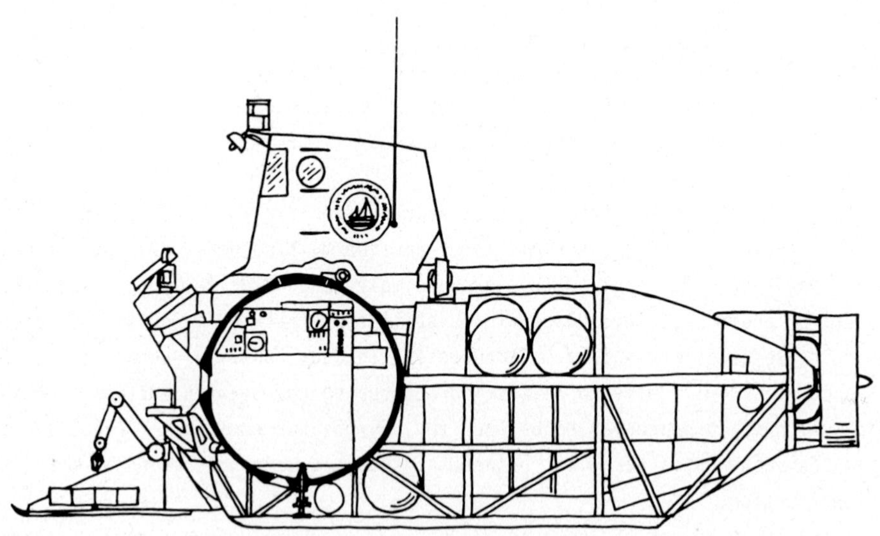

Fig. 3.2. The submersible Alvin of Woods Hole Oceanographic Institution (Moderated drawing after Oceanus, Vol.25, No.1, 1982).

wet submersibles on which the divers ride or are towed through the water are normally called diver transport vehicles.

The unmanned submersibles are usually referred to as remotely operated vehicles or ROVs, a rapidly growing class of submersibles.

TABLE 3.3
Tabulation of manned submersibles. The statistical material is not complete. Displacements and lengths are average numbers. Maximum depth capabilities and crew numbers are the lowest and highest values for each country. Most data are from about 1977.

Country	No.	Displ.(t)	Length(m)	Max.Depth(m)	Crew
U.S.A.	29	40	9	10 - 6060	1 - 7
France	15	64	12	244 - 11000	2 - 10
U.S.S.R.	8	18	7	250 - 2000	2 - 5
U.K.	8	11	7	335 - 916	2 - 5
Canada	8	45	9	76 - 2000	1 - 6
Germany,F.R.	3	9	6	250 - 300	2 - 4
Italy	3	10	6	300 - 365	2
Japan	2	49	11	300 - 600	3 - 4
Norway	2			500 - 1000	
Poland	1	1		200	2
Taiwan	1	14	7	300	
Colombia	1	9	6		2
World	81	35	9	10 - 11000	1 - 10

Twelve nations use manned submersibles in their ocean research work. Two of them, U.S.A. and France, own more than half of the world's total submersible research fleet. A typical research submersible is 9 m long, has a displacement of 35 t, a crew of 3, and can dive to 1000 m water depth.

Manned research submersibles have been used for many years in the exploration of the sea. They give the scientist an opportunity to see the objects of his study in situ with his own eyes, but manned submersibles are very expensive to build and operate.

During the 1980's small remotely operated vehicles, the ROVs, have taken over a very large part of the jobs that were carried out earlier by manned submersibles. An ROV is much cheaper than any manned submersible, and it can be operated by persons who are far less educated than submarine pilot. Besides a video camera for inspection work many ROVs are also equipped with manipulators that can carry out different mechanical jobs. The underwater video cameras are so sensitive that they can produce clear pictures under light conditions in which the human eye cannot see at all.

The ROV development is a result of massive research efforts in hydrodynamics and control theory made possible by the use of fast microcomputers.

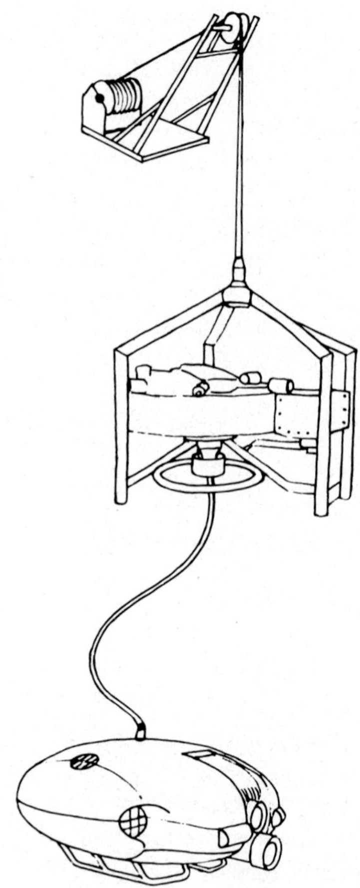

Fig. 3.3. An ROV with deployment cage and hoist.

A typical ROV is about 2 m long, it has a mass of 500 kg and a lifting capacity of 1000 N. It can operate at a maximum water depth of 1000 m with a 150 m tether connected to a deployment cage. It has a colour video camera and a 4-function manipulator. Its maximum speed is about 1 m/s. If not permanently installed on board the vessel, the ROV is usually transported in a standard 20

TABLE 3.4

Tabulation of remotely operated vehicles - ROVs. The statistics are based on data from "Offshore" April 1985. The numbers represent a growth of one third over the 1984 statistics.

Country	No.	Max.Depth(m)	Max.Load(N)	Manip.Functions
U.S.A.	110	460 - 2500	90 - 2600	0 - 7
Canada	80	120 - 2500	90 - 2600	0 - 7
U.K.	78	40 - 1800	90 - 4500	0 - 7
Norway	32	250 - 2500	220 - 2500	0 - 7
France	7	360 - 900	220 - 2600	0 - 5
Australia	4	360 - 900	- 130	0 - 3
Denmark	4	360 - 1500	90 - 2600	4 - 7
Singapore	2	- 250	-	0 - 3
World	317	40 - 2500	90 - 4500	0 - 7

ft container which is then used as the control center during operations. Some ROVs are so small that they can be carried as airplane luggage by two persons.

REQUIREMENTS OF RESEARCH VESSELS

The research vessel is a working platform for the scientists who carry out the ocean research work and for the equipment used in that research. The amount of time spent on scientific work at sea is a basic measure of the efficiency of the vessel. This should be borne in mind when comparing different ships or different solutions to a specific problem. An otherwise good research vessel will be inefficient if the work is interrupted frequently due to excessive roll motion which causes the scientists to be seasick, or if computer failures occur due to electrical disturbances.

Flexibility is a key requirement of any research vessel. Scientific objectives and the available equipment will change many times during the lifetime of a ship, and it should be designed to meet these changes.

In the following, a list of requirements to ocean research vessels is given with some comments. More detailed discussions of this subject can be found in Rosenblatt (1960) and in King & Kaysen (1984).

Main dimensions

The main dimensions of a research vessel are determined by the space required for the scientific work, such as laboratories, open deck areas, accomodations, storage space, etc. The upper limits are usually dictated by the available economic funds.

Generally speaking, a large ship will perform better than a small one. For given wave conditions the large ship will be quieter and therefore provide more working days per year. Its transit voyages will be faster, its fuel economy better, and the necessary crew relatively smaller compared to the scientific payload. On the other hand, there may be cases where two small vessels are preferred to one large one, because they can spread the activities geographically and scientifically. The size may also be limited due to the water depth and harbour facilities in some areas.

Special ship types such as catamarans, Small Waterplane Area Twin Hull (SWATH) ships and semisubmersibles may be superior to the conventional mono-hull ships in certain respects, e.g. ship motions and deck area, but they are usually too expensive to be considered.

General arrangement

In addition to all the installations that are standard in any ocean-going ship, a research vessel must provide adequate living space for the scientists and room for essential equipment. The laboratory facilities, computers, and the offices and accomodations for the scientists should be placed in the quietest areas, in terms of engine and propeller noise and in terms of ship motions. The optimum position with regard to ship motions will be around amidships and close to the waterline. With regard to noise the higher forward decks are preferred.

Deck space

Some of the most important functions of a research vessel require open weather deck space, preferably with free access to both sides of the ship and over the stern:

- Equipment deployment and recovery
- ROV and submersible handling
- Cable stowage, deployment and recovery
- Workboat launching and recovery
- Access to the ship and transfer to other ships
- Helicopter landing and takeoff
- Mooring and anchoring

In the design of a research vessel it is very difficult to cut back on laboratories and space for scientists and equipment, and it is impossible to cut back on the ship's crew, navigation bridge, engine room, etc. Therefore the research vessels often end up with unsufficient open deck space.

Seakeeping

Good seakeeping ability is crucial to the success of a research vessel. Scientists are not seamen. Their work requires more comfortable conditions than for the usual duties on board a ship, and they will not have years of training in enduring the motions of a ship in the open sea. Seaworthiness is a means of widening the opportunities available for the scientific work.

Good seakeeping involves many characteristics that are all functions of the basic shape of the vessel. The risk of shipping green water on the weather decks must be kept low; the motions and especially the accelerations in roll, pitch and heave should be small; speed loss in waves should be moderate; steering, also in following seas, should be safe; and the intact and damaged stability should be at least in accordance with existing rules.

Apart from sheer size, seaworthiness can be obtained through a combination of many different design features, such as:

- High freeboard, especially forward
- Deep draught
- Large beam
- Low block coefficient
- Large bilge keels
- Roll damping tanks
- Roll and pitch damping fins
- Sails

Sails have been used in several research vessels and should be considered seriously also in new ships. Sails dampen the roll motions and save fuel on transit voyages. Roll damping fins are very efficient, but unlike the roll damping tanks, they require a certain forward speed to be active.

Propulsion

The propulsion system of a research vessel serves three different purposes:

- Gives the vessel an adequate cruising speed
- Makes it possible to tow large loads
- Makes it possible to keep station or to move along a given path under computer control (dynamic positioning)

These requirements are logically fulfilled with a diesel-electric installation, controllable-pitch main propellers and 2-4 bow and stern thrusters, preferably of the azimuth type. Other types of machinery with e.g. mechanical gear couplings between main engines and propellers are also in use. Cycloidal propellers have been used because of their ability to deliver thrust in any direction, but their cruising performance is rather poor.

Research vessels generally require a low noise and vibration level inside and in the water. Special skew-back propellers, resilient mountings of the diesel engines, and other noise reduction techniques can be very effective in obtaining these goals.

Dynamic positioning

Research vessels will very often be required to keep station, to move along a given path, or to follow another vessel, e.g. a submersible. Modern control techniques and computers have made such operational modes possible with surprisingly great accuracy, even under the disturbing influence of waves, wind and current.

A central computer controls all propellers and thrusters of the ship . The computer receives position data several times each second from a reference system of transponders which can be acoustical, mechanical (taut wire), based on Radar or Decca, or it can be a combination of these. The computer compares the actual position, heading and speed of the vessel with the desired values, and orders necessary changes in the propeller settings in order to correct for any deviations.

The computer is programmed with a certain control strategy, and also with the characteristics of the different propellers and thrusters and of the vessel itself. On-line wind speed measurement makes it possible to use wind feed-forward which counteracts the influence of wind gusts before the vessel moves out of position.

The dynamic positioning system is very simple to operate by means of a joystick and simple scales for the desired course and speed.

Electric power

The computers, the data acquisition systems, and other scientific equipment on board the ocean research vessel all require very clean electric power, i.e. free of voltage and frequency disturbances. Therefore, the computer power lines should be protected from disturbances from other machinery, and the data transmission lines should be kept isolated from the ship's power lines and from machinery. Well-designed electric ground connections are also essential.

Large amounts of electric power are required also for navigation systems, dynamic positioning systems, seismic equipment, submersibles, ROVs, diving equipment, cranes, hoists, and for heating and air conditioning.

Ice strengthening

Strengthening against ice should be considered if there is a possibility that the vessel will be operating in cold waters during its lifetime. Not only the hull but also propellers and other appendages will require this.

NAVIGATION SYSTEMS

In ocean exploration work precise navigation is considered to be even more important than in ordinary shipping. All observations must be accurately located for obvious reasons. Systematic surveys must not be contaminated with position errors, important findings should be easily retrievable, and national or other border lines should be accurately observed.

All existing types of navigation systems may be found in use in a research vessel for different purposes depending on the location and the required accuracy. The navigation systems may be divided into five different groups according to their operational principles: Terrestrial, astronomical, radio, inertial, and satellite navigation. The first two are as old as navigation itself; the remaining three were introduced during the last three decades. Satellite navigation will probably be the dominant navigation system in the future.

Terrestrial navigation

Near the shore the ship's position can be determined by taking bearings of natural landmarks or lighthouses together with bottom soundings. The method is simple but not very accurate.

The magnetic or the gyro compasses are instruments belonging to the terrestrial as well as other navigation systems.

Astronomical navigation

Observation of the sun or the stars and planets has been the only accurate means of determining a ship's position in the open sea for centuries.

By the use of a sextant the angular height over the horizon of one or more celestial bodies is measured at a certain time given by a precise chronometer. By means of astronomical tables and some calculations based on spherical geometry the latitude and longitude (i.e. position) of the observer can then be determined from this altitude and time data.

The accuracy of astronomical navigation is within 2 km under ideal conditions. The method requires a clear vision of the sky and the horizon. High waves make it difficult to make the observation of both the object and the horizon.

Radio navigation

Since the turn of the century radio waves have been used to improve the safety and accuracy of the navigation at sea, first, simply by the broadcasting of precise time signals, and later, also in the form of radio beacons that can be located with a directional aerial. Today the following radio navigation systems are in normal use: Radar, Decca, Loran, and Omega.

Radar. Radar is a system in which a radio transmitter on board the ship sends out microwaves, and a receiver with a rotating screen records the energy reflected back from a coast or from other ships or seamarks. The time lag between transmission and reception of the radio signal defines the distance to the reflecting object. The direction is determined by the instantaneous angle of the rotating screen.

The radar image is usually displayed on a cathode-ray screen with the ship itself stationary in the center and the environment moving around it. But the reverse display, the true-motion system, is often used in dense traffic.

Decca. Decca is a radio navigation system in which the ship receives radio signals sent out from land based stations. A network consists of one master and three slave stations.

From each of the slave stations a radio signal is sent out in-phase with a master signal. When received by the ship there will be a phase difference between the two signals depending on the distances to the master station and to the corresponding slave. Points of a given phase difference are placed on a hyperbola with the master and the slave as foci. Similar hyperbolas are identified for the other two master-slave pairs. The ship's position is at the intersection of the three hyperbolas. The network of hyperbolas, the Decca lanes, are found on special charts for the areas covered by Decca.

The range of the Decca system is up to 500 km from the coast. The accuracy is between 100 m and 3 km depending on the distance to the land stations.

Loran. Loran is a system similar to Decca but intended for longer ranges. Instead of measuring phase differences between the received radio waves, time differen ces between radio impulses are measured. The cost paid for the longer range is lower accuracy.

Omega. Omega is a low-frequency system of the Loran type. The frequency is so low that the radio waves can effectively follow the curvature of the earth, which makes the system practically global. The accuracy is relatively poor.

Inertial navigation

The inertial navigation system measures the accelerations of the vessel, and finds by double integration the distance covered from a given reference point, and hence the position at any time. The system works without any information from the outside, which is a clear advantage in submarine navigation. Over a longer time period the double integration procedure may give rise to zero-drift errors. Therefore, the reference point is to be updated frequently, if possible, e.g. by means of satellite observations.

Satellite navigation

The satellite navigation system is in fact astronomical navigation using man-made celestial bodies. Instead of light the satellite sends out radio signals that inform the receiver about the exact speed, course, height, and position of the satellite. The relative speed between ship and satellite will influence the frequency of the received radio waves according to the Doppler principle, and this fact is used to determine the slant range from ship to satellite. Knowing the course and speed of the ship (which may be zero) the ship-borne computer can determine the ship's position very precisely after a few satellite fixes.

Until 1988 satellite navigation will be based on the 5 Transit satellites in circumpolar orbits. With this system position fixes can be made about once every hour.

In 1988 the Navstar GPS (Global Positioning System) will be in operation with 18 satellites in 6 circular orbits. This means that position fixes can be made at any time, anywhere on earth.

The Navstar GPS system operates at different precision levels, characterized by the number of operating channels. The single-channel system is intended for general marine applications and is free of charge. The two-channel system will be used in general aviation and in naval ships. The five-channel system is intended for supersonic aircraft and special applications.

The standard single-channel Navstar GPS system promises an accuracy of better than 100 m, conservatively estimated. Tests conducted offshore California in 1985 showed that 90% of actual fixes had an accuracy better than 25 m, and 65% were better than 15 m.

Even higher precision can be obtained by a differential measurement technique which compensates for position errors due to the uncertainty in the orbital parameters of a given Navstar satellite. When the Navstar position signals are received at a fixed known location it is possible to determine these small systematic errors precisely. The correction signals are broadcast from such stations to the ships which can then fix their position with a reliable accuracy of 10 m. This is actually close to the performance of the multi-channel Navstar systems.

In the future, satellite navigation will undoubtedly be the standard navigation system. The system will be an integrated part of a ship-borne computer that will also monitor the machinery and do all the calculations necessary for the navigator.

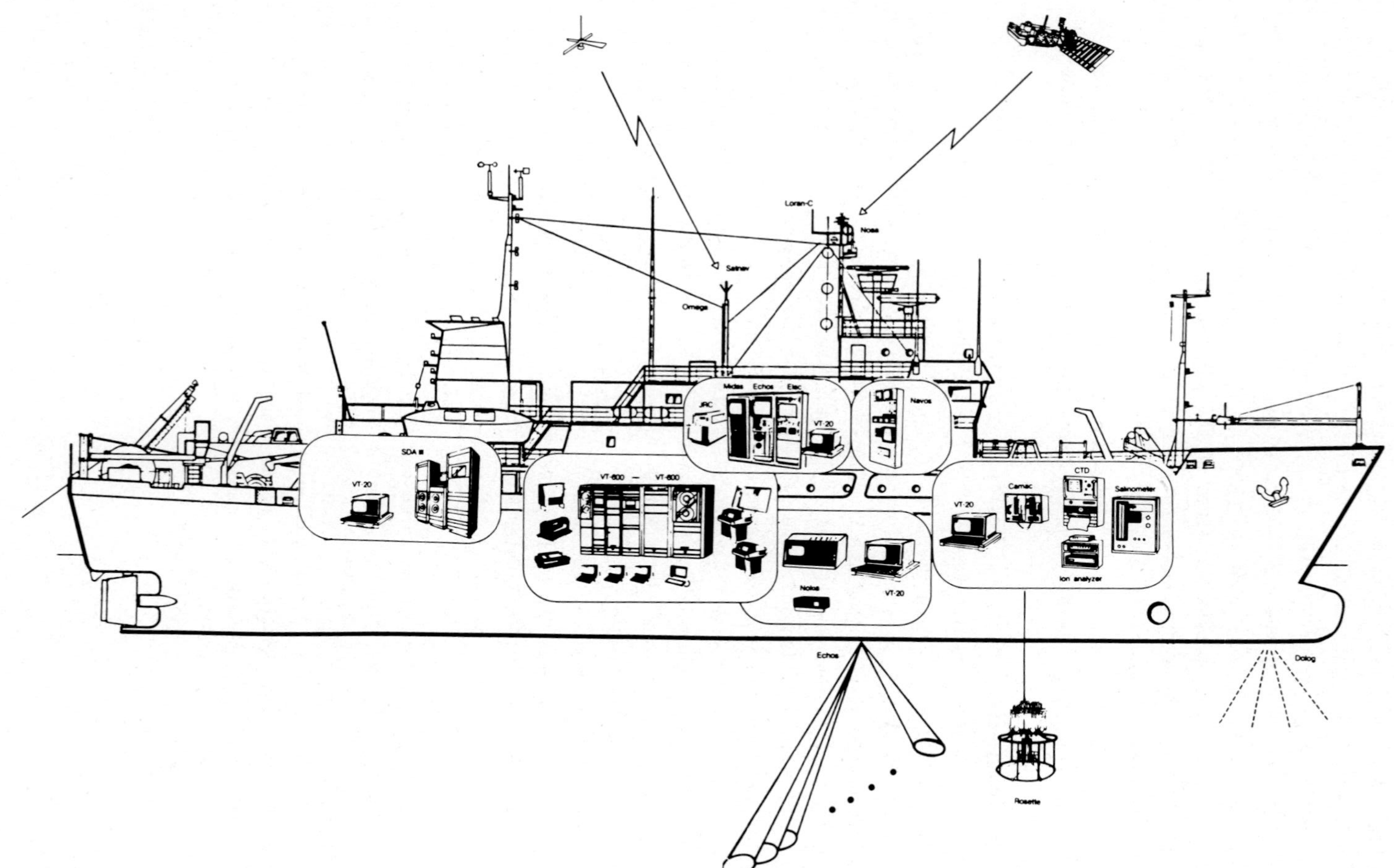

Fig. 3.4. The integrated computer system for navigation and scientific purposes in a modern research vessel (Courtesy Hollming Ltd.).

EXAMPLES OF RESEARCH VESSELS AND SUBMERSIBLES

The following section describes four different research vessels and submersibles that are typical examples of vessels in use or being built around 1985. The following vessels are shown:

- Research vessel RV "Sonne"
- Geotechnical research vessel, Hollming Ltd Yard Nos. 255-256
- Deep sea drilling vessel "Joides Resolution"
- Submersible "Perry PC-18"
- Remotely operated vehicle (ROV) "Super Scorpio"

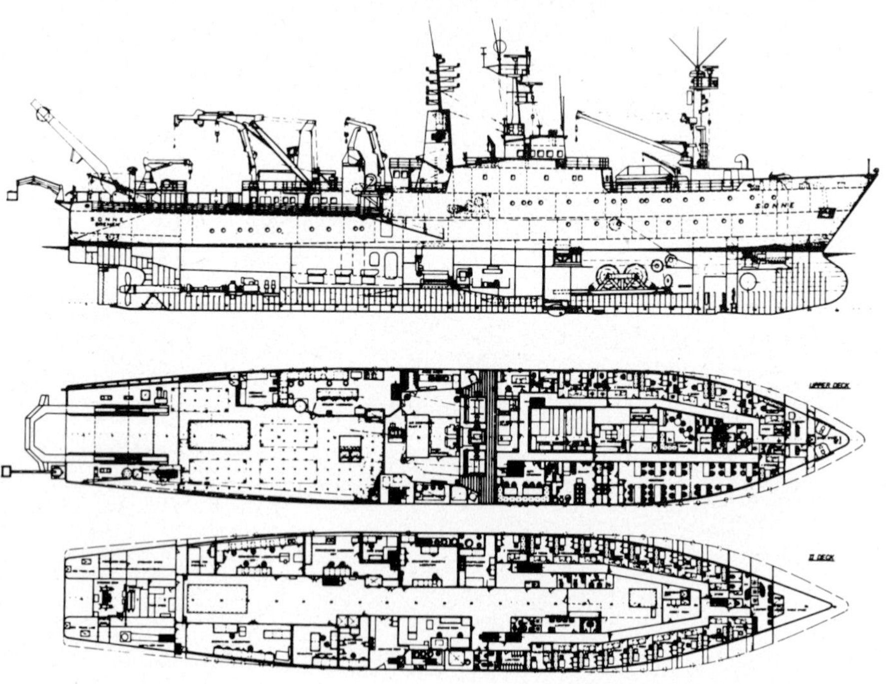

Fig. 3.5. Research vessel RV "Sonne" (Courtesy Reedereigemeinschaft Forschungsschiffahrt GmbH).

Research vessel RV "Sonne"

The RV "SONNE" is a multipurpose research vessel operated and owned by RF Reedereigemeinschaft Forschungsschiffahrt GmbH, Bremen, F.R. Germany. She was built originally as a stern trawler in 1969 and converted to a research vessel in 1977/78 (Fig. 3.5).

Technical data

Main dimensions: Length o.a.: 86.81 m, Length p.p.: 76.20 m, Beam: 14.20 m, Draught: 6.50 m, Displacement: 3834 t, Gross registered tonnage: 2607.

Personnel: crew: 26, scientists: 23.

Service speed: 13 knots.

Main engines: 4 x 735 kW diesel generators.

Propulsion: 2 x 1100 kW electric motors, 1 variable pitch propeller.

Manoeuvring: 1 x 588 kW bow thruster, 1 flap rudder.

Generators: 4 x 180 kVA, 380 V, 50 Hz.

Stabilized network: 40 kVA, 380/220 V, 50 Hz, voltage 1%, frequency 0.5%, dynamic 1.2%.

Bunker capacity: 920 t gas oil, 50 t fresh water.

Fresh water production: 30 t/day.

Maximum service duration: 90 days.

Navigation: 3 Radars, direction finder, Decca, Loran-C, autopilot, gyrocompass, Nautomat-navigation, satellite/Omega navigation system, Dopplersonar, EM-log.

Radio equipment: Short wave SSB 1.6 kW, radio telephony, teletype, VHF.

Acoustic equipment: Bathymetric multi-beam sonar system, 3 frequencies echo sounder 30/20/12 kHz, deep sea 12 kHz echo sounder in combination with pinger registration, sub-bottom profiler 3.5 kHz, shallow water survey echo sounder 30/210 kHz, acoustic transponder navigation system, depth indicator, horizontal echo sounder, fish finding echo sounder, navigation echo sounder, heave compensation for echo sounders.

Geophysical equipment: 6 x free piston high pressure compressors 150 bar.

Scientific equipment: X-ray spectrometer, deep sea TV-system, XBT-sonde, other measuring instrumentation.

Laboratories: Geological, water sampling, wet geochemical, dry geochemical, geophysical, gravimetry/magnetic, photo, registration, computer.

Workshops: Mechanical, electronic, chemical, streamer, seismic, mounting, gravimeter, echo sounder, drawing office.

Deck space: Upper deck: 220 m^2, space for three 20 ft. containers.

Lifting gear: A-frame at stern 12 t, central crane 8 t, derrick crane 5 t, jib boom 10 t, 4 winches 6000 - 8000 m rope each, plus other equipment.

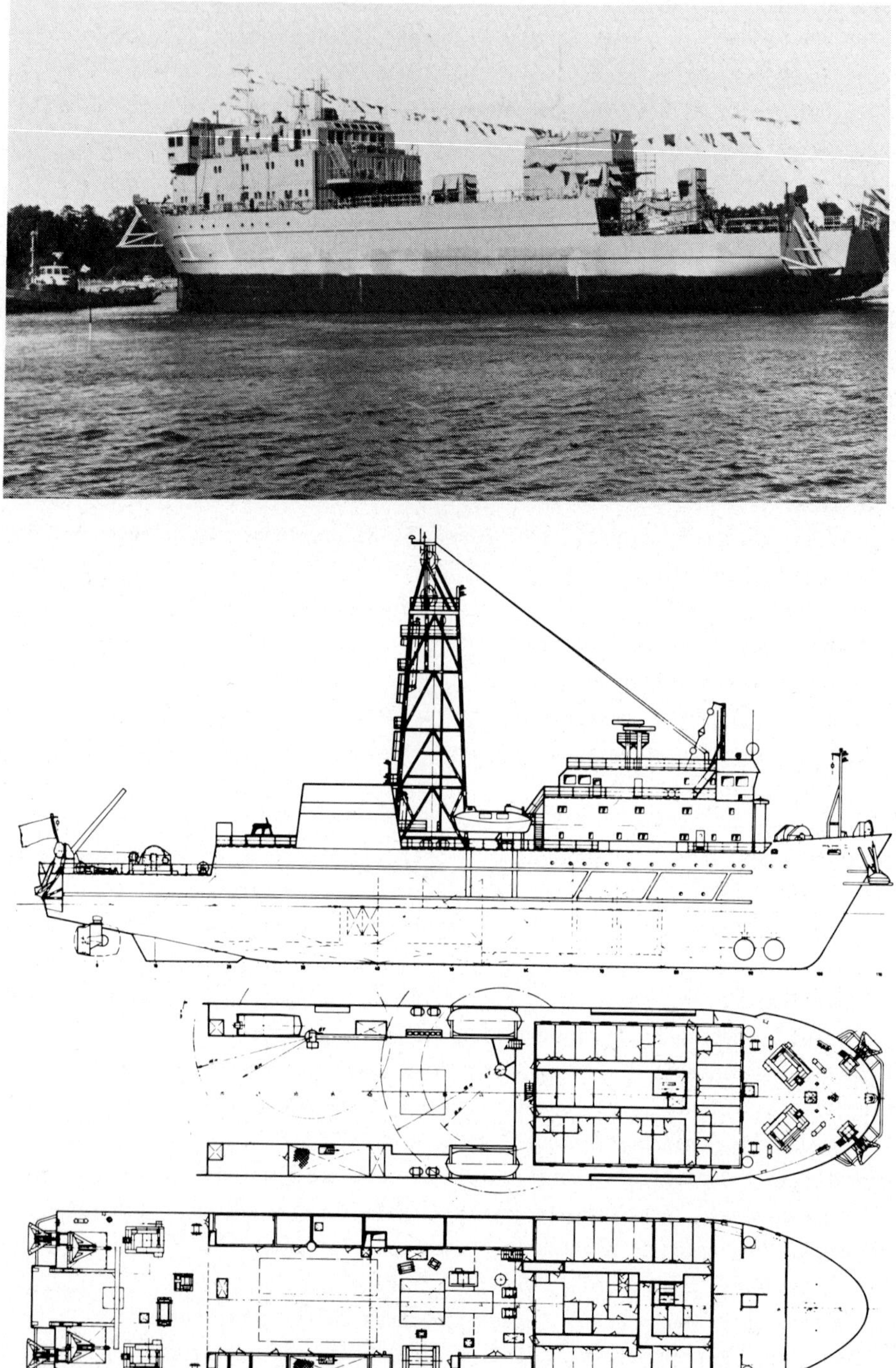

Fig. 3.6. Geotechnical research vessel, Hollming Ltd. yard Nos. 255-56, see also Fig. 3.1 (Courtesy Hollming Ltd.).

Geotechnical research vessel, Hollming Ltd. Yard Nos. 255-256

The geotechnical research vessels (see Fig. 3.6) from Hollming Ltd., Rauma, Finland are built for delivery in 1986-87 to the U.S.S.R. Ministry of Gas Industry. The vessels are intended for geotechnical research in water depths of up to 300 m. The drilling equipment is capable of drilling holes down to 200 m below the sea bed for core sampling and in-situ testing.

Technical data

Main dimensions: Length o.a.: 85.8 m, Length w.l.: 75.3 m, Breadth: 16.8 m, Draught: 5.6 m, Displacement: 5300 t, Deadweight: 2000 t.

Personnel: crew and scientists: 65 persons.

Service speed: 13 knots.

Main engines: 4 x 1150 kW diesel generators, engine room automation for unmanned operation.

Propulsion: 2 x 1500 kW electric nozzle rudder propeller units.

Manoeuvring: Dynamic Positioning system controlling the 2 rudder propellers and 2 x 850 kW bow thrusters. The DP system is equipped with two central computers and several parallel position reference systems, including a hydroacoustic system communicating with submerged buoys, and wind speed and direction sensors.

Maximum service duration: 50 days or 8000 nautical miles.

Navigation: Integrated computer navigation system, including satellite and radio positioning systems, plus a load calculator. In connection with the DP system the integrated navigation system can, for example, steer the vessel along a preprogrammed route.

Drilling equipment: 35 m derrick with heave compensation for up to 7 m vertical motion.

Drilling mud station: The dry bulk mud is stored in 5 pressurized tanks with a total capacity of 60 m^3. After mixing with water the mud is stored in 3 fluid tanks with a total capacity of 105 m^3.

Scientific equipment: Two different systems for cone penetrometer testing (CPT), a 30 t seabed unit and a wire line tool operating through the hollow drill string down to 200 m hole depth. Non-destructive sampling equipment using the push-in or core-drilling methods.

Lifting gear: A-frame and winches at the stern, 2 x 5 t cranes at upper deck.

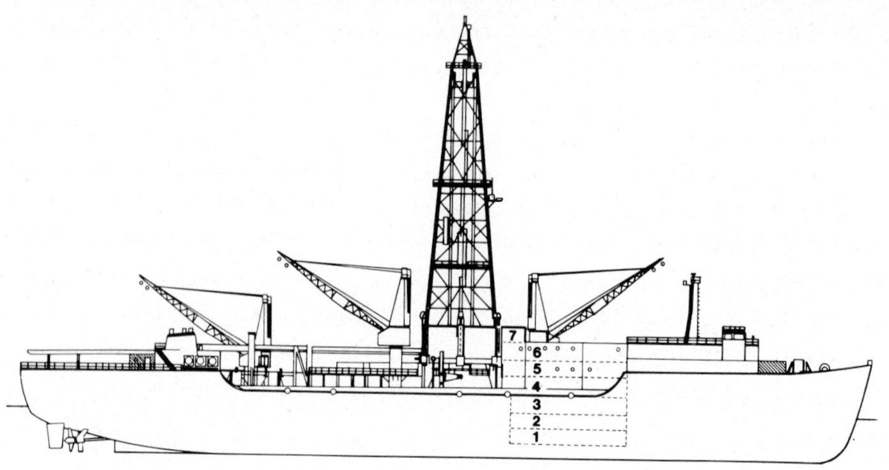

Fig. 3.7. Deep sea drilling vessel "Joides Resolution".

Deep sea drilling vessel "Joides Resolution"

The "Joides Resloution", built in 1977 as "Sedco/BP 471" and re-outfitted in 1984, is the drilling vessel for the Ocean Drilling Program (ODP), a 10-year scientific mission started in 1985 with main purpose to retrieve core samples from beneath the ocean floor. The ODP is sponsored through the Joint Oceanographic Institutions for Deep Earth Sampling (JOIDES), counting 14 universities and institutes in the U.S.A., Canada, F.R. Germany, France and Japan. Funding for drilling operations is provided by foundations in these countries; Texas A & M University is the science operator for the ODP.

Technical data

Main dimensions: Length o.a.: 143.3 m, Breadth: 21.3 m, Draught: 7.3 m, Moonpool 6.7 m diameter, Displacement: 16600 t.

Personnel: crew: 65, scientists: 50.

Service speed: 13 knots.

Main engines: 14700 kW diesel generators.

Propulsion: 2 x 3375 kW main propellers.

Manoeuvring: Dynamic positioning system controlling the 2 main propellers and 12 x 600 kW thrusters.

Bunker capacity: 3380 m^3.

Fresh water capacity: 110 m^3/day, Storage: 160 m^3.

Maximum service duration: 120 days.

Navigation: All modern navigation systems including satellite nav. with GPS.

Communication: State-of-the-art satellite communication with direct telephone, telex, facsimile, data transmission, and electronic mail. Normal radio communication as back-up.

Drilling equipment: Derrick 62 m above water line, electric top drive, 600 t lifting capacity, 400 t heave compensator. Maximum drill string is 9150 m. Maximum water depth 8230 m. Dry bulk mud and cement: 380 m^3, liquid mud: 610 m^3, drill water 2070 m^3.

Laboratory: Seven story lab. structure on starboard side, with underway geophysics lab on poop deck aft, and library on main deck (see Fig. 3.7):

Deck 1: Refrigerated core storage freezer.

Deck 2: Refrigerated core storage, cold storage, second lab.

Deck 3: Electronics shop, photo darkroom, photo finish room.

Deck 4: Computers, computer user room, science lounge, offices.

Deck 5: Paleontology lab, paleo-prep. lab, scanning electron microscope lab, chemistry lab, thin section lab, petrology lab, X-ray lab.

Deck 6: Core receiving, physical properties lab, paleomagnetics lab, core splitting and sampling, photo station.

Deck 7: Downhole measurements.

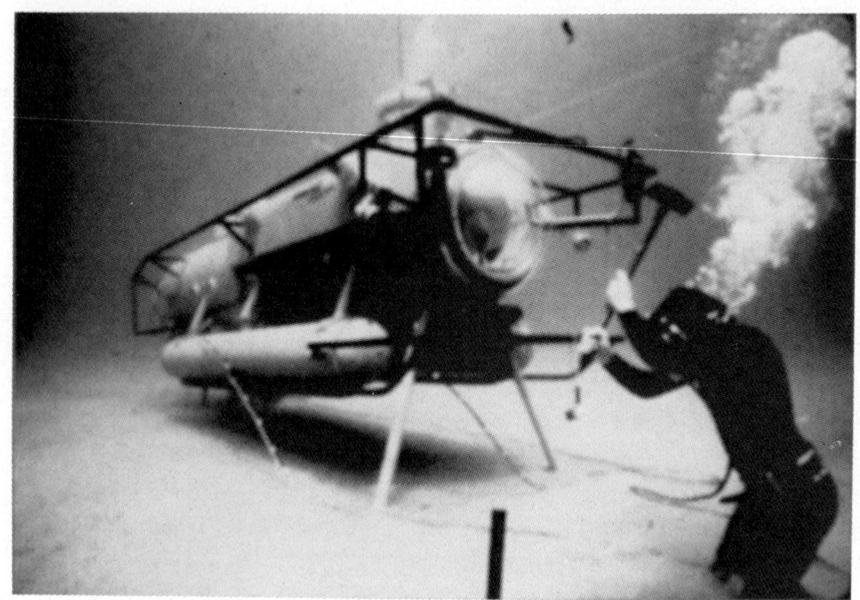

Fig. 3.8. Submersible "Perry PC-18" (Courtesy Perry Oceanographics Inc.).

Submersible "Perry PC-18"

The "Perry PC-18" is a 4-man diver lock-out submersible manufactured by Perry Oceanographics Inc., Florida, U.S.A. The submersible has been built in five different versions for research and offshore inspection work (Fig. 3.8).

Technical data

Main dimensions: Length: 7.7 m, Width: 2.4 m, Height: 2.6 m, Surface draught: 1.9 m, Displacement at neutral buoyancy: 10.9 t.

Depth rating: 300 m (observation), 200 m (diver lock-out).

Pressure hull: 1.4 m diameter cylinder with internal ribs.

Viewports: 1 x 0.89 m hemispherical viewport at front end, 8 x 0.20 m circular viewports around conning tower, 4 x 0.20 m viewports one in each hatch.

Personnel: 1 pilot and 1 co-pilot in forward compartment, 2 divers in diver lock-out compartment.

Speed: 1.3 m/s (2.5 knots).

Propulsion: Propeller with 7.45 kW electric motor 120 V dc can rotate 90 degs. port and starboard.

Manoeuvring: 1 horizontal and 1 vertical thrusters, each 314 N.

Batteries: 35 kWh 12 V lead acid batteries in droppable battery pods.

Ballast tanks: Forward tank 91 kg capacity, aft tank 227 kg capacity.

Trim system: 57 kg water capacity, high-pressure pump with overboard discharge and water transfer between tanks.

High pressure air system: 16.8 m^3 at 180 bar.

Ballast tanks: 0.4 t.

Oxygen system: 8.5 m^3 at 140 bar.

Endurance: 8 hours for observation missions, 4-6 hours with 2 hours saturation lock-out dive.

Payload: 454 kg in addition to standard equipment and 340 kg for crew.

Manipulators: 2 units, 4 degrees of freedom, 5 functions, horizontal, vertical, extend, grip, and rotate.

Lifting capacity: 667 N.

Exterior lights: 3 x 500 W 120V dc lights.

Navigation: Magnetic compass including an external transmitter, through-hull cable assembly, and remote indicator for pilot viewing, directional gyro, submarine sonar with sector scan and full down-tilt for depth sounding, auto pilot.

Communications: Sub-surface: a two-station underwater telephone system with a 15.2 m transducer cable assembly providing an operating range of 1400 m. Surface: 40 channel, 5 W radio telephone.

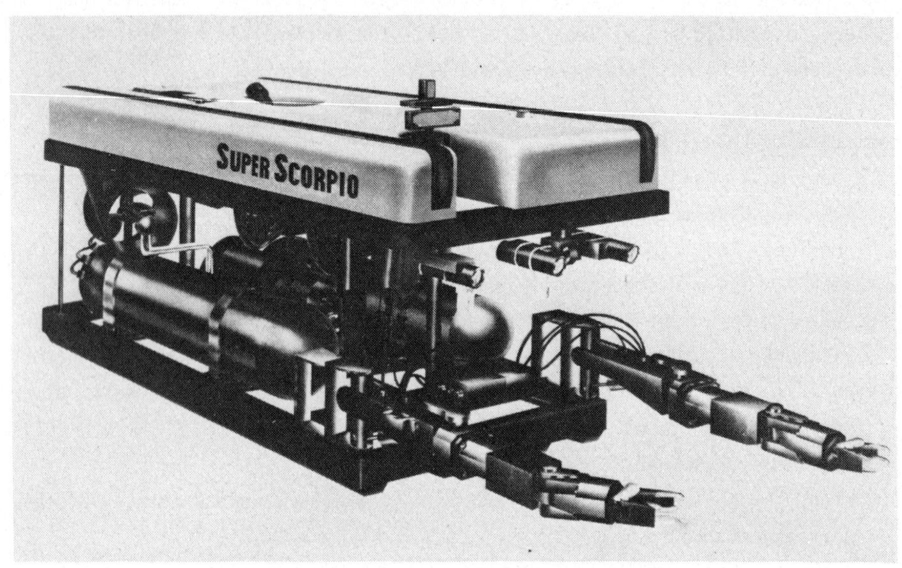

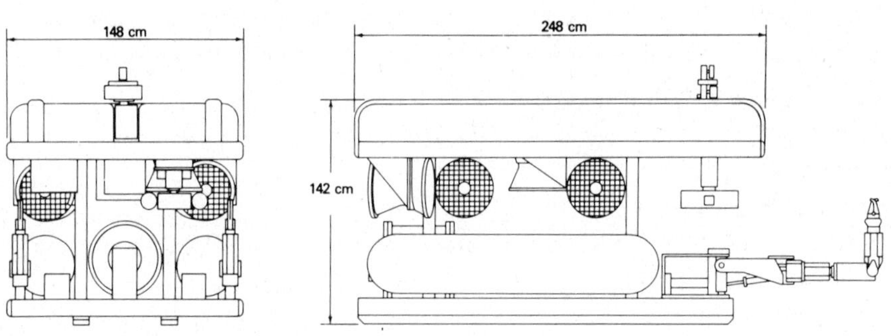

Fig. 3.9. Remotely Operated Vehicle ROV "Super Scorpio" (Courtesy Ametek Offshore Ltd.).

Remotely operated vehicle (ROV) "Super Scorpio"

The "Super Scorpio" is a heavy duty work and inspection ROV produced by Ametek Offshore Ltd., Aberdeen, Scotland. It was released in 1985, and its primary market is the offshore oil industry. With its powerful thrusters, strong manipulator/grabber system and great depth rating, the "Super Scorpio" is also typical of an ROV for marine mineral exploration work (Fig. 3.9).

Technical data

Main dimensions: Length (without manipulators): 2.48 m, Width: 1.48 m, Height: 1.42 m, Displacement mass: 1588 kg.

Depth rating: 1000 m (standard), 1500 m (optional).

Forward speed: 1.5 m/s (3.0 knots).

Propulsion: 5 x 2200 N hydraulic thrusters, 2 axial, 2 lateral, 1 vertical.

Payload: 1500 N (without additional buoyancy).

Lifting capacity: 2010 N vertical thrust.

Manipulators: 7-function manipulator, 5-function grabber, all hydraulic.

Hydraulic power unit: 46 kW unit serves thrusters and manipulators.

Umbilical cable: 1500 m of 36 mm armoured cable.

Control: Single joystick control.

Sonar: Range 1 - 610 m, optional scan converter.

Television: 3 TV camera coax channels.

Pan and tilt inspection: 270 degs. pan, 180 degs. tilt with rate control, 80 degs. pilot tilt.

Lights: 6 x 250 W variable control lights.

Depth sensor: Accuracy 1% standard, 0.1% optional.

Depth control: Auto/manual proportional type, accuracy 280 mm in auto mode.

Heading sensor: Gyro compass, accuracy 1% from magnetic reference.

Emergency locators: Acoustic beacon transponder, 100 hours strobe flasher.

Multiplexer: 64 channels.

REFERENCES

Ballard, R.D., 1982. Argo and Jason. Oceanus, 25: 30-35.

Mann Borgese, E. and Ginsburg, N. (Editors), 1980. Ocean Yearbook 2, Appendix G. The University of Chicago Press, Illinois, pp. 641-673.

King, J.H. and Kaysen, H.D., 1984. Trends in ocean research ships. Proc. Oceans'84, Marine Technology Society, pp. 758-767.

Mathisen, K. (Editor), 1985. Navigation/communication. Shipping News International, 41, 6: 10-23.

Redden, J., 1985. ROV contractors - ROV manufacturers. Int. J. Ocean Business, 45: 63-69.
Rosenblatt, L., 1961. The design of modern oceanographic research ships. In: J.O. Traung and N. Fujinami (Editors), Research vessel design. FAO, Rome, 71 pp.
Trillo, R.L. (Edit.), 1979. Jane's ocean technology 1979-80, fourth edition. Jane's Publishing Comp., London.

CHAPTER 4

MARINE GEOPHYSICAL EXPLORATION TECHNIQUES

H. RICHTER, R. HANSEN, W. POHL, H. ROESER and H. W. RIES

INTRODUCTION

In the course of offshore exploration mostly seismic methods are used. They are well-known and have reached a high standard. Apart from the seismic techniques, a wide spectrum of other geophysical methods were developed within the last years. The range of application extends from engineering problems (harbour construction, soil mechanics, etc.) to scientific investigations in the deep ocean. Accordingly, the operations vary from shallow water to deep ocean surveys (Lettau and Richter, 1985; Meyer et al., 1976).

Principally, the difference between shallow- and deep-water geophysical methods is small. In practical operation, however, there are so many different limiting conditions, either for deep or shallow water applications that special systems had to be developed. Survey differentiation starts already with the selection of the research vessel. It is obvious that small boats, having a minor draught, and used for near-shore surveys at lower charter prices, are unsuited to 30-day operation in the open ocean. Additionally, the exploration at the open sea requires fix-mounted equipment such as deep-sea winches, special echo sounders etc.

Another problem is the navigation of the ship and the positioning of the survey results. In coastal waters, within the range of modern radio navigation systems, the problem is solved and the attaching of the survey results to geographical coordinates is exact. On the other hand, in the open ocean, no exact procedure is available till now.

In most cases, marine exploration at sea starts with a topographic survey of the seafloor. Depending on the problem of further investigation, geophysical and geological methods are available, which either can be applied parallel to the topographic survey (seismic, magnetic, subbottom profiling) or on the basis of its results (TV, photo, sediment sampling, etc.). Most of the geophysical methods are described in this chapter. Because of the limited space available in the following sections, only brief accounts can be given to the techniques. Detailed information can be found in the relevant literature.

ACOUSTIC SURVEYING TECHNIQUES

Echosounder principle

Echosounders are the classic instruments for surveying the seafloor. They allow an exact and fast determination of the water depth from board of a cruising ship without major problems. The topgraphic survey of the seafloor is - besides safety of navigation for the ship - the typical application of echosounders (Ulrich, 1976). The principle is simple: A transducer transmits short acoustic pulses towards the seafloor. The seafloor reflects the pulse back to the ship. The time interval between the sending and receiving the reflected pulse is proportional to the water depth. The fairly constant sound velocity in water enables the displays of fathometers to be calibrated directly in meters or fathoms (1 fathom = 1829 m). Noise problems are easily controlled by means of narrow-band pass filters (Albers, 1965; Kinsler and Frey, 1962).

The simple procedure, but especially the handling advantages of the echosounders, supported experiments to extend the scope of application. The parameter most thoroughly investigated, is the pulse frequency (McQuillin and Ardus, 1977).

High frequencies (up to 500 kHz) permit transducers to be small and distances to be measured accurately. The short wavelength enables a transducer to be designed as a wavelength filter in order to generate narrow-beam signals.

On the other hand, the attenuation of high frequencies in water is high. The range of such systems is therefore between some 10 and several 100 meters at most (Kinsler and Frey, 1962).

As an example the absorption coefficients (Albers, 1965) in seawater at 5^{0} C for different frequencies are given below:

Frequency:		
	1 kHz	a = 0.00001 db/m
	10 kHz	a = 0.001 db/m
	50 kHz	a = 0.015 db/m

It is obvious that lower frequencies guarantee much longer ranges, and in case of the so-called sediment echosounders, a penetration and recording of the upper 50 meters of sediment or even more can be expected. However, the dimensions and the weight of the equipment becomes larger and focussing of the transmitted pulses is difficult within practical sizes.

In between these limiting characteristics special echosounders are applied. For practical purposes, the frequency of a special sonar system is selected so that on the one hand the desired range can be obtained and on the other, necessary shaping of the signal is still possible.

A new development are parametric echosounders (Konrad, 1975). They allow the signal to be focussed even at low frequencies. These systems generate the pulse as an interference product of two different frequencies. However, as the resulting parametric pulse contains only a small portion of the energy of the two generating frequencies, the resulting output power is low and enhancement is difficult.

A problem is the availability of special echosounders. The installation of additional transducers into the hull of a ship is a major effort and requires dry-docking. Therefore, the installation of fix-mounted special echosounders is a cost factor and can be tolerated only if the systems are used over a longer period. However, this definitely is not the case if - as usual - a ship is chartered for a limited survey time only. This restriction also applies in the case of a research vessel. Such a vessel, of course, is normally equipped with standard systems, but even so, it is impossible to cover all ranges of applications.

Successful attempts have been made to mount special transducers into two bodies (tow fishes) and tow them at the side or astern of the ship. Another solution, especially on board research vessels, are small moon pools (a pit through the ships hull), where special rigs enable transducers to be deployed even during the cruises.

Echosounders for shallow water

In the course of a shallow water survey, precise depth soundings can be performed with relatively small systems. The working frequency varies between 50 and 200 kHz, with the range of application of the high-frequency system limited to 10 - 30 meters of water depth (Richter, 1984).

The samll size of such transducers allows its attachment to a pole which can be mounted at the ship's rail. It is then possible to use the echosounders on board different ships. The systems are reliable and the accuracy of the depth measurements is on the order of a few centimeters. The influence of seawater waves can be corrected either by means of a special gyro platform or a system of accelerometers.

Echosounders for deep water survey

Depth measurements become more difficult when the bathymetric survey crosses the shelf slope and moves into the deep ocean. The pulse of a normal echosounder has a hemispherical shape. At a water depth of several thousand meters, such a signal hits the seafloor at a large area and is unable to record a detailed relief. It generates side echoes and diffractions which will bury all soundings (Fig. 4.1).

This problem was solved by the use of so-called narrow-beam echosounders (Honeywell Elac, 1966 and 1980). In this case, the transducer is constructed as a wavelength filter which focusses the transmitted signal to a beam with 1.5^{0}. However, the narrow-beam signal alone generates a negative effect. If such a

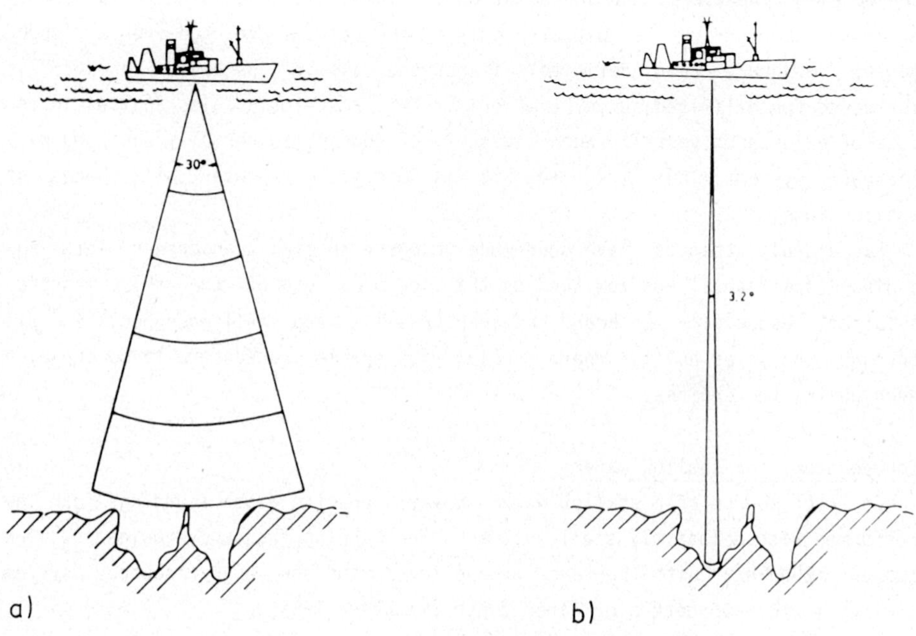

Fig. 4.1. Principle sketch for echosounder survey in deep water.
a) conventional system, b) narrow-beam echosounder

signal were to be transmitted from a rolling and pitching ship to a water depth of 4 or 5 thousand meters, the signal would poke at the seafloor at random and the results of the soundings would then become worthless. To avoid this effect, the narrow-beam transducer is mounted on a gyro-stabilized platform which guarantees that the signal always points exacly downwards, independently of the movement of the ship. Figure 4.2 exhibits the comparison of a profile recorded by a normal- and narrow-beam echosounder (Drenkelfort, 1970).

Multiple beam echosounders

For the topographic survey of the seafloor, computer techniques offer an

a)

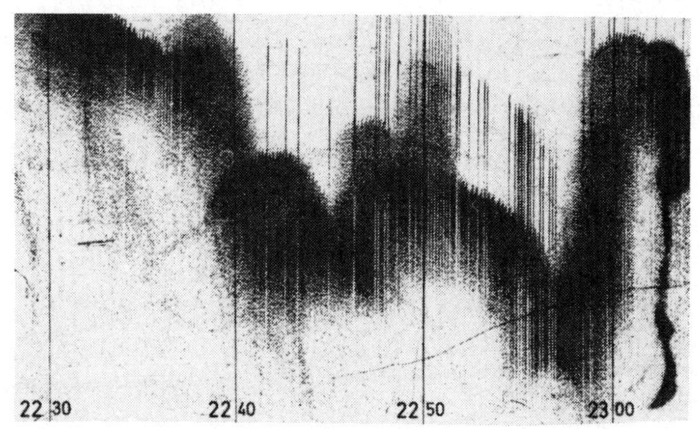

b)

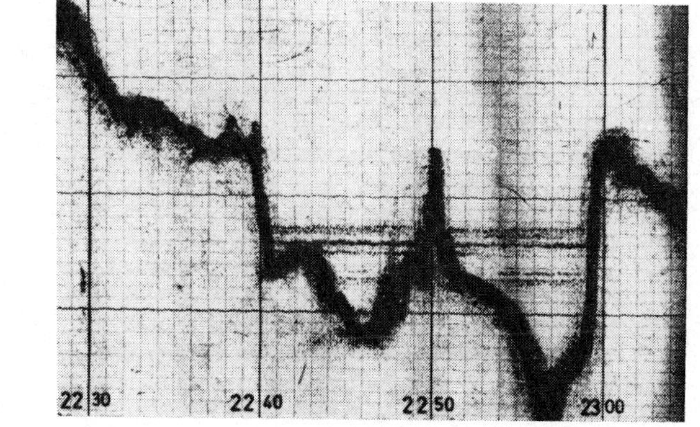

Fig. 4.2. Original echosounder display of a detailed bottom relief in the Red Sea (Atlantis II Deep).

a) Record of a normal echosounder

b) Record of a gyro-stabilized narrow-beam echosounder (30 kHz, 1.3^{0} beam width).

even better solution (Renard and Allenou, 1979; Smith, 1983). The seabeam system presents a surveying arrangement which records not only the trace beneath the ship, but surveys the seafloor in a strip by using a pattern of 16 sonar beams. The surveyed strip has a width of 3/4 of the water depth. Depth readings are processed on-line and displayed as contour lines (Fig. 4.3).

The working frequency of the seabeam systems is 12 kHz and the sonar pulses of the beam pattern are focussed to 2 2/3°x 2 2/3°. In the same way as for the narrow-beam echosounders, already discussed, focussed pulses are one of the basic requirements for mapping detailed reliefs (such as narrow valleys, faults, etc.) at the seafloor even in deep water. The second requirement, of course, is an exact position control of the transmitted and recorded signals.

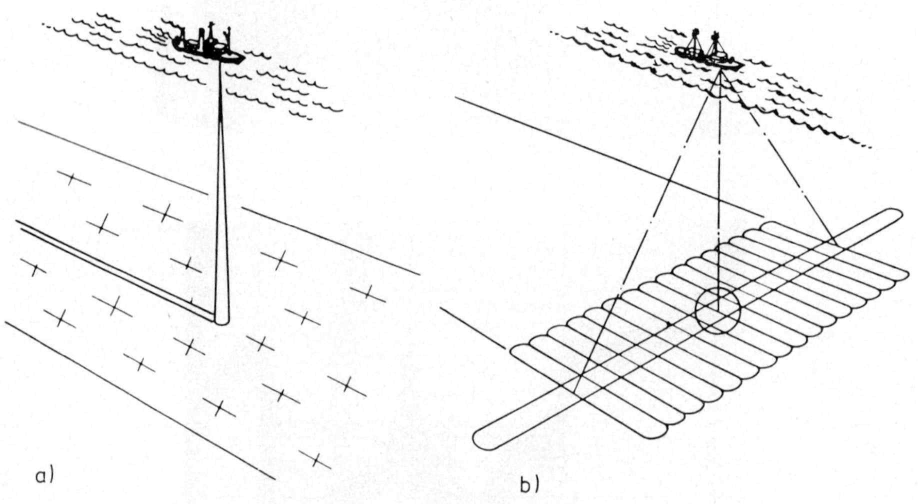

Fig. 4.3. Principle of echosounding.
a) Trace of a conventional narrow-beam echosounder
b) Swath of the seabeam system

In particular, the roll and pitch of the ship, as well as the refraction of the sonar pulses caused by temperature anomalies within the seawater, have to be controlled because otherwise a direction error of only 1° at a water depth of 4000 meters would cause a position error of about 70 meters compared to the nominal value.

To realize the sophisticated beam pattern of the seabeam system within technical limits, in spite of the comparable low frequency, transducer and receiver are separated and subdivided into single elements. These elements are fixed within two strips of about 10 feet length each and mounted at the ship's hull along and perpendicular to the ship's keel. The subdivision into single elements allows a directed receiving: Pulses (directions) which reach the elements at the same phases are added (increased), whereas all others are discriminated against by interference. This principle of amplification and discrimination is the basis of all directed systems and is valid for receivers as well as transmitters. The final characteristics of a special array depends on its geometrical shape. The seabeam array is arranged so that the resulting output forms a leaf-shaped beam pattern which is 2 2/3° in one direction and 54° in the other. Such beam pattern is called "fan-shaped".

As described above, the seabeam transmitters and receivers are mounted at the ship's hull, perpendicular to each other, and therefore, their fan beams are also perpendicular to each other. This means that the transmitter shines upon a strip of 2 2/3° perpendicular to the ship's heading and the receivers look at a strip of the same width but parallel to the ship's direction. As a result, the system gets replies only from the average of both beam patterns, i.e. the element of 2 2/3°x 2 2/3° directly underneath the ship (Fig. 4.4).

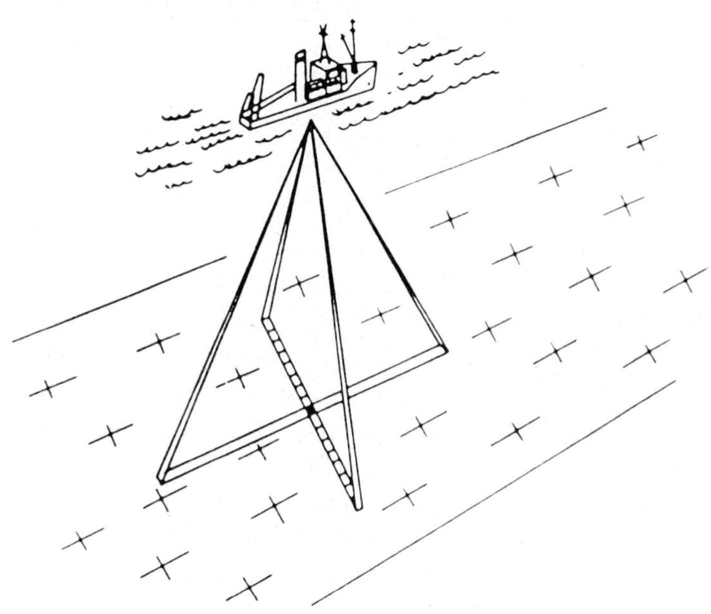

Fig. 4.4. The narrow-beam characteristic of multibeam echosounders.

However, the use of arrays has additional advantages: A series connection of the elements together with variable delay lines allows practically all directional effects within a circle beneath the arrays. In the case of the seabeam, the 16 beams of the pattern are therefore not generated by 16 transducers, but result from the output of the arrays, in connection with special calculated variable delay lines. Even the roll and pitch influence of

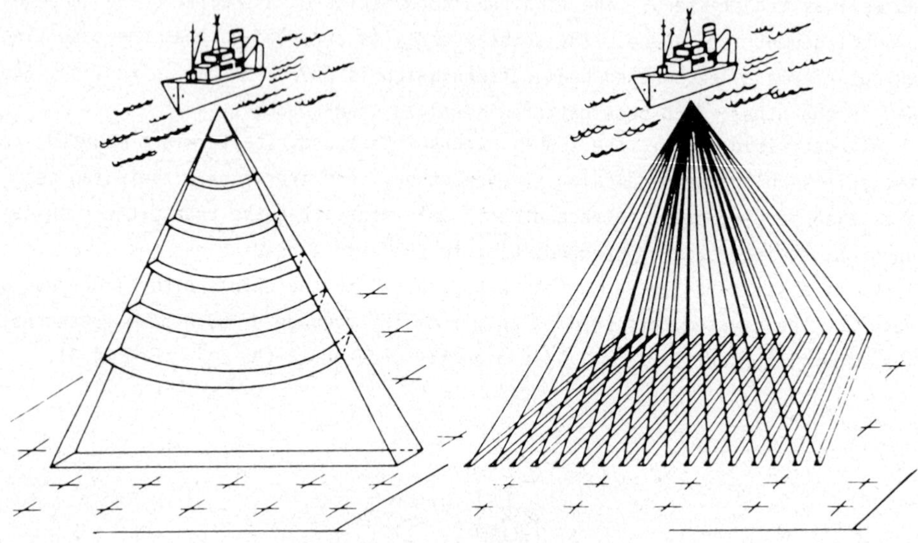

Fig. 4.5. The generation of the beam pattern.

the ship is corrected in this way. The reference is a gyro platform and the accuracy of the correction is $1/4^{0}$ relative to the horizon (Fig. 4.5).

Figure 4.6 shows a real-time plot of the seabeam system obtained in the course of a survey in the Pacific Ocean. For display, soundings obtained from pulses at a 6-sec interval were processed into contour maps with respect to the speed of the ship. Additional information such as heading, time, and depth at the center line are marked at the top of the plot.

Today, the multibeam echosounders are matured constructions and represent optimal systems for the topographic survey of the seafloor (Anonym, 1982; Anonym, 1985; Colvin, 1982). However, the construction of the topographic maps over larger area, out of several seabeam strips, requires additional corrections. As already mentioned above, up to now the exact navigation of a vessel in the open ocean is extremely difficult. The standard navigational methods used at present are based on the Integrated Transit Satellite Navigation Systems. However, especially in equatorial areas, the interval between subsequent satellite fixes is often more than 6 hours. As a result, the error in deadreckonning may accumulate to several nautical miles. Introduction of a Global Positioning System (GPS) in the near future may solve some navigational problems. Additionally, the curving and turning round of the ship are not taken into consideration in calculating the seabeam strips. The procedure for making the necessary correction is difficult and requires additional computer facilities (Richter, 1985). On the other hand, the direct display in many cases allows a correction of the ship's position on the basis of typical "landmarks" of the seafloor. It is possible for example to follow narrow valleys or likewise to survey a seamount up to the top only by means of these displays (this would be difficult with a conventional echosounder).

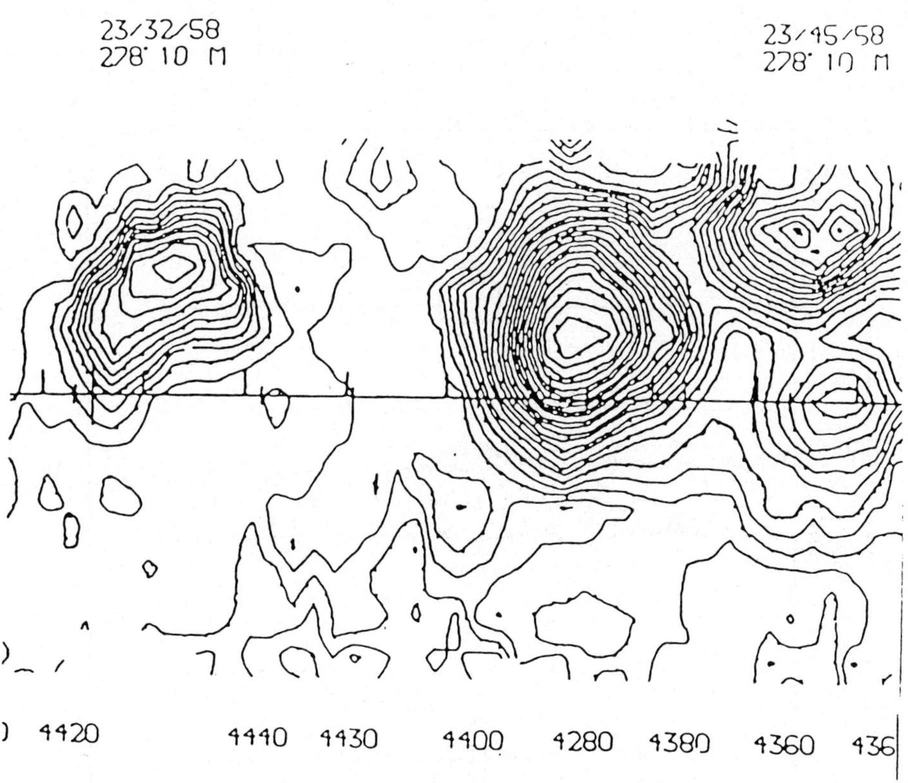

Fig. 4.6. On-line seabeam record from the Pacific Ocean. The display exhibits several seapeaks. For example, contour lines are computed in an interval of 10 meters. Tick marks at the contours are directed towards the dip of the structure. When a contour line hits the center line, its depth value is marked at the bottom of the display. In the upper part of the display, time, year, heading and the contour interval are noted.

Side-scan sonars

Besides the topographic mapping the aim of a survey often is a continuous inspection of the seafloor and in many cases objects in the range of only a few centimeters have to be mapped. Examples may be technical tasks, such as soil investigations for the deployment of an oil rig, or scientific studies for sediment phenomena. The instruments for such types of surveys are side-scan sonars (Prior et al., 1981; Luydendyk et al., 1983). Generally they are mounted in "tow fishes" and towed on a cable astern of the research vessel. Depending on the mission and the system used, the working frequency varies between 30 and 150 kHz. Similar to the seabeam the sonar pulses of the side-scan sonars are

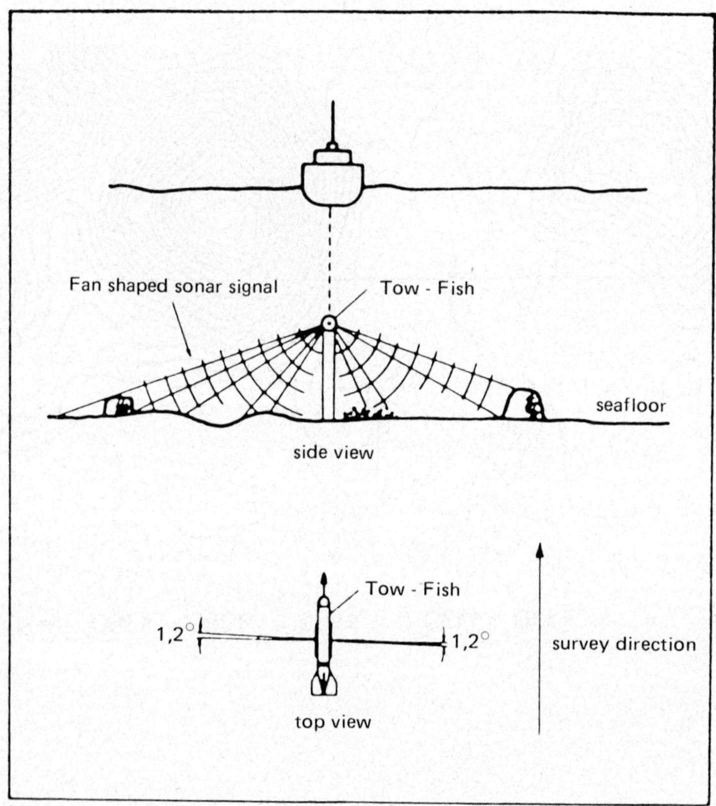

Fig. 4.7. Principle of side-scan sonar; fan-shaped sonar signal.

fan-shaped and accordingly the pulse hits the seafloor in a narrow swath of 100, 1000 or even more meters perpendicular to the heading of the towing ship. The towing altitude above the seafloor is almost 10% of the width of the surveyed swath. Due to the propulsion of the system the seafloor will be covered continuously by the fan-shaped pulses, i.e. the seafloor will be "sanned".

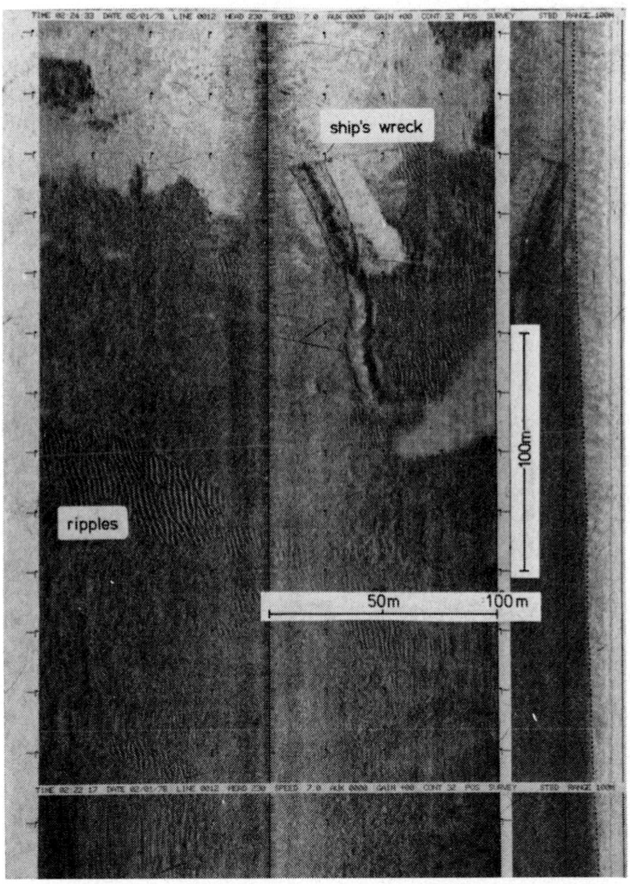

Fig. 4.8. Side-scan sonar section. The section exhibits a typical record of a shallow water survey. The display is on-line processed for slant range and scale. The tick marks at the lower part are printed in a distance of 25 m. The dotted line represents the water depth.

Unlike the seabeam, the transmitter array is used both as transmitter and recorder in the usual way as with conventional echosounders. The reflected signals are recorded and displayed with black-white photographs generated with a focussed flash light that shines inclined to the seafloor and marks in this way all unevennesses by light and shadow. Side-scan sonars, therefore, are sensitive instruments. High-frequency systems can reach a resolution of a few centimeters at a reduced swath width. For exact positioning of the detected phenomena, however, the records have to be corrected for the slant range of the sonar pulses. Normally, with conventional systems, the "slant range correction" is calculated from the flying altitude and swath width and applied manually. Modern systems are able to calculate the correction automatically and then the display is already a proper scale (Czarnecki, 1978).

Subbottom profiler

Another application of sonar techniques for surveying the sea are subbottom profilers. The working frequency varies between 3.5 and 7 kHz and the output power from 5 - 10 kW (Lettau, 1985; Drenkelfort, 1980; Lowell and Dalton, 1971). Depending on the type of sediment the pulses penetrate and map the layering within the upper 30 - 50 meters (Fig. 4.9). The results are similar to those of small seismic systems.

However, compared with seismic systems the wavelength is short (20 - 50 cm). Besides the advantages of easy handling, as is usual for echosounders, the subbottom profiler allows highly resolved detailed interpretation of the recorded sediment layers. The range of applications extends from soil investigations for underwater constructions to academic studies such as determination of sedimentation rates, and related problems. Another typical use is pipeline tracing. In many cases, pipes or cables have to be buried into the seafloor sediment for protection reasons and frequent inspection is necessary. Most of the subbottom profilers transmit their signals unfocussed. In other words, the shape of the pulse is more or less a sphere or cone. The high density of the pipeline material relative to the surrounding sediment, generates strong reflections at once in case the outer rim of the wave head hits the pipe. As the subbottom profiler approaches the pipe, the reflection time becomes shorter and reaches a minimum, when the profiler is perpendicular above the pipe. The recorded reflection pattern is a typical hyberbola. The zenith represents the pipe and the travel time recorded at this point is proportional to its depth (Fig. 4.10).

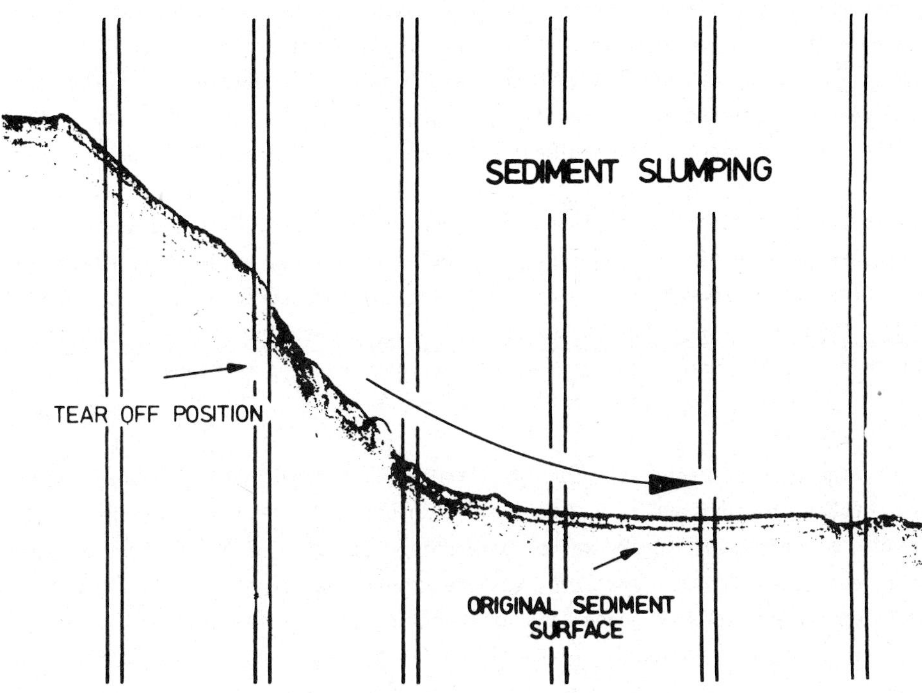

Fig. 4.9. Subbottom profiler record from the Red Sea.

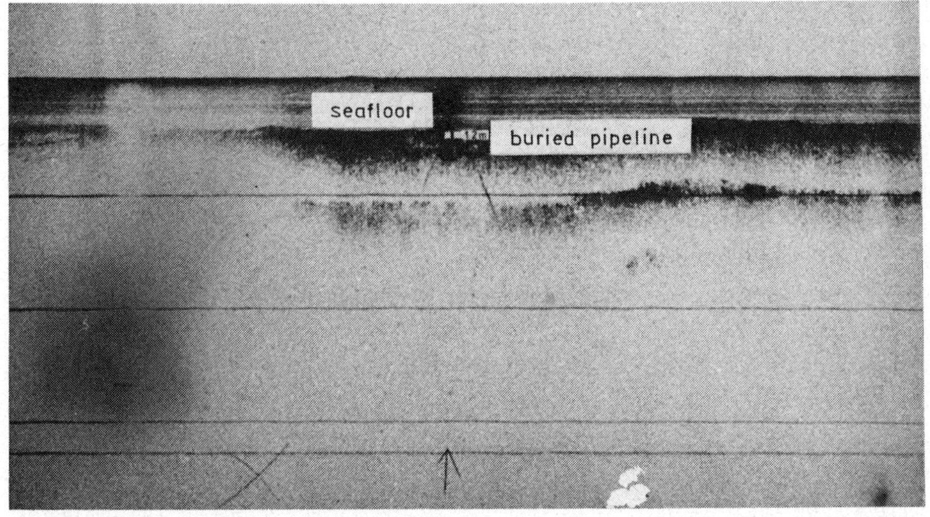

Fig. 4.10. Trace of a pipeline obtained by a subbottom profiler record.

Subbottom profilers represent matured systems and they are one of the standard tools for oceanographic research. Transducers are often mounted into tow fishes and can be used with any ship available. However, results are generally better when the system is fix-mounted in the hull of the ship.

The use of narrow beam sediment echosounders is of interest in deep water surveys. However, for wavelength of 20 - 50 cm it is difficult to design a transducer as a wavelength filter within reasonable sizes. Existing systems are therefore parenthetic echosounders, i.e. the narrow beam pulse is generated as an interference effect of two frequencies tuned off resonance (Konrad, 1975). That, however, limits the output energy and consequently, the application of narrow beam subbottom profilers has been insignificant up till now.

Deep tow system

In the case of deep water surveys, the application of the side-scan sonars and subbottom profilers is not possible without modifications. As mentioned above, a side-scan sonar works at a bottom distance of about 10% of the surveyed strip width, and the unshaped sonar pulse of normal subbottom profilers generates side echoes and diffractions in case the seafloor is rough in deep water. The records are then usuallly buried in noise. To utilize the advantages of such systems even in deep water and in spite of the mentioned limitations, tow fishes and towed platforms with data link systems are developed to carry the exploration tools to the seafloor.

Several of these "deep tow systems" are in use (Fagot et al., 1980; Blidberg and Porta, 1976; Huntec; Osborne, 1984; Richter, 1980). For exploration purposes, they generally include subbottom profiler, side-scan sonar and oceanographic probes such as temperature, pressure, salinity, light attenuation measuring devices, sonar speed instruments, etc. Additionally, so-called house-keeping equipment is installed: gyro, position control, accelerometer, bottom distance and obstacle avoidance sonar (Spiess et al., 1978; Hutchins et al., 1982). Electronic instruments and cable connectors are state of the art. The selection of the tow cable is more difficult. Its task is not only the power supply and data link but also the traction of the tow body. In deep water, the cable itself represents a considerable surface which generates buoyancy even at low speed of the towing ship. The diameter of the cable therefore becomes critical. For instance, an 18-mm diameter coax cable corresponds to a surface of 108 m^3 at a water depth of about 4500 m when normally 6000 m of cable are paid out. The resulting buoyancy must be compensated for in relation to the speed of the ship, by the weight of the towing body itself or a suitable depressor weight. It is obvious that due to the large surface of the cable the speed of the ship has the strongest influence (Fig. 4.11).

A weight compensation is possible only when the ship is travelling slowly.

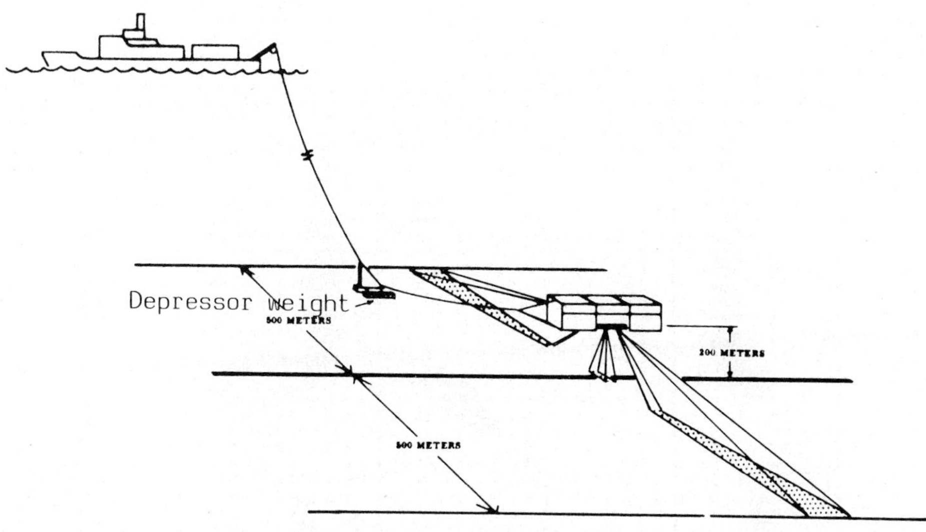

Fig. 4.11. Principle of a deep tow system using a depressor weight.

As an empirical result, the buoyancy of the 18-mm coax cable can be compensated for by a weight of about 2 t for the tow body at the water depth discussed and a ship speed of about 1 - 1.5 kn. But again, the speed of the ship is critical. Small variations already cause rapid descent or ascent of the vehicle.

Finally, the use of deep tow systems requires sophisticated technical efforts and the progress of the survey cannot be compared with shallow water missions. Therefore, the application normally will be limited to the clarification of special questions, that rose in the course of a more general survey (Fig. 4.12).

Consequently, attempts are made to speed up the procedure. The physical conditions for the buoyancy of the system cable-towfish are constant. The efforts, therefore, are directed to widen the surveyed strip.

These efforts were successful especially for the side-scan sonar. Within so-called High Speed Exploration Systems (HSES), the side-scan sonars cover a strip of more than 5 km at a bottom distance of 500 to 1000 m. The larger bottom distance has several advantages: The danger that the towed vehicle runs into an unexpected obstacle is reduced and the smaller towing depth also

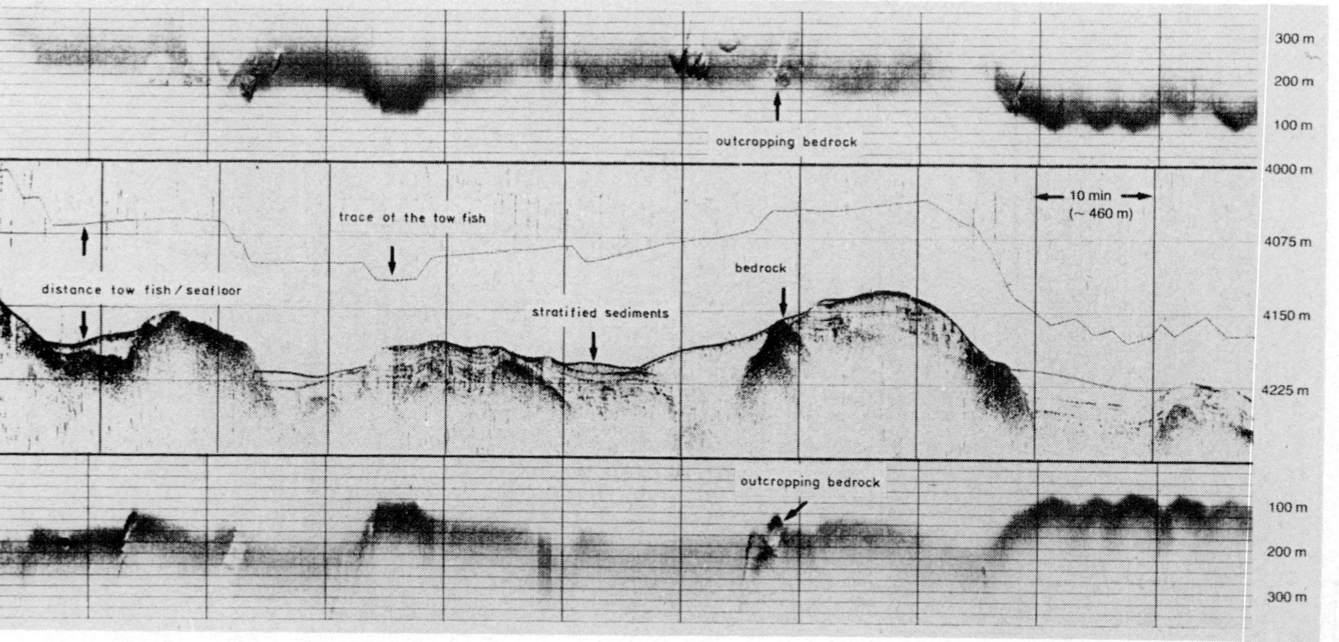

Fig. 4.12. Results of a deep tow survey: subbottom profiler and side-scan sonar records (AMR Deep Tow, Pacific Ocean).

decreases the buoyancy of the cable.

On the other hand, increased bottom distance decreases the effectiveness of the subbottom profiler due to side echoes, especially in case the seafloor topography is rough. The use of narrow-beam systems is difficult, as described above. Attempts were made to apply parenthetic systems. But even such systems suffer from low energy and stabilization problems within a towed carrier become critical.

SEISMIC EXPLORATION METHODS

Seismic methods are well-known and one of the main tools in petroleum exploration. Consequently, the methods are highly developed and there is a wide scope of applications and interpretation procedures. A discussion - even as an outline - will definitely be beyond the scope of this book. For detailed information numerous publications are available. The following discussion therefore will be concentrated on principles, systems and methods used in the course of marine hard mineral exploration. Generally, seismic refraction and reflection methods are in use (Anstey, 1970; Dohr, 1981; Dix, 1952; Bullen, 1963; Evenden et al., 1971).

Refraction seismic

The classical task for refraction seismic measurements during marine exploration surveys is the investigation of the deeper earth formations, especially the basement problem. The method is based on sound velocity determinations within larger geological sequences. The efforts necessary are comparatively high. The required deep penetration of the seismic energy and the large distances between the sound source and the recording points need high energy levels for the sound source, which normally can be reached only with explosives. This, on the other hand, causes problems not only from the environmental standpoint, but also from security considerations.

The field techniques and the interpretation methods are sophisticated, and described in more detail in special publications (Dohr, 1981; Anstey, 1970).

Reflection seismic

Single-channel method. Apart from the petroleum seismic instrumentations small systems are available and adapted to the requirements of oceanographic research. The principal components of such systems are almost identical to those of an echosounding or subbottom profiling system (Fig. 4.13).

A wide range of types of equipment is available and there are many possible combinations of component parts thus allowing great flexibility in adaption of the particular exploration problem. The geophysicist must carefully define the zone of penetration needed as well as the required resolution of the system

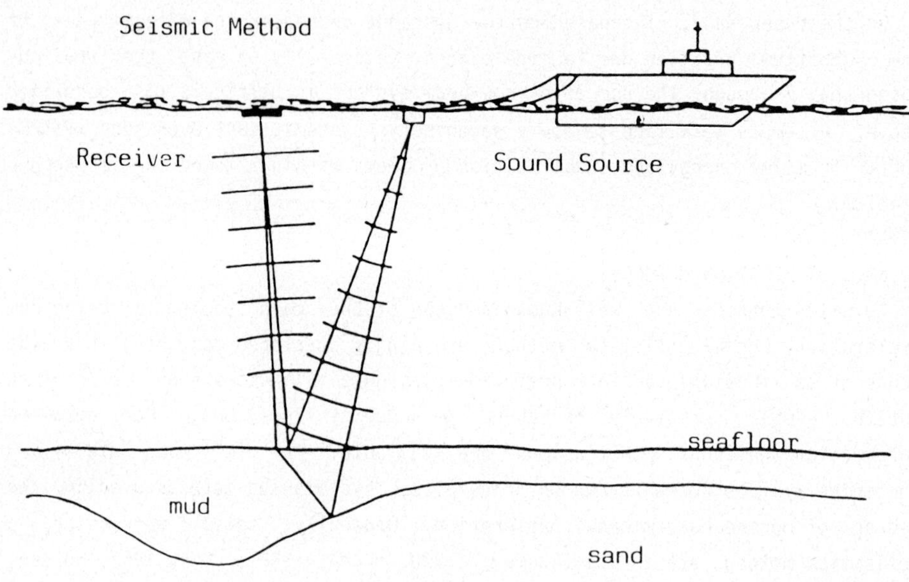

Fig. 4.13. Principle sketch for single-channel seismic surveys.

(McQuillin and Ardus, 1977).

Basic physical conformities may be summarized as follows:

High frequency - Low power - Low penetration - Good resolution
Low frequency - High power - High penetration - Low resolution

Sound sources. Most energy sources for the single-channel seismic systems depend on converting electrical into acoustical energy though systems powered by compressed air (see Table 4.1) are also in use.

Recording. To receive the reflected acoustic energy, so-called hydrophones are in use. The sensing elements are mostly piezoelectric units mounted in groups inside a flexible tube (streamer) mainly to minimize the generation of noise. It should be mentioned that also single elements are employed.

The electrical signals produced by the hydrophone are generally recorded analogue, amplified and displayed on-line on electrosensitive paper. Some special accessory units may enhance the quality of the record considerably (Hansen, 1984). These enhancing units may be:

- Frequency filters
- Automatic gain control
- Time-varied gain

- Wave filters
- Stacking units
- Heave compensation.

TABLE 4.1
Compilation of sound sources.

Name	Type	Pulse length (ms)	Acoustic energy (kWs)
Pingers	Electromagnetic	0.5	0.005
Boomers	"	1	0.1 - 1
Sparkers	Spark-gap	1 - 20	0.1 - 100
Airguns	Compressed air	10 - 20	1 - 200

Digital recording and processing techniques lead to further enhancement.

The resolution of analogue systems is 80 - 100 cm and 20 - 30 cm for digital recording and processing of the data respectively (Richter and Hansen, 1976).

Seismic profiling records provide the user with a detailed picture of geological structure beneath the seabed (see Fig. 4.14). Having defined the form of the superficial sediment layers and having constructed maps with layer thicknesses or depths, the seismic profiles can now be used to identify structures and lithological variations. Muds, clays, sands, gravel, glacial till, homogeneous or weathered rock, all produce characteristic profile records. With adequate control from drilling or sampling, a three-dimensional picture of the geology can be developed. A combination with side-scan sonar data then provides the best means of mapping variations in the type of sediment exposed at the seafloor.

The number of engineering applications of the seismic profiling method rose as offshore constructions and development increased. Engineering site surveys have two basic functions: Firstly, they allow the engineer to avoid (if possible) geological conditions that complicate a project or result in higher costs. Secondly, they provide the basic information that is necessary for planning future drilling, piling, dredging, excavation, mining or whatever other operations the project demands. High resolution surveys make it also possible to explore for man-made objects like buried pipelines, ship wrecks etc.

Multichannel seismic surveys. The basic principles of seismic reflection have been discussed already. Multichannel seismic methods are nearly exclusively used within hydrocarbon exploration. The search for offshore oil and gas

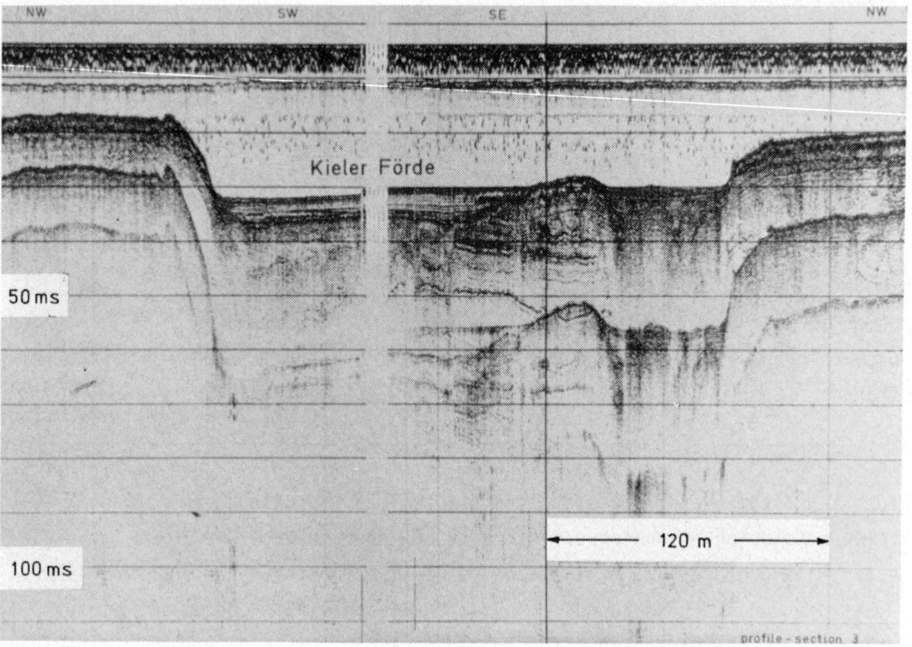

Fig. 4.14. Single-trace seismic Boomer profile, analogue recorded and on-line displayed on electrosensitive plotter paper.

considerably promoted the development of this technique.

Multichannel seismics will be applied if a deeper sounding of the sediment sequences is necessary, i.e. 3000 - 5000 m or more sediment thicknesses (Anstey, 1970; McQuillin and Ardus, 1977). One of the main disturbances in this case is the appearance of multiple reflection caused by seismic energy that travels several times between the surface and a strong reflector or even between reflectors, and then, partly to completely, buries deeper events. The most powerful method to avoid such effects is "common- depth-point stacking". It is based on the increase in sound velocity generally with depth in general, i.e. the differences in travel time for a distinct seismic event recorded at a geophone close to the shotpoint and at the farthest one of the array, decrease with deeper events ("Normal Move Out", NMO). This, however, is invalid for multiple reflections. The repeated travel within the same sediment sequence prologates the travel time but it will not effect the move-out (average sound velocity). This difference in time can be used to identify and discard multiple reflections.

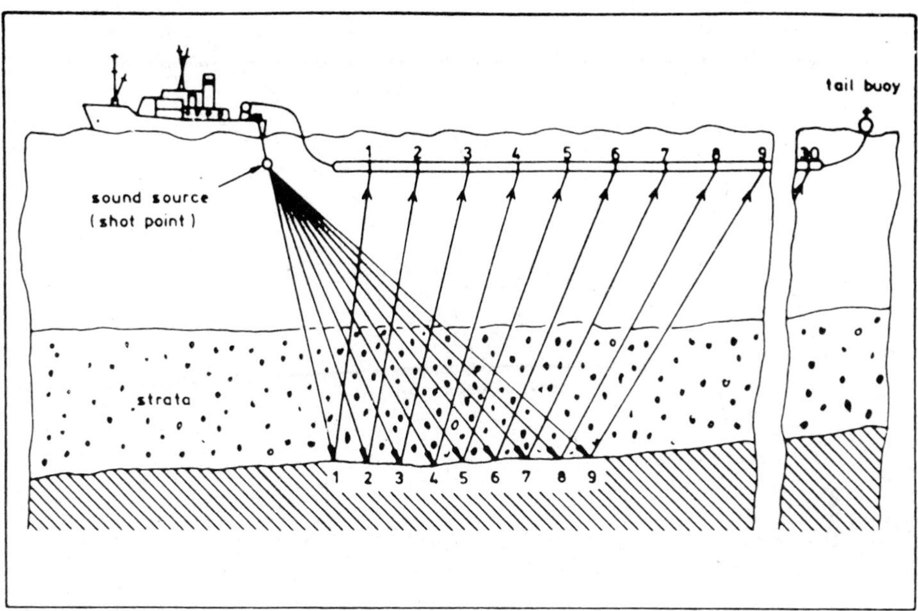

Fig. 4.15. Principle sketch for the multichannel seismic method.

The method of common depth-point stacking is based on repeated recording of the same subsurface point using different travel paths for the seismic energy. A schematic sketch is given in Fig. 4.16 (Dohr, 1981; Mayne, 1962). For the final display and interpretation, the time differences caused by the different travel paths will be corrected (NMO correction) and results will be added up (stacked). The resulting output enhances the true events and discriminates against multiple and random effects. For practical purposes field arrays similar to that in Fig. 4.17 are in use.

Only digital techniques made it possible to record and process the large amount of data. Additionally, the development of sound sources being able to operate at time or distance sequences that are electronically controlled, make marine seismics more efficient. In the early days of marine seismic surveys explosives were used as sound sources. Nowadays, air guns powered by compressed air (of around 150 bars) are in common use. Other types of sources are sparkers, gas exploders and water guns. The frequency range of these sources is 5 - 300 Hz, but the frequency band 10 - 100 Hz is preferred.

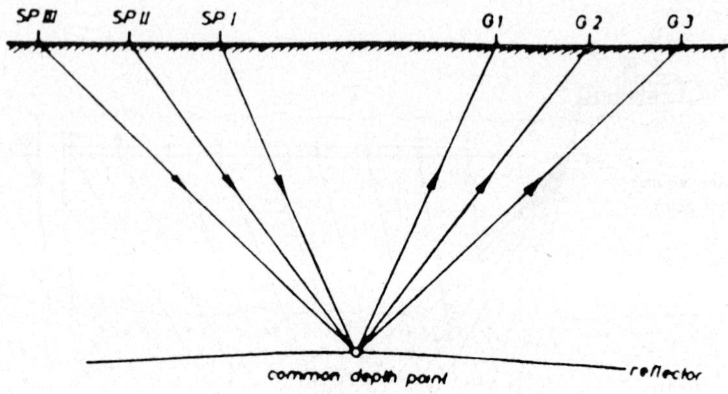

Fig. 4.16. Basic principle of multiple coverage.

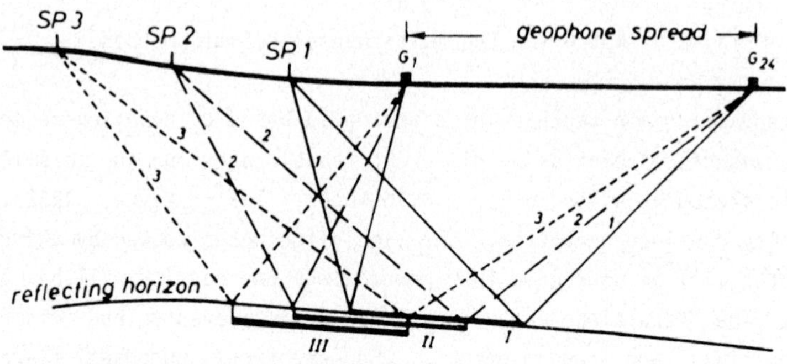

Fig. 4.17. Multiple coverage of reflecting horizon moving only the shot point.

Hydrophone streamers are made of a number of equally spaced active sections. Each contains a number of single elements in serial or parallel connection with the recording unit. The number of active sections is 12, 24, 48, 60, 96 or even 192. Several geophysical companies have systems with higher numbers of channels under development.

To maintain constant diving depths along the length of a neutrally buoyant streamer of up to 5 km length, a number of depth controllers must be fitted in. Each controller is clamped onto the streamer and constant depth is maintained by the action of fins, servolinked to a depth sensor. Several depth sensors in the streamer are used to monitor cable depth for operational control.

The position of the streamer relative to the ship's axis can be controlled by use of a transmitter on a tail buoy placed at the end of the streamer. Continuous control over the streamer is necessary to retrieve exact geometrical conditions for the later processing.

The recording unit controlls all processes during profiling on board of the seismic survey vessel. It accepts data from the positioning system, sends trigger signals to the sound source controller which releases the shot, and a record of several sec duration is recorded after each shot for all channels simultaneously. The single-channel data will be amplified and frequency-filtered, and the data of all the channels will then be multiplexed, digitized and stored on magnetic tape.

Due to the sophisticated methods and because of the theoretical considerations in multichannel seismics, data have to be processed prior to the final display. This digital processing requires larger computer capacities and will therefore generally be performed on-shore in a special computing center. On board ship, in most cases only displays are available that control the proper performance of the survey.

The aim of seismic processing in exploration (see Fig. 4.18) is to produce a visual display, showing a highly resolved reflection pattern so that interpretation may result in an unambiguous geological section of the strata (Robinson and Treitel, 1964; Silvermann, 1967). The processes of transforming the digital equivalent of the seismic signals into the final seismic section are not simple and involve a wide range of advanced mathematical techniques. A selection of the essential processing steps and their main aims can give only a rough impression of seismic processing. These steps include:

- True amplitude recovery (TAR). Due to its energy loss of sound waves running in seawater and sediment, deeper reflections are weaker than shallower ones. TAR compensates for this.
- Dynamic correction (normal moveout correction, NMO correction). Time delay for the travel time of a distinct seismic event with increasing shotpoint distance of the geophone.
- Stacking. The process "stacks" the NMO-corrected common depth point records. True seismic events will be added (enhanced), multiple and noise discriminated.
- Filtering.
  a) convolution, bandpass filter, notched filter
  b) deconvolution

The process depresses the frequency spectrum of the recorded signals to shorten the pulse- and light-reverberating seismic energy.

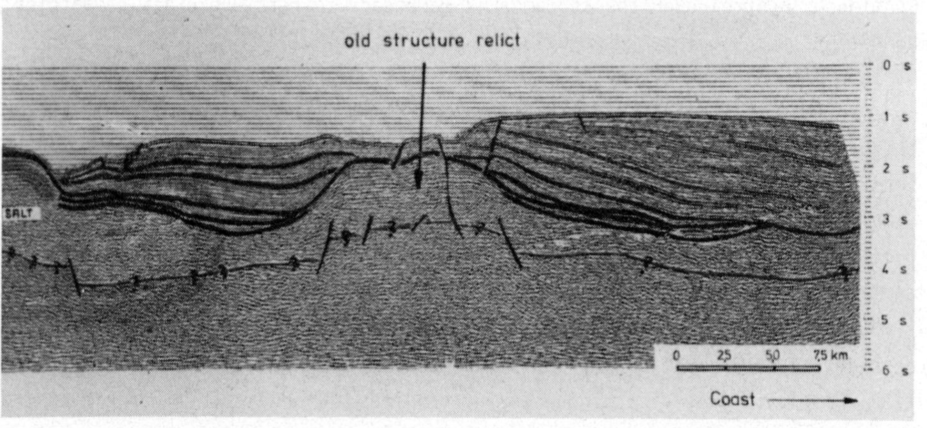

Fig. 4.18. Seismic cross section with salt dome and sediment basin.

MARINE GRAVITY MEASUREMENTS

Gravity measurement is one of the potential field methods used in marine geophysical prospecting, but it is meaningful only when an accuracy of better than 1 mgal is achieved. This is a severe limitation because the horizontal and vertical accelerations on a cruising ship can reach amounts of 50 000 or even 100 000 mgal. These accelerations are least near the intersection of the axes of roll and pitch (Dehlinger, 1978; Lange, 1961). The gravity sensor, mounted on a gyro-stabilized platform (see Fig. 4.19), should be as close as possible to the center of gravity of the vessel, in an compact air-conditioned container while the electronic part can be located up to 30 m away in an existing room or in an additional office container.

During a survey, the gravimeter is continuously levelled on the gyrotable. Levelling is checked in the harbour at the start of a cruise and corrected if necessary. Here, too, a gravity tie with a land station is recommended in order to determine the drift of the meter and to obtain all readings in a given system. Digital recording of the data is carried out in a constant rhythm, for instance the seismic "shot sequence" along with continuous analog recording, which is primarily considered as a control for the gravimeter operator during the survey. It has been sufficiently proved that all necessary values at a

Fig. 4.19. Gyro, platform and gravity sensor.

recording interval of about 10 seconds should be recorded.

On board, the duty of the operator of a gravimeter is limited to the regular control of faultless functioning of all gravity and navigation instruments. The exact determination of the ship position, the observance of a straight course, and a constant ship speed are necessary to achieve reliable gravity results.

The observed gravity is influenced by the Eötvös effect, which depends on ship speed and course. This effect is caused by the rotating earth. The centrifugal force resulting from the rotation of the earth is increased when the ship moves to the east, and is decreased when it moves to the west. The Eötvös effect can be as large as $\pm$ 40 mgal.

Other corrections to be applied are: time constant or time-delay filter, latitude correction, Bouguer correction, and two- or three-dimensional seafloor correction.

An important value in marine gravimeter surveying is the RMS error, calculated from "discrepancies" (differences in gravity) at line intersections. After adjustment, this error nowadays is between ± 0.2 and ± 0.5 mgal. Basically, the achievable accuracies are those of the applied corrections.

For interpretation purposes, the gravity values are interpolated on a regular grid and afterwards, the single isolines are calculated and plotted. The result is either a line presentation or an isogal map (Fig. 4.20).

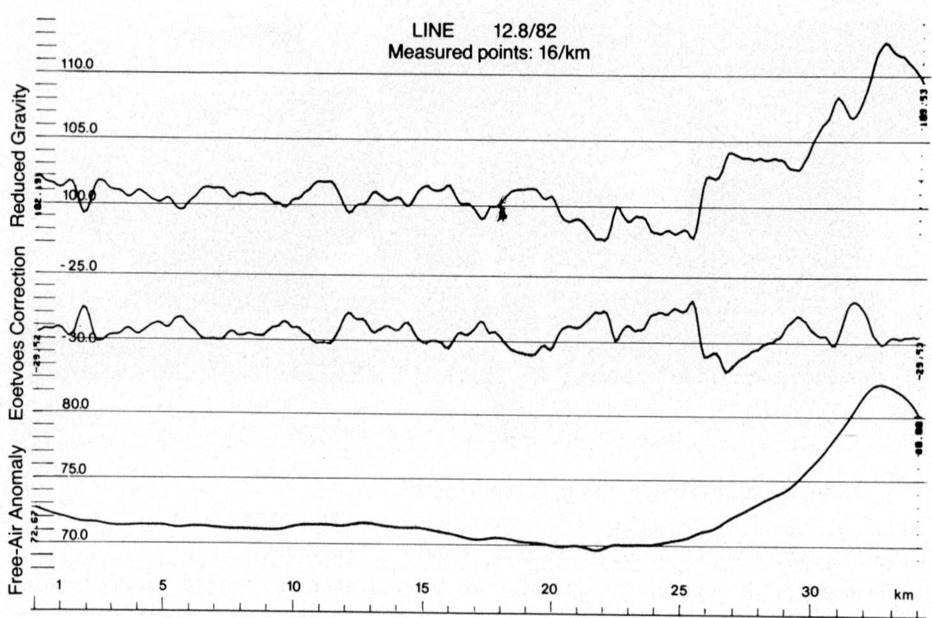

Fig. 4.20. Line presentation of onboard-processed gravity data (windforce: Bft 4; swell: 2 feet).

Today, gravity surveys are often carried out simultaneously with marine seismic surveys. The additional costs for marine gravity and its interpretation is about 5 to 10% of the costs of the seismic survey.

A typical gravimetric map is shown in Fig. 4.21.

MARINE MAGNETICS

Measurements of the magnetic field at sea have become a very important research area. After development of the proton-precession magnetometer and its marine configuration, the measurement of the total intensity got very simple, and extensive activity started about 1960.

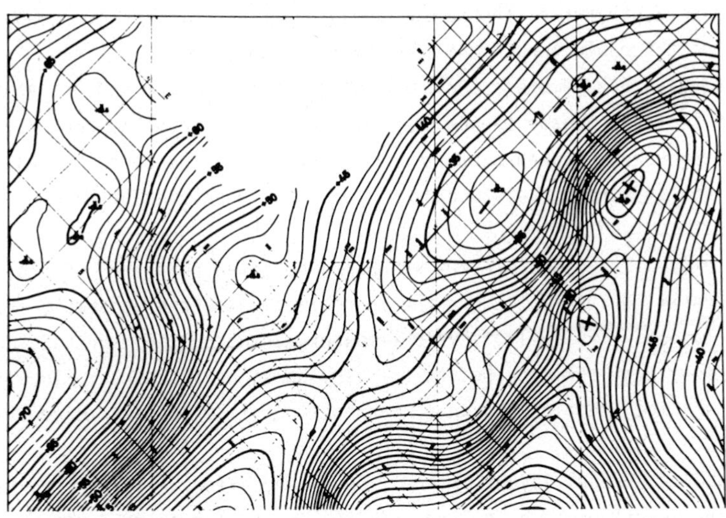

Fig. 4.21. Seagravity Bouguer map of a shelf area in the North Sea. The central part shows a typical gravity minimum over a salt dome.

The magnetic surveys demonstrated the existance of a pattern of lineated geomagnetic anomalies in large parts of the oceans (Vine, 1966). Their interpretation by the hypothesis of seafloor spreading confirmed Wegener's (1922) idea of drifting continents (1912), which was later rejuvenated by Hess (1962).

Geomagnetic measurements have not only greatly contributed to our understanding of the ocean floor, but are also of commercial value. Objects like pipelines and other iron bodies, can be detected with a magnetometer from a surface vessel. Further, they can assist the exploration for placer deposits, hydrothermal deposits, hydrocarbons etc., mostly by indirect deduction.

Execution of marine geomagnetic surveys

Instrumentation. For marine surveys, by far the most important instrument is the proton-precession magnetometer (Vacquier, 1968) which measures the total intensity of the geomagnetic field. Normally it is towed by a long cable behind the surveying ship which simultaneously can carry out seismic, gravimetric and bathymetric measurements. The cable length should be about three time the length of the ship to reduce the influence of the magnetic field produced by the ship (Bullard and Mason, 1961). In shallow water buoyant cables may be used; otherwise the sensor would be at a depth of 10 - 50 meters, depending on the cable length and the speed of the ship.

In some cases gradiometers are used which consist of two sensors at the same cable with the rear sensor typically 150 m behind the front sensor. By summing up the differences between sensors one can calculate the total intensity along the line and thus reduce disturbances from the time-dependent changes of the field.

There exist also magnetometers which can be towed near the seafloor. They give more detailed information on the magnetization of the uppermost layers.

Survey layout. Typically, cycle times of marine proton magnetometers are a few seconds. Thus, the data points are very dense on the survey lines, whereas the line-spacing must be chosen as large as possible to save expensive ship time.

Therefore, a careful layout of the lines is necessary which has to take into account the questions to be solved and problems of logistics. Data processing during the survey can give hints for further optimization of the grid. Generally, a net of parallel lines with some tie lines is best; the lines may be equally spaced or in groups which sometime give better information of the character on local anomalies. If a special feature (e.g. a pipeline) is to be followed, zig-zagging with two line directions may be very economic.

Data reduction. Nowadays, in most cases the International Geomagnetic Reference Field (IGRF) is the main field (IAGA, 1981); it is then subtracted. When necessary, small corrections are applied to it. These may be even more important when surveys from a longer time interval are combined because the local secular variation can differ considerably from the more regional representation in the IGRF. In most cases, the distance between the sensor of the magnetometer and the ship is not large enough for the magnetic field of the ship to be considered negligible. The remaining influence depends mainly on the magnetic properties of the ship, the course, the cable length and the magnetic inclination (Bullard and Mason, 1961). This so-called heading effect can be determined with sufficient accuracy by a test cruise with about eight different directions.

By far the greatest problems bring the short-term variations of the geomagnetic field. They can be larger than anomalies searched for and they may be strongly dependent on latitude and longitude. Sometimes it is possible to use nearby observatories or to establish special ground stations. In other cases buoys with recording magnetometers are used. The problems can also be reduced by gradiometers; they however have problems with the magnetic field of the ship which may severely disturb the integrated gradient.

Because of the problem inherent in the removal of the variations it is often best to make no corrections at all and mark or delete only those parts of the lines which contain no large disturbances. The problems expected must be carefully considered before making a survey; the best solution depends on many factors (Whitmarsh and Jones, 1969).

Data representation. Because the points on the survey lines are very dense and the line spacing often is too large for a full measurement of the anomalies, automatic contouring programs or manual contouring seldom can be used. Instead, it is now usual to plot the anomalies perpendicular to the lines or to project them into appropiate directions.

Coloured representations are just coming into use; the standard methods do not give satisfying representations.

Magnetic anomalies

Seafloor spreading and magnetic properties of the oceanic crust. Large parts of the oceans are covered by a pattern of lineated magnetic anomalies which are offset at sometimes very long faults (Roeser and Rilat, 1982). They are generated by the processes of seafloor spreading which was explained by Vine and Matthews (1963) and independently by Morley and Larochelle (1964) (see Fig. 4.22). Along mid-oceanic ridges the crustal plates on both sides of the ridge move away from one another thereby giving room for ascending magma. When magma cools through the Curie temperature the magnetic minerals contained in it become magnetized parallel to the magnetic field of the earth at that time.

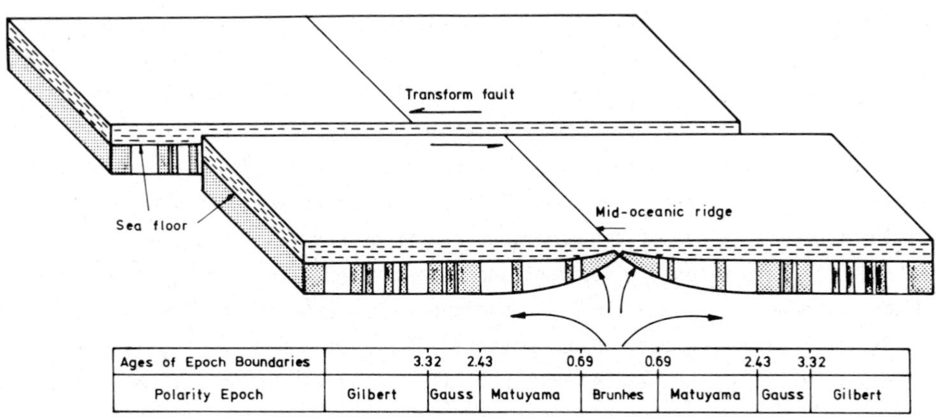

Fig. 4.22. The ocean floor as a huge tape recorder. In this model a transform fault cuts the ridge and the magnetic anomalies. Ages of polarity epoch boundaries are given in millions of years.

After some time the plates have separated again so much that hot magma can rise into the gap and cool. This process can repeat over millions of years. As the direction of the earth's magnetic field reverses in irregular intervals (Ness et al., 1980) stripes with two anti-parallel directions of magnetization are generated. The well-known reversal pattern made it possible to determine the age of a large proportion of the oceanic crust with remarkable accuracy.

The concept of seafloor spreading was confirmed by rock samples obtained by the Deep Sea Drilling Project (Denham and Schouten, 1979).

The greatest gaps in the linear pattern coincidence with plateaus which typically rise between 1000 and 5000 m from the surrounding deep sea (Nur and Ben-Avraham, 1982). Many of them are known to be continental fragments, whereas others are huge volcanic complexes. Important features are the so-called "magnetic quiet zones" where only weak magnetic anomalies exist for many different reason.

The magnetization of deep sea sediments is mostly weak but very stable (Opdyke et al., 1966). The origin of this magnetization remained a matter of debate until a few years ago when it was found that extremely small magnetic particles in the sediment were produced by bacteria (Kirschvink et al., 1985).

Exploration for natural resources. There are no straightforward methods available to predict the existence of natural resources below the seafloor from magnetic anomalies. A chance for a direct proof may exist in the case of placer deposits where the heavy mineral assemblage is often accompanied by magnetite or other magnetic minerals. Even in these cases it is equally important to delineate the source rocks magnetically (e.g. granites) (Sujitno et al., 1982).

The hydrocarbon potential of an area can be assessed only indirectly. The seafloor spreading concept allows the exclusion of large areas as non-prospective. In some cases magnetic anomalies may indicate intrusions which could have improved the maturity of possible source rocks.

Manganese crusts and nodules cannot be found by magnetic measurements. No relation is known between the magnetic signature of an area and the composition or the economic values of the manganese crusts or nodules found there.

Magnetic measurements give important clues to the assessment of hydrothermal deposits because they are closely connected to the exhalation of igneous rocks. A direct connection between the magnetic anomalies and the place where hot brines rise out of the sea bed does not exist. This may be caused by hot brines that come out of igneous complexes that are not yet cooled below their Curie temperature.

Artificial anomalies. Magnetic surveys in shallow water or with deep-towed magnetometers in deeper water provide a fast and cost-effective means to locate objects containing iron which may even be buried in the sediments. Pipelines, well heads and casings, mines, telephone cables, anchors, sunken vessels and submarines are such common items, and sometimes a magnetometer is the only tool capable of detecting them.

The search for pipelines is a special case. Although their shape is the same for long distances, their magnetic signature may vary considerably and even change its sense, because the remaining part of the pipe's magnetization depends on the orientation during the time of cooling through the Curie temperature or during strong deformations.

Interpretation techniques

When looking on marine magnetic data the first step is to define magnetic lineaments and to delineate areas with uniform magnetic signature. Afterwards it must be decided which anomalies should be modelled in order to get a feeling for the range of possible bodies.

Two-dimensional modelling plays the main role in marine geophysics because normally the surveys are not dense enough to justify the use of more sophisticated methods. Processes like Fourier transformation, upward and downward continuation and transformation on the pole suffer from the problem that the field is not known between the lines. The assumption of two-dimensionality which is often made brings errors into the result that are not easy to assess.

The seafloor spreading anomalies are a special case, and for their interpretation methods have been developed that are adapted exactly to this task. They work in the frequency domain (Parker and Huestis, 1974) or in the complex plane (Roeser, 1985).

Three-dimensional modelling is a must for the investigations of magnetic anomalies at fracture zones (Searle and Ross, 1975) or the anomalies of seamounts. For seamounts it has been tried to determine the magnetization direction, although often the available data are not sufficient to do that.

HEATFLOW MEASUREMENTS

Investigations of the thermal conditions within the earth's crust and the upper mantle are based in most cases on measurements of the heatflow through the surface of the earth. In the first instance investigations are undertaken for scientific reasons (Haehnel, 1972; Langseth, 1965; Von Herzen and Maxwell, 1959). Nowadays, the method developed is employed in offshore exploration as an important tool for the exploration for hydrothermally influenced ore deposits. Good results, for example, were obtained in the rift areas of the Red Sea, on the East Pacific Rise and in the Galapagos Rift.

Offshore heatflow investigations are less difficult than similar procedures on land (Birch, 1966). For determining the heatflow values, the temperature gradient and the thermal conductivity are measured within seafloor sediment. In most cases, the instruments are mounted along normal sediment coring devices (Fig. 4.23).

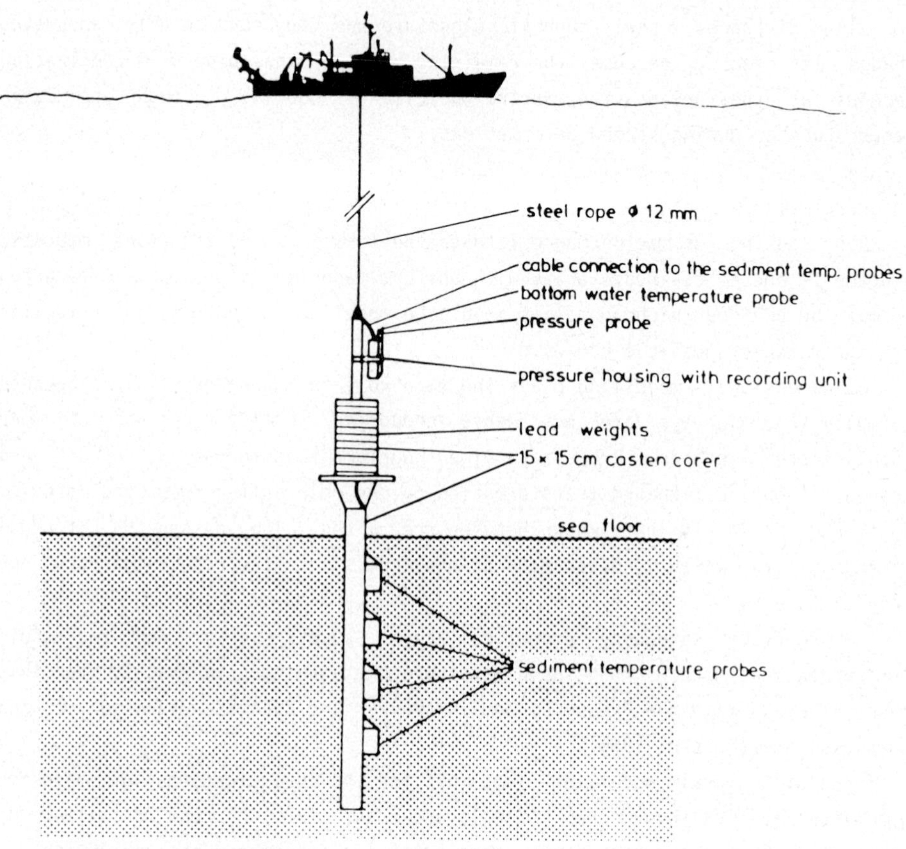

Fig. 4.23. Principle of heatflow measurement.

As shown in the figure, up to 5 temperature probes are mounted along a conventional box corer, at fixed distances. Via cable they are connected to an instrument housing at the topweight of the corer. The housing contains the measuring devices and a tape recorder for data storage. The heatflow measurements are performed parallel to the sediment sampling. After penetration of the corer into the sediment, the corer remains in place for 5 to 6 minutes. This period of time is necessary to guarantee a constant temperature and avoid possible frictional effects. For control purposes, the bottom water temperature, pressure (water depth) and the inclination of the probe are measured and also recorded on tape. The second parameter, the thermal conductivity of the sediment is measured on board from the recovered core. For this purpose, a thin needle-like cylinder is pushed into the sediment at a number of equidistant points along the core and warmed up by an internal heater (Fig. 4.24).

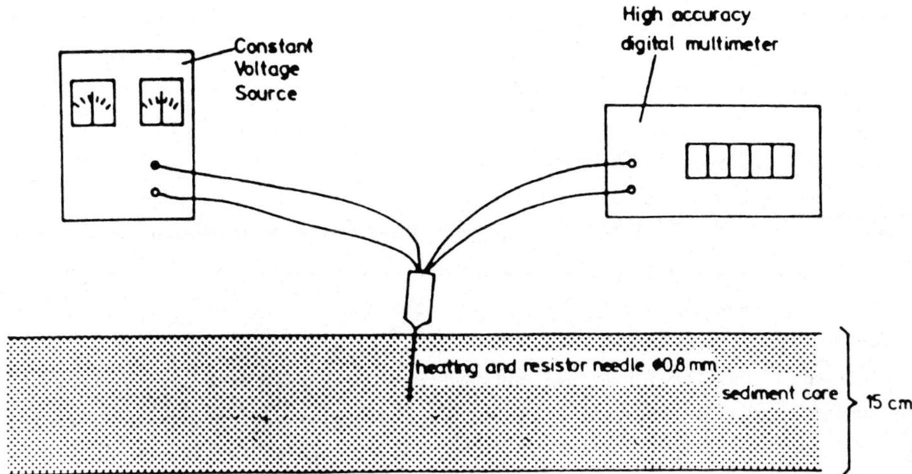

Fig. 4.24. Measurement of the thermal conductivity of a sediment core.

The rate of the temperature rise is measured by a small thermistor also inside the needle (Von Herzen and Maxwell, 1959; Langseth, 1965). The measured values are recorded from the beginning (before heating) over a period of about 2 minutes in a 10-second interval.

Both values, the average temperature gradient and the thermal conductivity are then used to calculate heatflow values. The use of heatflow measurements as an exploration tool has shown that a meaningful evaluation requires a large number of single measurements. Therefore, a combination of heatflow measurements and sediment sampling would be too time-consuming. To overcome this problem, a special probe was developed. Similar to a coring device the temperature probes are mounted along a solid metal 5-m pole. The recording unit is placed again within an instrument housing at the top weight. In this way it is possible to repeat the measurements at up to 6 different locations within an area without the necessity of taking the system onboard in between. The system measures even thermal conductivity in situ. This and the ability to make several measurements as a series gives advantages to the new system.

VISUAL INSPECTION OF THE SEAFLOOR

The exploration methods described above measure physical parameters of the sediment and substrata or the interface sediment/seafloor. Interpretation and evaluation are usually based on practical experience. The advantages of these methods are a fast survey progress and areal overview. For detailed analyses, however, a direct inspection of the phenomena is necessary. Adapted to the assignments, special probes were developed for visual inspection of the seafloor (Spiess, 1985). Most of the systems are operated from a deep sea wire or cable on board of a research vessel.

Photographic survey, still cameras, photo sled

Optical methods allow a direct visual inspection of phenomena at the seafloor. The interpretation is easy and uniquivocal. On the other hand, due to the limited light propagation in water the areas which can be inspected are limited to several square meters, and consequently the survey speed is low (Richter, 1980).

A common practice for a fast and more or less random overview in the course of normal sampling programs is the use of photo cameras attached to free fall grabs (see Chapter 5). However, the cameras are simple and the illumination is not optimal. Therefore the quality of the pictures is not always sufficient and qualified for sophisticated interpretation.

Much better results can be obtained by the use of photo sleds. These systems are designed to retrieve photographic pictures of high quality from the seafloor (see Fig. 4.25). The camera equipment is mounted inside a stable metal frame, the sled. The main components of a photo sled are: a deep-sea photo camera of 70-mm or nowadays even 35-mm frame size, a flash-light array and the triggering unit. Often, two cameras are used to permit taking of stero photographs. The film reserve and the battery capacity for the flash lights allow up to 600 shots within an application cycle. Besides the photo system, the sled allows the installation of additional oceanographic probes such as water samplers, thermometers etc. The photo sled is deployed from a deep-sea wire or cable on board the vessel, and its handling is easy. A first estimate for the bottom distance is the wire payed out. The final approach to the seafloor can be controlled by means of pingers. Pingers that are mounted at the frame transmit sonar pulses in a fixed interval of 1 second. When the sled approaches the seafloor, the research vessel is able to record this pulse on two transmission paths: the direct path and shortly later, the echo reflected at the seafloor. The time difference is proportional to the bottom distance and decreases with the approach. In this way, an exact determination of bottom distance is possible.

The pinger signal can be coded with additional information such as water temperature, pressure or event markers used to confirm the functioning of the system. A typical application is the message for the successful operation of

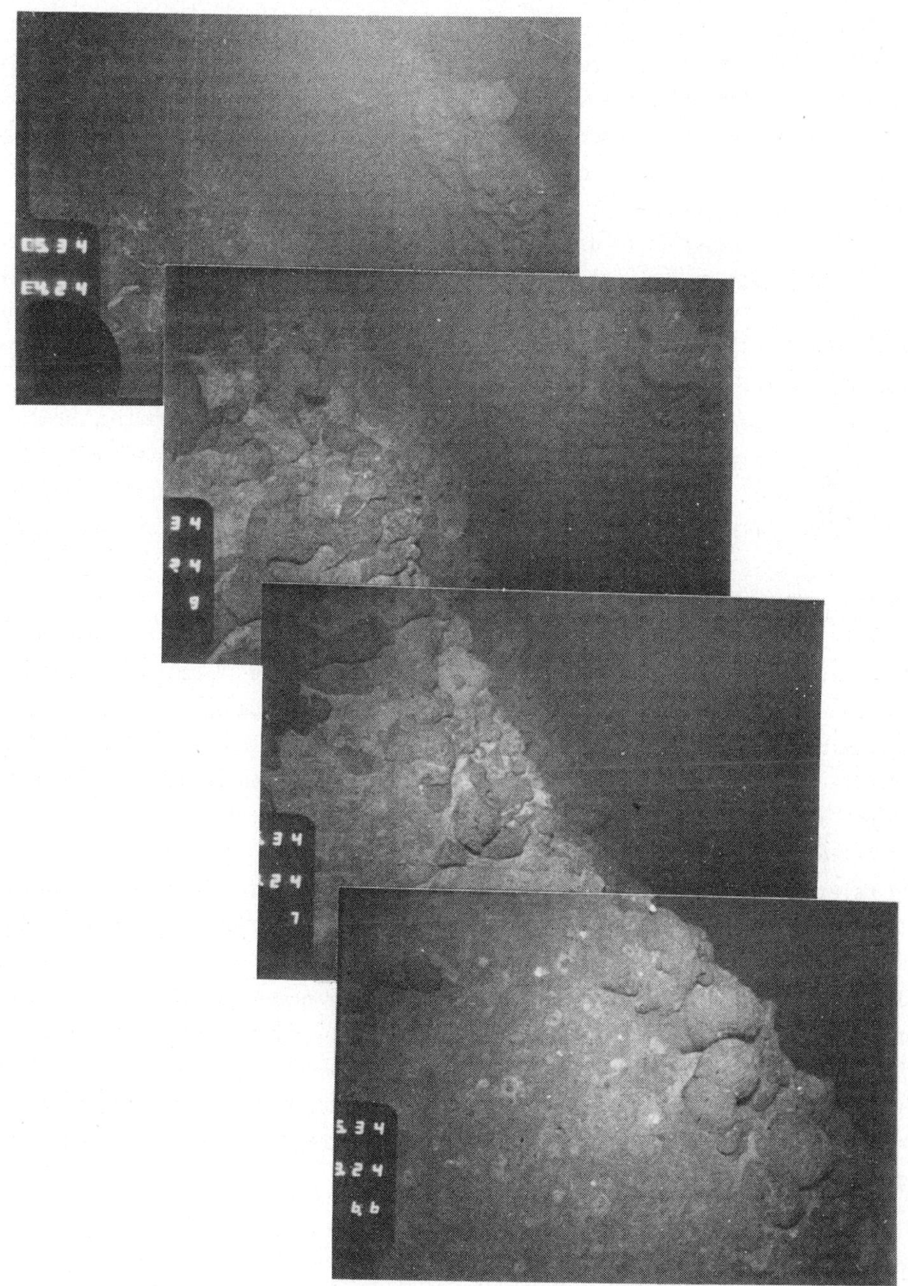

Fig. 4.25. Series of deep sea photos of the Pacific Ocean. The set exhibits pillow lava in the neighbourhood of a seamount (water depth about 3500 m). Note that the pictures originally were produced in colour.

water samplers often used together with a photo sled.

In most cases in using photo cameras, photographing is triggered by weights, keeping the system in a more or less constant bottom distance, i.e. slightly more than the length of the trigger rope and deployed always when a picture is taken. Normally, the operation is controlled with the deep-sea winch; in many cases, however, the motion of the ship generated by the swell is sufficient to trigger photographing.

When it is possible to use a cable instead of a simple wire the photo sled can be designed in a more sophisticated manner. The main advantage is the use of additional TV cameras. Such TV cameras allow the inspection of the seafloor at a monitor and the optimal placing of the photographs. In this case, of course, the triggering of the cameras will be performed via the cable.

The use of a cable makes the photo sled more efficient. On the other hand, a cable generally reqires a more sophisticated winch system and this limits the application to only a few modern research ships. In general, however, application of both cable or deep sea wire restrict the mobility of the ship.

Deep sea TV

All problems connected with the cables are also valid for deep sea television systems. As in the case of the photo sled, the TV cameras, light and trigger units are mounted in a steel frame and connected to the ship via a transmitting cable. But as the TV inspection of the seafloor is performed along profile lines, the buoyancy of the cable must be compensated for. In most cases, a so-called depressor system is used (see Fig. 4.11).

A weight of about 2.5 t is mounted at the cable 80 - 100 m in front of the TV sled and the weight of the sled itself is compensated to zero by buoyancy. This arrangement causes two effects: The depressor weight compensates the buoyancy of the cable, and the more or less weight-neutral area between the depressor and the TV sled attenuates the up- and down-movement of the depressor generated by the swell, transforming them into a to and fro motion of the sled. This transformation keeps the cameras at a constant distance above the seafloor and in this way produces the preconditions for a continuous inspection. However, with this sophisticated depressor array the operating speed is generally limited to 1 - 2 kn.

REFERENCES

Albers, V.M., 1965. Underwater acoustic handbook - II. The Pennsylvania State University Press, Library of Congress Catalog Card No. 64-15069.

Anstey, N.A., 1970. Seismic prospecting instruments. Gebrüder Borntraeger, Berlin, Vol. 1, 156 pp.

Anonym, 1982. Multibeam bathymetric survey system. Sea Technology., pp. 28-31.

Anonym, 1985, 1985. GLORIA surveys the sea floor. Mining Magazine, August, pp. 83-84.
Birch, H., 1976. Heat-flow measurements in the Atlantic Ocean, Indian Ocean, and Red Sea. J. Geophys. Res., 71: 583-586.
Blidberg, R.D. and Porta ,D.W., 1976. An integrated acoustic seabed survey system for water depth to 2000 feet. OTC, 2655: 447-464.
Bullard, E.C. and Mason, R.G., 1961. The magnetic field astern of a ship. Deep-Sea Res., 8: 20-27.
Bullen, K.E., 1963. An introduction to the theory of seismology. Cambridge, 350 pp.
Carnetzki, 1978. Pattern recognition for side scanning sonar. Univ. of Rhode Island, 110 pp.
Colvin, D.W., 1982. Wide-swath bathymetry surveying of the UK continental shelf by sector-scanning sonar and the possibilities for on-line contouring. Oceanology International, paper 0182 5.1.
Dehlinger, P., 1978. Marine gravity. Elsevier, Amsterdam, 294 pp.
Denham, C.R. and Schouten, H., 1979. On the likelyhood of mixed polarity in oceanic basement drill cores. In: M. Talwani and C.G.A. Harrison (Editors), Deep drilling results in the Atlantic Ocean crust. Amer. Geophys. Union, Washington, pp. 160-165.
Dix, C.H., 1952. Seismic prospecting for oil. Harper and Broth., New York, 414 pp.
Dohr, G., 1981. Applied geophysics - introduction to geophysical prospection. In: Geology of petroleum, Vol.1. Enke, Stuttgart, 222 pp.
Drenkelfort, H., 1970. Darstellung von Schichtungen im Wasser und im Sediment mit Ultraschall-Echoanalagen. Interocean'70, Düsseldorf, Vol. 2, pp. 269-274.
Drenkelfort, H., 1980. Sediment Echolote. Information 1/80, Honeywell- Elac-Nautic GmbH, Kiel, pp. 1-15.
Evenden, B.S., Stone, D.R., Anstey, N.A., 1971. Seismic prospecting instruments. Gebrüder Borntraeger, Berlin, Vol. 2, 195 pp.
Fagot, M.G., Cholson, N.H. and Moss, G.J., 1980. Deep-towed geophysical array system development. NORDA-TN-75, 70 pp.
Haehnel, R., 1972. Heat-flow measurements in the Ionian Sea with a new heatflow probe. Meteor Forsch. Ergebnisse, Reihe C, No.11, pp. 105-108.
Hansen, R.D., 1984. High resolution seismic results of the detailed Sonne- 17 survey areas (Chatham Rise New Zealand). Geol. Jb. D65: 57-67.
Hess, H.M., 1962. History cf ocean basins. In: A.E.J. Engel, H.L. James and B.F. Leonard (Editors), Petrologic studies. Geol. Soc. Am., New York, pp. 599-620.
Honeywell Elac, 1966. Narrow beam echsounder for survey and research, Electroacustic GmbH, Kiel. Intern. report.
Honeywell Elac, 1980. Narrow beam echosounder. Electroacustic GmbH, Kiel, internal report.
Huntec, 1970. Hydrosonde deep tow system. Huntec Ltd, Scarborough, Ontario, Canada, internal report.
Hutchins, R.W., et al., 1982. Characterization of seafloor sediments geoacoustic scattering methods using heigh resolution seismic data. Oceanology International, paper no. 0182 5.6.
IAGA, 1981. International geomagnetic reference fields. EOS, 62: 1149.
Kinsler, E.L. and Frey, A.R., 1962. Fundamentals of acoustic. Wiley, New York, 524 pp.
Kirschvink, J.L., Jones, D.S. and MacFadden, B.J., 1985. Magnetic biomineralization and magnetoreception in organisms. Plenum, New York, 438 pp.
Konrad, W.L., 1975. Design and performance of parametric sonar systems. Naval Underwater System Center, New York, 24 pp.
Lange, W., 1961. Seegravimetrie. Freiberger Forschungshefte, Reihe C, 83 pp.
Langseth, M.G., 1965. Techniques of measuring heat flow through the ocean floor. Geophysical Monogr., Am. Geophys. Union, No. 8, 276 pp.

Lettau, O. and Richter, H., 1985. Explorationsverfahren in Flachwasser- und Tiefseegebieten. 5. Mintrop Seminar, pp. 255-290.
Lettau, O., 1985. Subbottom Profiler Messungen im Roten Meer. Deutsche Hydrogr. Gesellsch., 16 pp.
Lowell, F.C. and Dalton, W.L., 1971. Development and test of a state-of-the-art sub-bottom profiler for offshore use. Offshore Technology Conf., Vol. 1: 143-158.
Luyendyk, B.P., Hajic, E.J. and Simonett, D.S., 1983. Side-scan sonar mapping and computer-aided interpretation in the Santa Barbara channel, California. Mar. Geophys. Res., 5: 365-388.
Mayne, W., 1962. Common reflection point horizontal stacking techniques. Geophysics, 27: 952-965.
Meyer, K., Richter, H. and Hansen, R.D., 1976. Grundlagen und Beschreibungen neuer geophysikalischer, geologischer und bodenmechanischer Untersuchungen bei offshore-Bauvorhaben. Interocean 1976, Düsseldorf, pp. 354-368.
McQuillin, R. and Ardus, D., 1977. Exploring the geology of shelf seas. Graham and Trotman, London, 234 pp.
Morley, L.W. and Larochelle, A., 1964. Paleomagnetism as a means of dating geological events. Roy. Soc. Canada, Spec. Pub. 8: 512-521.
Ness, G., Levi, S. and Couch, R., 1980. Marine magnetic anomaly timescales for the Cenozoic and Late Cretaceous. Rev. Geophys. Space Physics, 18: 753-770.
Nur, A. and Ben-Avraham, Z., 1982. Ocenaic plateaus, the fragmentation of continents and mountain building. J. Geophys. Res., 87: 3644-3661.
Opdyke, N.D., Glass, B., Hays, J.D. and Foster, J., 1966. Paleomagnetic study of Antarctic deep-sea cores. Science, 154: 349-357.
Osborne, L.C., 1984. Canadas seabed 2 program: reliable high-speed mapping heads out for the deep seas. Sea Technology, October, pp. 16-19.
Parker, R.L. and Huestis, S.P., 1974. The inversion of the magnetic anomalies in the presence of topography. J. Geophys. Res., 79: 1587-1593.
Prior, D.B., Coleman, J.M. and Roberts, H.H., 1981. Mapping with side scan sonar. Offshore, 4: 151.161.
Renard, V. and Allenou, J.P., 1979. Sea beam multi-beam echo-sounding in "Jean Charcot". Intern. Hydrograph. Rev., LVI(1): 35-67.
Richter, H. and Hansen, R.D., 1976. Digital recording and processing of high resolution seismic for offshore soil investigation. Offshore Technology Conf., OTC 2612: 953-960.
Richter, H., 1980. Explorationstechnische Entwicklungen. Geol. Jb., D38: 157-182.
Richter, H., 1984. Lotungen (Echolote, Side Scan Sonar u.a.)- Ein Überblick. Deutsche Hydrograph. Gesellsch., 15 pp.
Richter, H., 1985. Fächerlote- Kartierung und Auswertung. Deutsche Hydrograph. Gesellsch., in prep.
Robinson, E.A. and Treitel, S., 1964. Principles of digital filtering. Geophysics, 29: 395-404.
Roeser, H.A. and Rilat, M., 1982. A world map of the magnetic sea-floor spreading anomalies. Geol. Jb., 23: 71-80.
Roeser, H.A., 1985. The rapid calculation of sea-floor spreading anomalies in the complex plane. In press.
Searle, R.C. and Ross, D.A., 1975. A geophysical study of the Red Sea axial trough between 20.5^{o} and 22°. Geophys. J. Roy. Astr. Soc., 43: 555-572.
Silvermann, D., 1967. The digital processing od seismic data. Geophysics, 32: 988-1002.
Smith, S.M., 1983. Sea beam operator manual. Scripps Institute of Oceanography, La Jolla, CA.
Spiess, F.N., Lowenstein, C.D. and Boegeman, D.E., 1978. Fine grained deep ocean survey techniques. Offshore Technology Conf., OTC 3135: 715-721.
Sujitno, S., Boujo, A., Cressard, A. and Robach, F., 1982. A preliminary approach for offshore tin prospecting by high sensitivity magnetisms: a new tool. Note by Indonesia, CCOP (XIX), 70, 25 pp.
Ulrich, J., 1976. Anwendungsmöglichkeiten moderner Echolotsysteme in der

ozeanographischen Forschung. Electroacustic GmbH, Kiel.
Vacquier, V., 1972. Geomagnetism in marine geology. Elsevier, Amsterdam, 185 pp.
Vine, F.J. and Matthews, D.H., 1963. Magnetic anomalies over oceanic ridges. Nature, 199: 947-949.
Vine, F.J., 1966. Ocean floor spreading: new evidence. Science, 154: 1405-1415.
Von Herzen, R. and Maxwell, A.E., 1959. The measurements of the thermal conductivity of deep-sea sediments by a needle-probe method. J. Geophys. Res., 64: 1557-1563.
Wegener, A., 1922. The origin of continents and oceans. Methuen, London, 212 pp.
Whitmarsh, R. and Jones, M., 1969. Daily variations and secular variations of the geomagnetic field from shipboard observations in the Gulf of Aden. Geophys. J., 18: 477-483.

CHAPTER 5

MARINE GEOCHEMICAL EXPLORATION METHODS

W.L. PLÜGER and H. KUNZENDORF

INTRODUCTION

Geochemistry is widely accepted in marine mineral exploration and forms an integral part of nearly every marine exploration campaign. Its primary function is to back up marine geological investigations and the marine geophysical surveys described in Chapter 4.

Marine geochemistry includes the chemical analysis of the minerals and samples by means of direct and indirect methods. Although many exploration activities do not include geochemical exploration explicitly in their working schemes, there actually is nearly no modern mineral exploration without geochemical work. More or less extensive geochemical investigations are quite often hidden under exploration activity headings like sampling, analysis and data evaluation.

During the past 20 years, geochemical exploration has proved to be an important technique in terrestrial mineral exploration and there are numerous publications in the form of textbooks (e.g. Hawkes and Webb, 1962; Levinson, 1974; Rose et al., 1979) and individual articles. The Association of Exploration Geochemists (AEG) has published an excellent bibliography on the subject (Hawkes, 1982; 1985). Furthermore, the methods of geochemical exploration have been refined considerably and geochemical methods are now used increasingly in the search for hidden ore bodies.

Geochemical exploration surveys are generally composed of:

- Sampling
- Chemical analysis, and
- Interpretation of analytical data

In geochemical exploration, the primary and secondary dispersion halos of mineralizations are determined and interpretated in order to locate these.

The general philosophy behind this technique is that elements are dispersed from their original site, either in solution, in particulate form or in detrital phases. By determination of the concentration gradients of suitable chemical elements or mineral phases, the location of the primary mineralization can probably be detected.

In principle, two geochemical exploration steps exist:

- Reconnaissance surveys
- Detailed surveys

During a reconnaissance survey, prospective and non-prospective areas are distinguished usually on a large scale. The detailed surveys concern the follow-up exploration of prospective target areas on a smaller scale.

Marine geochemical exploration techniques are of a relatively recent origin. They are generally not as refined as the terrestrial methods, but apparently they have a great potential in several phases of marine mineral exploration.

Sampling the seafloor is many times more expensive than sampling in terrestrial mineral exploration. This has an imperative effect on the marine sampling strategy, i.e. grid sampling has to be planned very carefully.

As described in Chapter 2, the nature of the various marine mineral commodities, their formation and therefore also their geochemical behavior are quite different. As a consequence, there exist no uniform geochemical method for exploration of the different underwater mineral deposits; each type requires its own exploration strategy, involving sampling (choice of sample grids and sampling devices), chemical analysis (choice of sample type and analytical instrumentation) and data interpretation.

The principle methods in use in marine mineral exploration have been divided into direct and indirect methods (Cronan, 1980). Indirect methods include geophysical techniques and geochemical methods, which in turn deal with analysis of recovered material and in-situ analysis. Direct methods comprise visual observation of the seafloor by various instrumentations (see Chapter 4) as well as sampling of the water column, the seafloor surface and the seafloor substrata.

This chapter describes the standard methods for sampling seawater, sediments, rocks and minerals. Analytical methods to be used on board an exploration vessel and in land laboratories are then briefly outlined. Finally, two marine geochemical exploration examples are given.

MARINE EXPLORATION SAMPLING DEVICES

For the purpose of marine geochemical exploration a wide range of sampling equipment able to collect water, sediment, rock and mineral samples exits. Although sophisticated indirect geophysical exploration methods have been developed in recent years (see Chapter 4), sampling is still an unalterable prerequisite for nearly every exploration campaign. By sampling and subsequent analysis of the recovered material onboard ship, exploration projects can be guided saving time and costs. Recovered samples are also of considerable importance to characterize the marine environment, to draw conclusions for genetic interpretations and to evaluate marine mineral deposits.

As a result of the enhanced activity in marine mineral exploration during the past two decades the development of sampling equipment has gained new impulses, resulting in the design of new sampling tools. Nevertheless, several proven sampling devices, designed many years ago, still are in operation.

Marine sampling apparatus can be divided into:

- Water sampling devices
- Seafloor surface and upper substrata sampling apparatus
- Seafloor substrata sampling instruments

In the following, a brief description is given of selected devices for water, sediment, mineral and rock sampling.

Water sampling devices

In marine mineral exploration water sampling and subsequent analysis of these samples is especially important in the search for active hydrothermal sites on spreading ridges and adjacent areas. Also, during exploration for other mineral commodities, water samples are taken often from a scientific point of view in order to compile, e.g., the inventory of chemical and/or (micro-) biological composition of the water column and/or to draw genetic conclusions with regard to the formation of the deposits. For example, the latter is of interest for the formation of manganese nodules in areas with high biological productivity in surface waters, and with regard to the flux and uptake of elements by deep-sea sediments.

There exists a wide variety of water samplers, ranging from a single wirebound sampling bottle, via hydrographic chains consisting of an optional number of bottles, to sophisticated sampling systems which contain up to 24 water samplers able to collect water from every desired depth. To avoid contamination, the majority of the water samplers are manufactured from hard plastics or stainless alloys.

Fig. 5.1 shows an example of a small-sized Nansen water sampler and its mode of operation (Turekian, 1985). The sampler is connected to the wire at a discrete distance from the wire end, which is equipped either with a heavy dead weight or with a sediment sampling device. The water sampler has two fixed connections to the wire, whereby the upper one can be unlocked by a messenger weight that is sent down along the cable. If the upper connection is unlocked, the sampler turns around and the former open lids are closed, resulting in the enclosure of the water sample from the discrete depth. At the same time, the ambient water temperature is fixed by the tear-off of the mercury column of the attached thermometer.

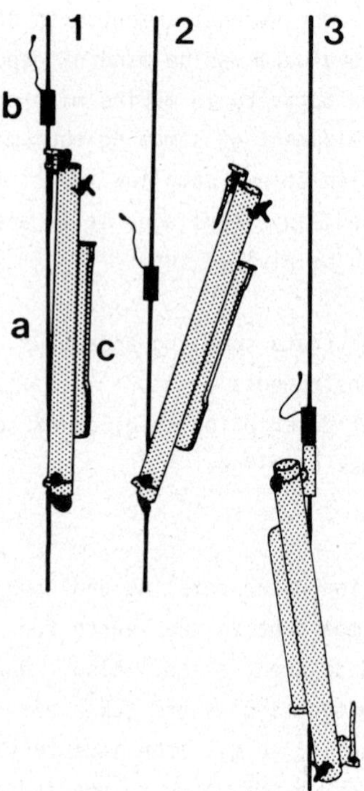

Fig. 5.1. Stylized illustration of a Nansen water sampler (a). Release of the device is initialized by a messenger weight (b), which descends along the wire. Ambient water temperature is measured by two thermometers, installed in a thermometer rack (c). Closure of the initially open water bottle (1) is obtained by rotation (2). By heaving, the rotated sampler (3) is recovered from a pre-defined depth. Approx. length of the water sampler is 80 cm.

This type of water samplers may be combined in a hydrographic chain by adapting a series of instruments to the wire at discrete distances. The complete water sampler chain often is positioned by the use of bottom distance pingers and release of water samplers is again by messenger weights. After touch of the first sampler by the first messenger weight, the second messenger is released and in turn activates the second sampler; messenger weight of the second water sampler activates the third water sampler etc.

A special method to sample bottom water with Nansen water samplers has been described by Sipos et al. (1980) where Nansen water samplers were mounted on free-fall grab samplers. In this way, water sampling is independent of the hydrowire and contamination caused by the wire is avoided. According to these authors, a comparison of the Pb contents of water samples taken by wire-bond samplers and by free-fall samplers in the Pacific Ocean clearly shows contamination from the hydrowire.

The Nansen water sampler only collects 1.2 l of seawater. In many cases, larger water samples are required and therefore, larger samplers have to be used. Standard equipment of many research vessels are Niskin water samplers being able to collect up to 60 l of seawater from the water column. This type of sampler is also equipped with thermometers to register the water temperature at the sampling site. Closing of the water sampler is again facilitated by messenger weights but no turning around of the sampler bottle is necessary due to a special closing mechanism.

While lowering the sampling equipment through the water next to the ship's hull, initially open systems like Nansen and Niskin bottles may be contaminated by surface water, i.e. by oil slick and metallic contributions from the ship's hull. To avoid these contamination sources, so-called "Go-Flow" water sampling bottles (General Oceanics, Miami, U.S.A.) were developed which are immersed in closed state and opened by a pressure-activated release at approximately 10 m depth, altered if necessary. During lowering, the sampler is flushed until sampling initiation. Closing the sampling device at a preselected water depth is facilitated by a rotating valve which is activated by a messenger weight. When the sampler is retrieved, an inert gas can be injected into the bottle to force the sample through a filter out of the bottle valve. This type of water sampler is available with bottle capacities of between 1.7 and 60 l. Bottles may also be combined to a hydrographic system along the deep-sea wire as previsously described or they may be in the form of a sampling system containing up to 24 bottles mounted on a rack, which then is lowered by a single or a multiconductor electromechanical cable (Fig. 5.2). Such multiple-sampler systems are controlled by a deck command unit, which triggers the sampler at every desired depth and provides a system status and a confirmation that the sample was collected properly.

A further development of this kind of multi-samplers is a miniaturized version, provided with teflon or polyethylene bags instead of the rigid polyvinyl chloride tubes for collecting water samples. Configurations of 1 to 24 samplers, in sizes of .5 to 30 l are available. The advantages of this system are the sealed contamination-free bags composed of special material,

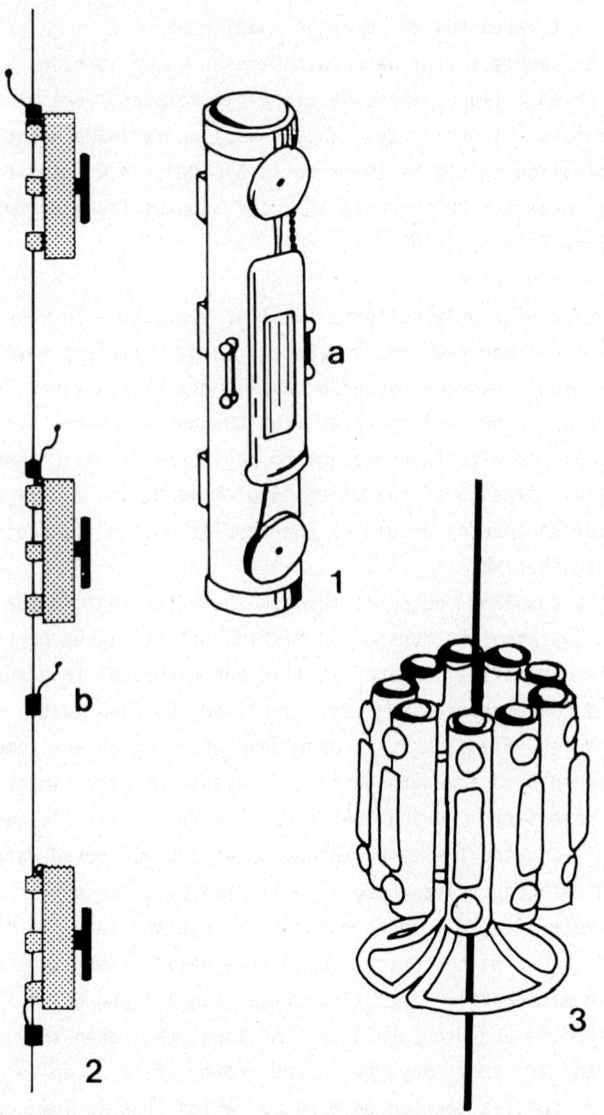

Fig. 5.2. Sketch of a "Go-Flow" water sampler (1) with an adapted thermometer rack (a) as developed by General Oceanics Inc. Various water samplers may be combined either to a hydrographic chain (2) which is released by messenger weights (b), or to a rosette water sampler device (3) which is released at different depths by a deck-command unit, using a electromechanical cable. Water samplers are available in size bewteen 60 and 130 cm.

which are opened, filled and resealed at the sampling site at discrete depths. The samples may then be preserved, stored and shipped without opening the bags.

Seafloor surface and upper substrata sampling apparatus

The choice of the equipment to sample the seafloor and/or the underlying sediments, rocks and minerals depends highly on the practical or scientific goal of the research project. Since sampling, especially in deep water, is time-consuming and ship-time is loaded with high costs, a careful selection of sampling equipment has to be made to minimize time spent and to maximize sampling efficiency.

Since the beginning of the research of the seabed a hundred years ago a wide range of sampling devices has been developed. Besides highly sophisticated instruments, rather primitive sampling devices are still in use. Table 5.1 summarizes sampling equipment which at present is standard gear onboard an exploration vessel. Additionally, new developments like TV-controlled, electro-hydraulic grabs are also presented.

Dredges. Dredges are rather primitive devices that were in use since seabed sampling started in the last century. They are normally employed when larger volumes of hard rock, manganese nodules or other coarse components have to be collected from the seafloor. Dredges are constructed either as a rigid steel box of various dimensions and capacities or as so-called chain-bag dredges, which consist of a rectangular steel mouth with a flexible bag of steel chains (Fig. 5.3).

For operation, dredges are connected to the ship's wire and they are then towed along the seafloor by the moving ship. Before a dredge haul is started, the seafloor topography in especially rough terrain has to be known to optimize the sampling route. Ship speeds during dredging operations are between 0.5 and 1.5 knots and the wire paid-off may reach twice the water depth in order to keep bottom contact. Sometimes a heavy dead weight (compensator) is adapted to the wire at a distance of about 100 m in front of the dredge and a bottom pinger is installed (a few meters above the dead weight) to control bottom contact. Bottom contact is also evidenced by tensiometer readings.

To localize the dredge track, a continuous recording of water depth, ship position and wire paid-off is necerssary. If time is at disposal, the drege track may be recorded by using an acoustic transponder navigation system. In any case, the position of the true sampling site cannot be determined exactly; the recovered material is derived from any point along an approximately 3 km long pathway at the seafloor.

TABLE 5.1.
Summary of seafloor sampling devices for marine exploration work.

Device	Size	Application	Remarks
Dredges:			
Box dredge & chain bag	Up to 120x60 cm mouth opening	Rock fragments, manganese nodules, crusts	Sampling sites not accurately known
Grabs:			
Van Veen, Smith-McIntyrne, Shipek	Cross section up to 500 cm^2 or more	Surface sampling of sediments and placers	Usually disturbed samples, often loss of fine-grained material
Free-fall	Cross section up 1500 cm^2	Manganese nodule recovery	Still photo camera adaptable
Corers:			
Spade	20x30 cm	Surface sampling up to 50 cm depth	Undisturbed samples; peneliquid layer may be lost
Box	15x15 cm 30x30 cm	Sampling of surface and deeper stratigraphic layers up to a depth of 15 m	Undisturbed samples combined heatflow measurements possible
Piston	Diameter 4 to 10 cm	Sampling of sediment layers up to 30 m depth	Surface samples slightly disturbed
Vibro-corer	Diameter 5 to 9 cm	Sampling of placers and coarse sediments	Disturbed samples from defined depths
Free-fall	Diameter 4 to 8 cm	Sampling of sediment up to 1.2 m	Sampling along transits

Fig. 5.3. Box dredge with recovered manganese nodules (upper photograph). Chain bag dredge with basalts being recovered at the RV "Sonne" (lower photograph).

<u>Grabs</u>. Various types of grabs exist to provide surface samples from the seafloor. All devices, with exception of the free-fall grab, are deployed on a wire from the research vessel. Opposite to dredge samples, the locations of the samples taken by grabs are fairly accurately known.

One of the oldest devices is the grab developed by van Veen (1936), which is available in sampling sizes ranging from 0.25 to over 1 m. By the closure of two quarter-cylinder buckets, which depends on the strain from the hoisting wire, a sample from the sediment surface is collected (Fig. 5.4).

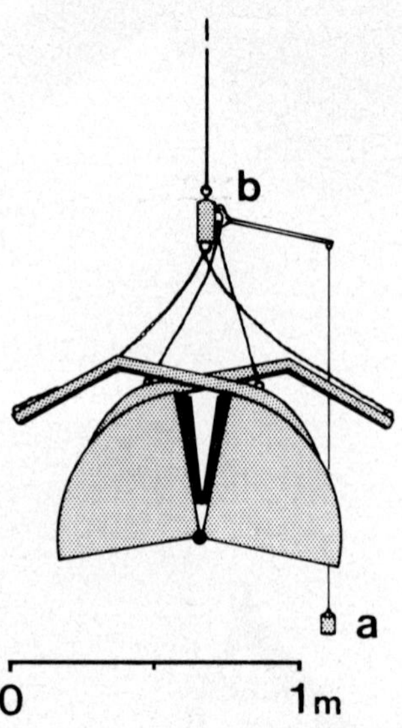

Fig. 5.4. Schematic illustration of the van Veen grab. After tipping the seafloor by the trigger weight (a), the release hook (b) is unlocked and a sample is taken by closing of the jaws, when heaving the device.

In case of the grab after Smith and McIntyrne (1954), the closure of the jaws is spring assisted, and a frame is mounted around the grab to prevent tipping over on contact with the seafloor. The frame acts also as a counter mass to the springs, resulting in a penetration of the buckets into the seafloor before closure. Both grabs, van Veen and Smith-McIntyrne, are not designed to be watertight and there is a loss of fine-grained sediments during the raise of the grab. If pebbles or rock fragments are held between the jaws, no entire closure will be achieved, resulting in the loss of parts or even the total sample.

A further type of grab, in which samples are protected from washing during retrieval, is the Shipek grab (Fig. 5.5). Complete closure is obtained by a half-cylindrical bucket, rotated by two external helical torque springs. This grab can be provided with a camera and flash to obtain additional information from the seafloor.

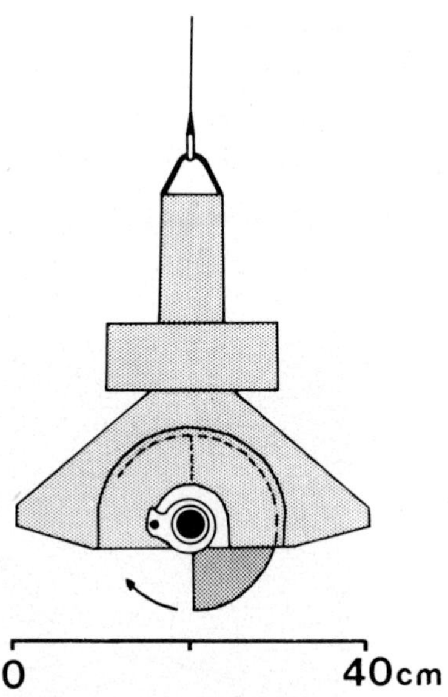

Fig. 5.5. Schematic presentation of the Shipek grab. Sampling enclosure results from rotation of the cylindrical bucket by external helical torque springs in the center of the device.

Since a few years, new generations of grabs are available. A first development was a pneumatic grab designed by Preussag (see Chapter 7) to obtain large samples from the seafloor in water depths up to 1000 m. The system is operated by compressed air at 200 bar from two tanks, which may be easily exchanged in less than a minute aboard the ship between two sampling sites. A special configuration of the jaws enables an almost undisturbed sediment sample to be taken. A filled grab with completely closed jaws is guarenteed by a grab closing force of 1.5 tons. Total weight of the system is 1.85 tons and the maximum penetration into the sediment reaches 70 cm (Anonym, 1983).

Exploration devices, like dredges and grabs commonly used until now, show considerable disadvantages in sampling massive sulfide deposits and ferromanganese encrustations grown on hard-rock substrata. Dredges are not recommended as a sampling device, because only blind sampling is achieved.

Often, especially in the case of rough submarine topography, equipment may be lost. Grabs are often not able to operate in depths of several thousand meters and are not capable of breaking-off mineralizations or rock fragments. Therefore an electro-hydraulic grab, equipped with a TV-system for on-line operation has been developed recently. The system deploys a coaxial cable for carriage through the water column and for transmitting of the video and telemetry data. The battery-powered hydraulic unit, installed at the device itself, is remotely controlled from the surface vessel and ensures a grab closing force of 3.6 tons, sufficient to collect any kind of sample. Sampling site selection is guided by a TV system and there is a possibility of emptying

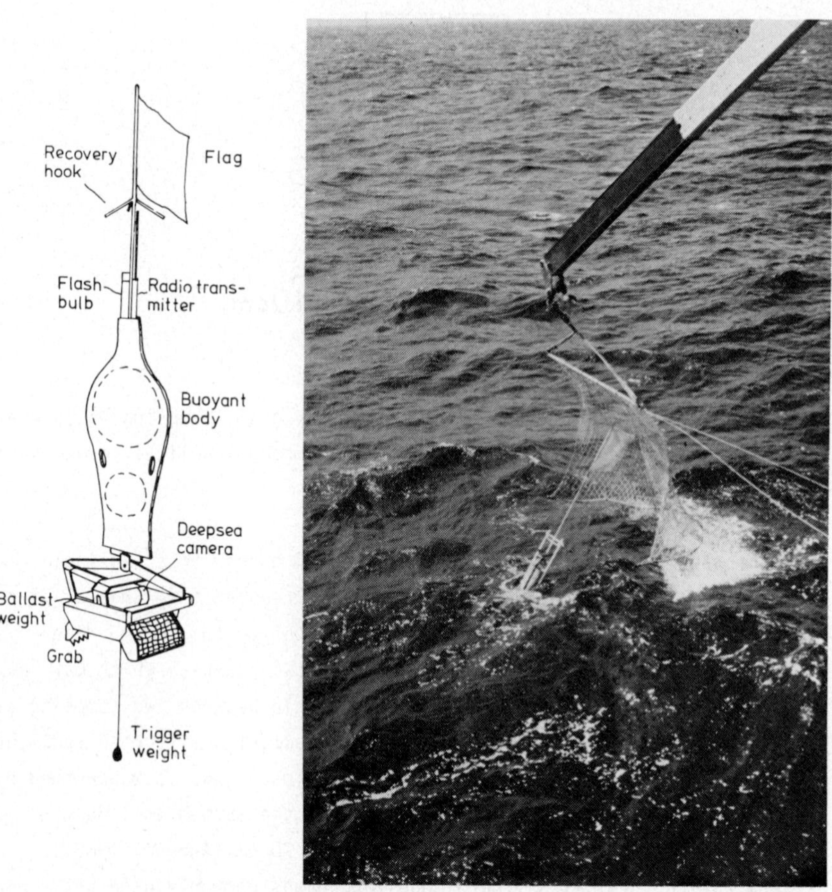

Fig. 5.6. Sketch of a free-fall grab (left side) and recovery of the grab by a towed net aside the ship's hull (right side).

the grab directly on the seafloor, followed by repeated sampling until the desired sample has been caught. The application of this new sampling device in the recent exploration for massive sulfides in the Pacific Ocean as well as a photographic presentation can be found in Chapter 7.

In manganese nodule exploration a special kind of grab is widely in use. Since manganese fields cover large areas on the seafloor, sampling has to be carried out on numerous sampling sites in order to obtain statistically well-defined data on coverage and metal content in relatively short time. To fulfill these requirements, free-fall grabs were developed. Free-fall grabs are launched from the exploration vessel, descending to the seafloor by means of ballast weights. After bottom contact a sample is collected from an area of approximately 0.15 m^2, the ballast weihgts are thrown off and the free-fall grab returns to the sea surface (Fig. 5.6). The average speed during descending and ascending the water column is about 80 m/min resulting in approximately 2 hours of sampling operation at 5000 m water depth. Refinding the devices at the surface is supported by special equipment mounted at the sampling devices, i.e. flags, flash and VHF radio beacons. Generally, a still camera and a flash are adapted to the free-fall grab, which takes a photograph of the undisturbed seabed before sampling. For special purposes, a modified Nansen water sampling bottle may be included in the device to sample near-bottom water.

Seafloor substrata sampling instruments

The sampling devices (dredges and grabs) discussed above are only able to collect either coarse material or more or less disturbed samples from the surface or surficial deposits of the seafloor.

In many studies, however, sampling from greater depths and/or preservation of the original sediment structures is required. Especially for stratigraphical control, interstitial water studies, measurement of physical properties and detailed geochemical studies, undisturbed sediment cores of suitable length are obligatory. For these reasons various sampling devices were developed, that, with the exception of the free-fall corer, all are wire-lowered from the ship.

Easy to handle and frequently used to core the seafloor is the spade corer (Fig. 5.7). A section of 20x30 cm sediment is recovered from depths up to 45 cm by use of standard equipment. The sediment surface remains undisturbed, although the peneliquid layer of the sediment-seawater interface is lost by the heaving of the device. Bottom approach is usually remotely controlled by a pinger. After bottom contact, the box is pressed into the sediment by the heavy weights which are on top of the central piston. Hoisting the device results in a 90° turn of the spade, which cuts the sediment column and closes the box at its underside.

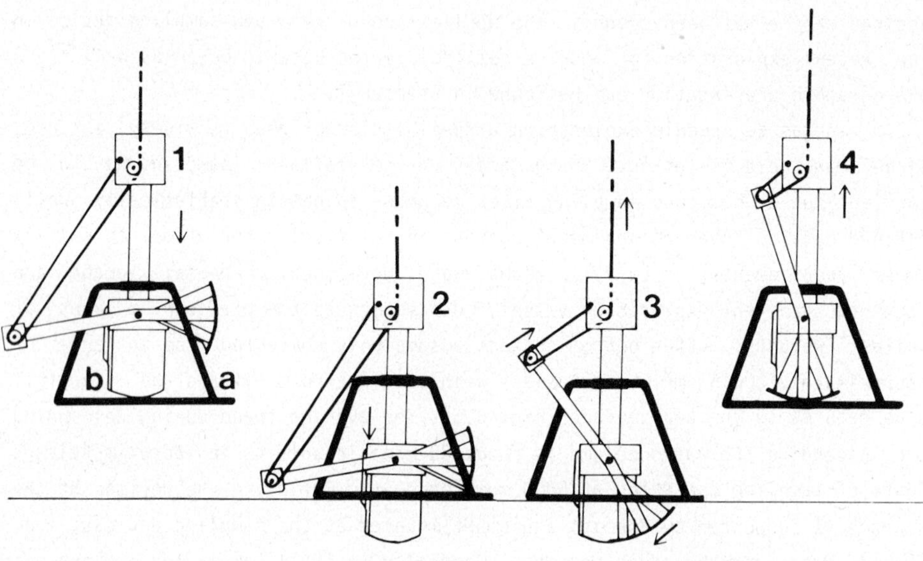

Fig. 5.7. Spade coring sequence (1 to 4) demonstrating sampling of the uppermost sediment layers. Size of the sampling frame (a) is approximately 1.5x1.5 m; total height of the sampling equipment is about 2 m. The sampling box (b) covers an area of 20x30 cm and its maximum penetration depth is 45 cm.

If a longer sediment core is required, box corers with variable lengths (between 3 and 15 m long) amd different box sizes (generally 15x15 cm, or 30x30 cm) are in use (Fig. 5.8). Box corers are equipped with adjustable heavy weight stands, between 2.5 and 5.5 tons, depending on the length of the box and the type of sediment to be sampled. The box itself is formed by two rectangular pieces of stainless steel or aluminium, longitudinally joined by screws. At the lower end, a core nose containing two spring-loaded flaps is installed. Descending the device with open flaps is again remotely controlled by a pinger. Sediment penetration results simply by the heavy weight of the sampling system; normally the speed of the system at the moment of penetration is between 40 and 70 m/min. When pulling out, the flaps of the core nose will be closed by the turn of two external tabs. Heat flow measurement may be combined with the coring; various sensors are installed along the outer side of the box and a magnetic tape registration unit, adapted to the wire is employed to store the in-situ heat flow measurement data.

Fig. 5.8. Box core (30x30 cm) with recovered sediment, partly opened, and sampled plastic containers (100x15x5 cm) for documentation purposes. Remaining material is used for on board analysis.

Another widely accepted device to obtain long cores (up to 30 m) is the piston corer (Fig. 5.9). Normally, piston corers contain clear plastic liners inside a cylindrical steel barrel, which is driven into the sediment by a heavy weight stand of several tons. The diameter of the barrel and its inserted liner is variable and ranges between 4 and 10 cm. Remotely controlled by a pinger, the equipment is wire-lowered to the seafloor, where it is triggered by a pilot corer or pilot weight a few meters before contact of the main device. This results in a free fall of the corer and in a gain of momentum which increases the penetration depth. The free-fall height is determined by the difference in length between trigger wire and total length of the piston corer. The mode of operation of the piston corer device is shown in Fig. 5.10. The piston reduces the pressure in the liner and sucks the sediment into the core. A stainless steel spring-finger or flap-type core catcher inside the nose cone of the barrel prevents loss of sediment from the liner during heaving. After removel of the nose cone onboard ship, the sediment-filled liner is pulled out of the barrel, cut and sealed.

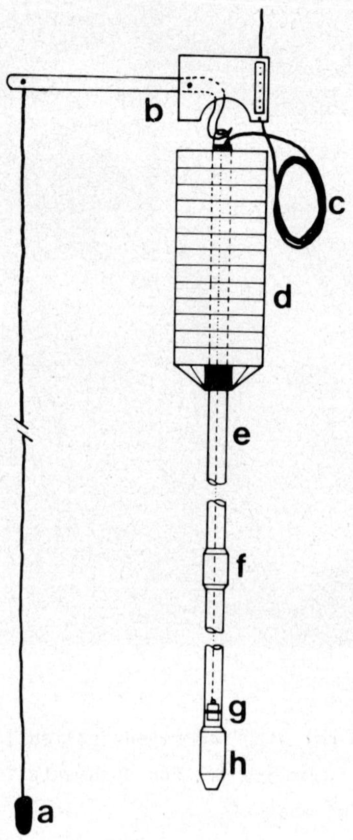

Fig. 5.9. Piston corer with pilot corer or pilot weight as a trigger (a), release mechanism (b), free-fall wire (c), adjustable weight stand (d), core barrel (e), connection joint (f) to extend total length, piston (g), and nose cone (h) with inserted core catcher.

Spade, box and piston coring often fail if the sediments, which have to be sampled are very hard or have intercalated coarse fragments. Also, sampling of placer minerals from greater depths may be rather difficult due to incomplete coring by box or piston corers. To sample them more sufficiently, vibrocorers with electric, pneumatic or hydraulic power sources inducing vibration or hammering at the nose of the corer are applicated. The corer itself consists of a metallic tube, which can also carry a plastic liner. Length of the barrel is variable and in some versions of vibrocorers, a piston is adapted inside the barrel to lower internal wall friction and to create a low pressure above the core during bottom penetration. Sometimes, a complete undisturbed sediment core is not required. Then, samples may be collected by vibration coring in conjunction with an air-lift system. In this way, compressed air is forced into the core nose and an ascending water/air stream transports the sediments through a flexible plastic tube to the ship. The sediment/water mixtures of different depths are collected in several containers, where sediments are settled and collected for further investigations.

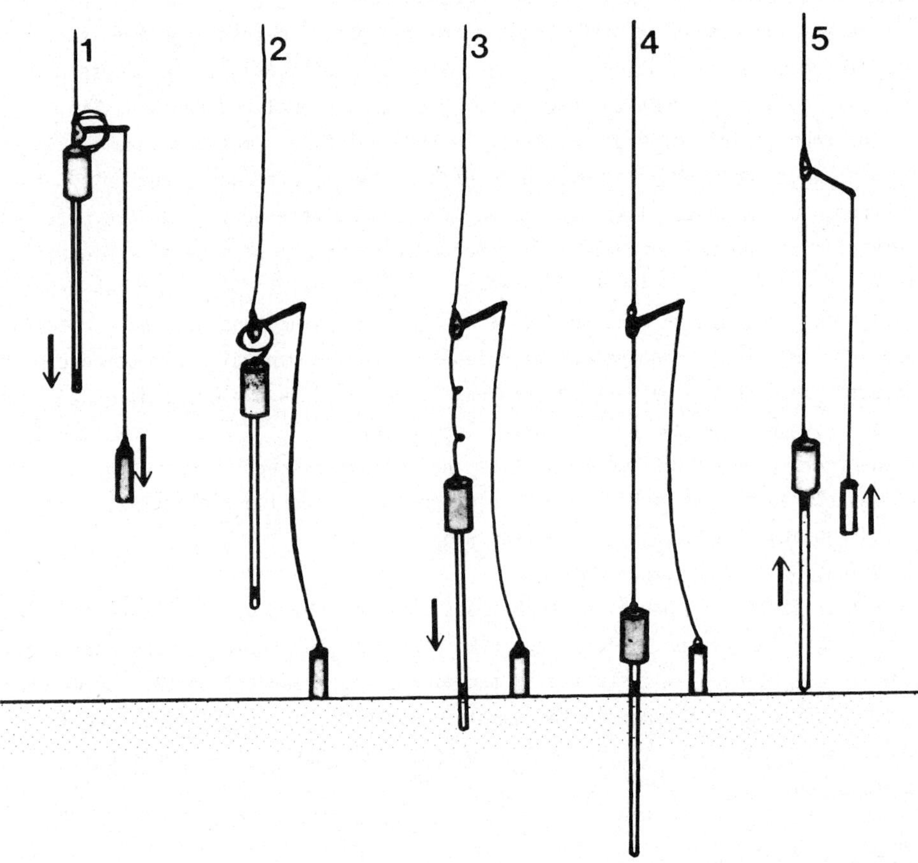

Fig. 5.10. Schematic illustration demonstrating the coring sequence by a piston corer.

To collect sediment cores of limited length in a rather short time, a free-fall sediment sampler was designed by Benthos Inc., U.S.A., in cooperation with the Woods Hole Oceanographic Institution. The corer has an ascent rate of about 450 m/min, which is induced by a ballast part including a nose cone, pilot weight, core barrel, ballast and a protective shroud for the glass float. After coring of up to 1.2 m sediment, the ballast part is expanded and the floating part, consisting of two glass spheres and the core liner assembly descends with a speed of approximately of 75 m/min to the sea surface. Usually a series of corers is dropped from the ship along a transit and retrieval of the devices is by ship or by longboat. During night-operations the free-fall corers are equipped with a flash beacon.

ANALYTICAL METHODS IN MARINE MINERAL EXPLORATION

There is no essential difference in the analytical methods applied in marine and terrestrial mineral exploration, although modifications are necessary if analysis of samples from the marine environment are analyzed onboard ship.

In terrestrial mineral exploration the analytical methods available at present were thoroughly reviewed by Fletcher (1982), although a number of other instrumental methods like neutron activation analysis were left out. It is convenient to adapt the analytical concepts of terrestrial mineral exploration. There are, however, some differences in that relatively fast ship-board analytical techniques are needed at the reconnaissance and at some detailed exploration steps. Analytical techniques in marine mineral exploration can be grouped into those that are adapted easily on board an exploration vessel and those that are used at the land-based laboratory.

Generally, analytical work involves the following steps:

- Sample preparation, including dissolution and extraction procedures,
- Instrumental analytical procedures, and
- Evaluation of analytical data

While sample preparation techniques consider the physical state of the sample material (water, sediment, rock, minerals) the choice of the analytical procedures (generally instrumental techniques) is dependent on the occurrence and content of elements in the sample. Evaluation of analytical data and presentation of data in the form of geochemical maps is an important step in exploration.

Sample preparation

Marine mineral exploration samples such as seawater, sediments, rocks and minerals generally need sample preparation steps prior to analysis. This section summarizes procedures which are used onboard ship to prepare samples for subsequent ship-borne analysis.

Seawater contains inorganic particulate matter and marine organisms or their decomposition products in the form of suspended matter. These products are able to adsorb or to liberate trace metals rather quickly. For that reason, suspended and particulate matter is separated from the seawater either by filtration through micropore filters (0.45 μ pore size) or by centrifugation of the seawater directly after recovery. If no separation is carried out onboard ship and samples have to be stored and shipped afterwards to the land-based laboratory, seawater samples often are immediately froozen after recovery.

Prior to the determinations of trace elements (e.g. Fe, Co, Ni, Cu, Zn, Pb) in seawater and interstitial waters by Atomic Absorption Spectrophotometry (AAS), these elements are complexed by a 1% APDC (Ammonium-Pyrrolidine-Dithiocarbamate) solution and extracted with MIBK (Methyl-Isobutyl-Ketone) in acid solution (Brooks et al., 1967).

Sediment, rock and mineral samples are carefully registered after recovery. They are then classified and often photographed for later investigations (see also Chapter 6).

As already described, sediment samples may be collected by various devices. Plastic liners from piston corers are normally cut to convenient size, e.g. one meter length. A first sampling of the sediment takes place at the cutting edges. If necessary, the whole core may be cut into longitudinal sections. One half of the liner is stored as a reference sample in the ship's freezer room, the other half is used for onboard analysis. Representative samples from spade corers are taken by insertion of short liners into the core. These samples are usually saved as reference pieces; the remaining material is used for ship investigations. Sediment profiles from box cores of various length are obtained by pushing plastic containers of one meter length into the sediment core. The containers are closed, sealed and stored in the freezer room. Since box cores have relatively large cross sections (15x15 or 30x30 cm) sufficient sediment material for further investigations remains onboard ship. Chemical changes in the sediments may still take place even during storage at low temperatures in the freezer room, and especially also during the transport to the land-based laboratory. To avoid these reactions, dry-freezing procedures often are applied.

Sediment samples that have to be analyzed chemically onboard ship are dried at 105° C for 6 to 24 hours. Drying time depends highly on the analytical accuracy required. If, for certain reasons no high accuracy is needed, shorter drying periods may save considerable time. For special purposes, i.e. determination of the Hg contents in sediments, drying temperatures have to be reduced to 60° C because of Hg volatility.

Rock and sediment samples also have to be dried according to the above mentioned procedures. Special attention is paid to the hygroscopic behaviour of the ferromanganese nodules and crusts after drying. Storage of these samples in desiccators is obligatory.

Next step in sample preparation is crushing and grinding. Rocks and minerals are crushed by use of a hammer, or in case of larger quantities by a jaw crusher. Crushing requires special care to avoid contamination from crushing equipment. To obtain fine-grained material (analytical size, -100 mesh), different types of mills may be employed. Here again, contamination may occur from the use of non-adequate grinding equipment. Highly recommended is the use of agate components, which do not contribute metals to the sample powders as do steel or tungsten carbide components (Fletcher, 1982).

Analysis by X-ray fluorescence (XRF) does not require any further sample preparation, since the -100 mesh powders have to be pressed to stable pellets only, using a hydraulic or a manually operated pressing tool with pressure capacity of 5 to 40 tons. More accurate XRF analysis may be obtained by using a glass bead instead of a pellet, because enhanced sample homogeneity is achieved, no mineralogical effects are observed, and there is little need for correction of the obtained data. Glass beads are prepared by fusing the sample with a flux at defined sample-to-flux ratios. The disadvantage of this kind of sample preparation is the need of an analytical balance onboard ship. On the other hand, samples which have to be analysed by AAS need usually a complete or partial dissolution, or a fusion followed by the dissolution of the prepared glass bead. In exploration work various dissolution techniques have proved to be adequate and a summary of these methods is given in Table 5.2. If the samples to be analyzed contain larger amounts of organic matter or sulfides, special dissolution techniques are needed (Fletcher, 1982).

TABLE 5.3

Decomposition techniques for mineral exploration samples. After Fletcher (1982).

Method		Chemical compounds used
Digestion		HNO_3, HCl, HCO_4, HF other combinations possible
Fusion	acidic	$KHSO_4$, $K_2S_2O_7$
	halide	NH_4I, NH_4Cl
	alkaline	Na_2CO_3, NaOH, $LiBO_3$ and others
Partial decomposition		0.1-1.0 N HCl and various others

Analytical techniques

A number of methods are applied to determine the elements (Table 5.3) that are present in exploration samples and in the mineral deposit itself. The analytical methods described below are in routine use at the land-based laboratories, but some of them, e.g. AAS and XRF, are also adapted in the shipboard laboratory.

While atomic absorption spectrophometry (AAS) is the most frequently used analytical method in mineral exploration, X-ray fluorescence analysis (XRF) is clearly dominating in marine work. This is partly because XRF needs only simple sample preparation. Accurate weighing is very difficult to accomplish on board a research vessel, especially under harsh weather conditions. Therefore, many of the methods used in terrestrial mineral exploration are invalidated in certain stages of marine exploration. Nowadays as a result of detailed analytical experiments conducted mainly in the 1970s, most of the leading research vessels have installed computerized XRF instrumentation.

TABLE 5.3
Selected elements to be determined in marine marine exploration samples.

Mineral commodity	Metals to be determined
Placers	Sn, W
	Au, Cr, Pt
	Fe, Ti, Zr, Th
Manganese nodules and crusts	Mn, Fe, Co, Ni, Cu, Zn
	Mo, As, V
	P
Metalliferous sediments and massive sulfides	Cu, Zn, Fe, As, Au, Ag

The most important analytical techniques in marine mineral exploration, are:

- X-ray fluorescence analysis (XRF)
- Atomic absorption spectrophotometry (AAS)
- Inductively-coupled plasma emission spectroscopy (ICP-ES)
- Instrumental neutron activation analysis (INAA)

Summarized characteristics of these methods are given in Table 5.4. Besides these methods, a number other special methods adopted to the specific purposes are in use.

TABLE 5.4

Analytical methods used in marine mineral exploration.

Analytical method	Characteristics
X-ray fluorescence (XRF) (A) Wavelength-dispersive (B) Energy-dispersive	Excitation of characteristic X-rays of elements in the sample; dispersion according to wavelength (A) or energy (B); energy detection by radiation detectors. Expensive, but reliable multielement method with high sample throughput; detection limits at the 1 to 10 ppm level; simple sample preparation procedures; easy installation on a research vessel.
Atomic absorption spectrophotometry (AAS)	Absorption of element specific light by atoms at the ground state, which are produced by vapourising sample solution in a high-temperature flame or graphite atomizer. Cheap and simple equipment; easy to operate; low detection limits, often at the ppb levels; multielement analysis not possible; difficult to use onboard ship.
Emission spectroscopy (ES)	Excited outer electrons emit characteristic line spectra when transiting from excited to ground states; complicated line spectra; direct-reading line intensity instruments exist; energy sources are d.c. arc or inductively coupled plasma (ICP); multielement analysis possible; low detection limits (ppb level); dissolved samples for analysis; complicated instruments.
Instrumental neutron activation analysis (INAA)	Irradiation with thermal neutrons in a nuclear research reactor; nuclear reactions produce gamma-emitting isotopes; detection by semiconductor detectors; complicated gamma-ray spectra treated by computer for intensity calculations; multielement analysis; low detection limits, varying according to element under considerations; skilled personnel required for routine analysis.

X-ray fluorescence analysis (XRF). XRF is a well-established analytical method frequently used in routine geochemical analysis. In fact, major element analysis of geological samples is almost exclusively carried out by this method. A number of textbooks describe physical principles and applications in great detail (e.g. Jenkins and De Vries, 1970; Kunzendorf, 1973). The basic components of XRF instrumentation are shown in Fig. 5.11 (A).

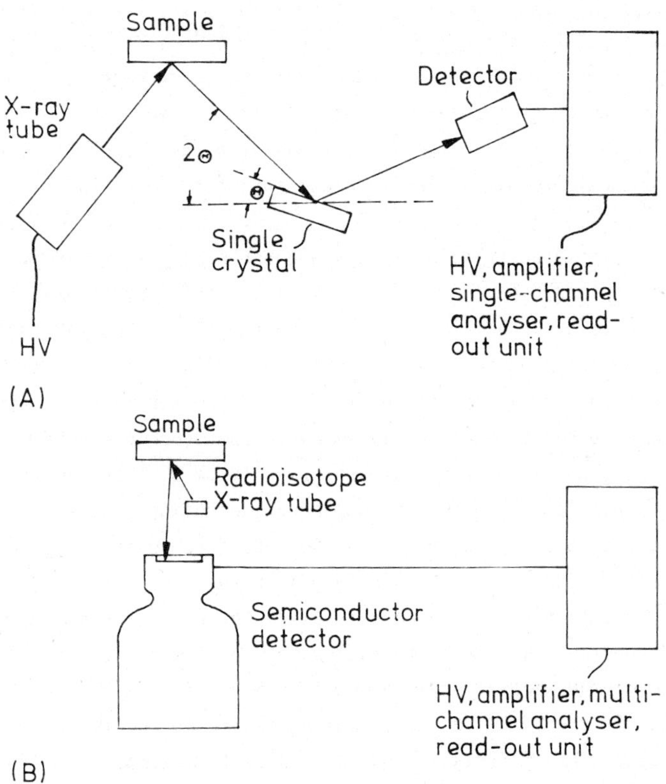

Fig. 5.11. Schematic principles of XRF instrumentation. (A) Wavelenght-dispersive system, (B) Energy-dispersive system.

Briefly, X-rays generated by a cooled X-ray tube strike the sample which usually is in the form of a pelletised powder or a bead, but the sample may also be analyzed as a powder. The primary X-rays in turn excite characteristic X-rays of elements in the sample. The fluorescent yield for these elements is highest for incoming radiation being only slightly higher in energy than the characteristic X-ray energy. X-ray tube energy spectra are therefore generated by tube targets that have to be changed depending on which series of elements has to be analysed.

Characteristic X-rays from the sample are then sorted according to their wavelengths by a single crystal using the principle of Bragg reflection (n = 2d sin O, where is the wavelength of the characteristic X-rays, and d is the crystal reflecting plane spacing). Characteristic X-radiation may be collimated after leaving the sample and before reaching the surface of the radiation detectors (scintillation and/or proportional flow counter). Detector pulses are amplified and counted in a single-channel analyser. Data read-out nowadays is via microcomputer, including necessary corrections for X-ray interference effects (line interference, X-ray absorption and enhancement). The procedures for conversion of X-ray intensities into element contents are complex and not within the scope of this section. For calibration purposes, international reference standards are available (Flanagan and Gottfried, 1980; Govindaraju, 1984).

Because many characteristic X-rays and especially those from the light elements have low energies (less than 10 keV), most automatic XRF instruments operate under vacuum conditions, i.e. the sample is irradiated in a vacuum chamber. The lowest Z-number element analysed routinely is Na. Although some XRF investigations report detection limits at 1 ppm for certain elements, in general, these are at or slightly below 10 ppm. For elements with atomic numbers of less than 19, they are greater than 10 (e.g. Leake et al., 1969). It is, however, not possible to define general detection limits for the method because these are strongly dependent on the matrix composition. A simple matrix correction procedure uses Compton-scattered X-rays.

Major oceanographic research vessels are now equipped with modern wavelength-dispersive XRF instrumentation to meet the analytical requirements of systematic marine mineral exploration.

A somewhat different XRF technique was developed in the 1960s. According to the mode of detection of X-rays (Fig. 5.11 (B)), the method is known as energy-dispersive X-ray fluorescence (EDXRF). While sample preparation methods are essentially the same as those for the conventional technique, excitation of characteristic X-rays in the sample may be produced by small radioisotopes emitting low-energy gamma- or X-rays in addition to the X-ray tube (e.g. Kunzendorf, 1972; Plüger, 1975). Also, detection of X-rays is directly via a semiconductor detector with much better energy resolution than that of the detectors in conventional systems. At low energies (low-Z number elements), XRF is superior to the semiconductor detectors but at higher energies ($>$ 30 keV) the EDXRF systems are better. The main advantage of this technique lies in the short distances between source, sample and detector permitting the construction of portable instruments involving scintillation detectors and X-ray filters, weighing generally less than 10 kg. Commercially available instruments have been used in the shipboard analysis of manganese nodules (Lüschow and Kraft, 1973).

Semiconductor-based systems, on the other hand, are characterised by detection limits that are comparable to the conventional wavelength-dispersive systems, whereas portable EDXRF systems seldom achieve detection limits of better than 1000 ppm.

Semiconductor EDXRF systems were used in manganese nodule exploration (e.g. Friedrich et al., 1974; Friedrich and Plüger, 1974). Although these systems produce simpler characteristic X-ray fluorescence spectra than conventional systems, they have a major drawback in that the X-ray detector systems need to be cooled to the temperature of liquid nitrogen for reasons of energy resolution. New electronically based cooled systems may become available in the near future so that EDXRF systems may become more compatible with conventional wavelength-dispersive systems.

Atomic absorption spectrophotometry (AAS). The method most often used by far in mineral exploration is AAS. The main reason for this is that rather inexpensive, relatively small instruments are on the market and also, for many elements rather low detection limits (at the ppb-level) can be achieved. The method has been described in numerous publications (e.g. PE, 1976; Price, 1972).

The physical principle of the method is resonance absorption of light. A sketch of AAS instrumentatioon is given in Fig. 5.12. A light source generates a spectral line of the element to be analysed. Light is then focussed into the vaporised sample (flame) and the intensity of line absorption is a measure of the element content in the sample. The flame is often generated by an air and acetylene mixture. Temperatures in the flame reach more than 2000^{0} C which reduce the ions to neutral atoms in the ground state. A reference beam may be included in AAS instrumentation (double-beam instruments).

Flameless AAS (heated graphite atomizer technique) has been developed for the determination of extremely low element contents.

As with most analytical techniques, also AAS is not absolutely free of intereferences, although this is often not mentioned when suggesting the method. Fletcher (1982) lists spectral, chemical, ionisation and background interferences. Elements that normally can be determined without difficulties after digestion with hot acid are for instance Li, Na, Mg, K, Ca, Mn, Fe, Co, Ni, Cu, and Zn. A large number of other elements has been determined in exploration samples by AAS. Important elements that cannot or are difficult to determine by AAS are, e.g. Zr, Nb, La, Ce, Th and U.

AAS has been used extensively in the analysis of manganese nodules, metalliferous sediments and massive sulfides.

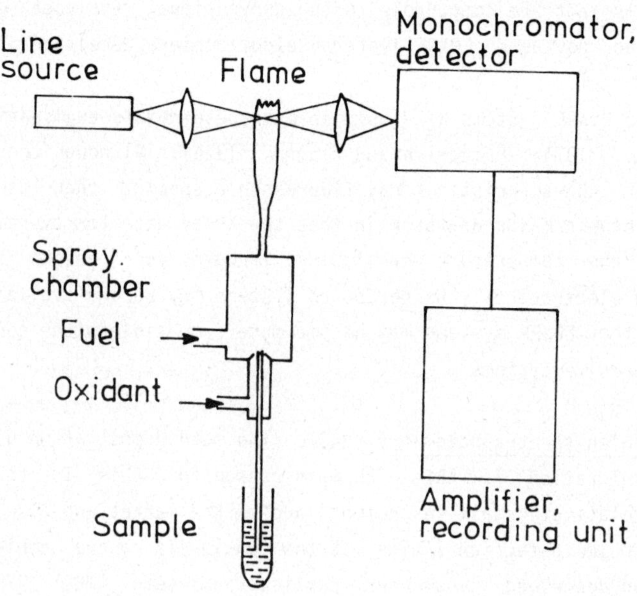

Fig. 5.12. Schematic presentation of AAS instrumentation.

Emission spectrography (ES). This method has often been used in the early years of geochemical exploration. It was very popular because of the possibility of multielement analysis, although interpretational techniques for complex emission spectras did not permit the method to be more than semiquantitative.

While AAS uses neutral atoms at ground state for analysis, ES makes use of excited states of outer electrons, which, after excitation, emit characteristic radiation during transition to ground states. Complicated characteristic line spectra are generated for each element in the sample so that complex spectra have to be handled for analysis (e.g. Fletcher, 1982). Intensities of preselected lines are measured directly by electron photometers but in early times, photographic plates had to be interpreted.

Much higher temperatures (5000-7000^{0} K) are needed for vapourising and ionising flames which can be achieved only by e.g. a d.c. arc. In this case the sample is in the form of a powder deposited on tape which then is fed directly into the arc. Often a d.c. arc is difficult to operate at stable conditions. Also, the special sample preparation technique is a serious drawback. However, thousands of analyses have been performed usually giving an estimate for up to 30 elements.

Relatively new is the implementation of an inductively coupled plasma (ICP) as a more stable energy source. Much higher temperatures can be produced in the plasma. While d.c. arc spectrometers operate on powdered samples, ICP-ES works with dissolved samples. A description and evaluation of the method can be found in Boumans (1979).

Whereas the introduction of ICP has stimulated the use of ES, this method shows also spectral and other interferences. As is with most of the analytical methods requiring more or less complicated decomposition schemes, analysis will never be faster than the decompositional procedures, which is valid also for ICP-ES. Instruments are expensive and require operational skill. Being a relatively new analytical method, further refinements of procedures and techniques are expected.

Instrumental neutron activation analysis (INAA). INAA requires usually a nuclear research reactor for routine analysis, although a number of neutron generators and accelerators have been proposed for analytical purposes. The method was developed with a view to analyse geological samples without advanced sample preparation or dissolution schemes that are applied in radiochemical neutron activation analysis (RNAA). Because of its relatively high cost, INAA is often neglected as an important analytical method in geochemical work. However, some nuclear research centers, during the past 10 to 20 years, have offered inexpensive multielement analysis of geological samples by INAA to mining companies involved in mineral exploration. At present the method is predominantly applied in the analysis of precious metals and the rare earth elements. Many textbooks were published on the subject or special INAA techniques are included in the treatment of analytical methods (e.g. Fite et al., 1971, Wainerdi and Uken, 1971). The rather complicated technique can be described only briefly in this section.

Generally, 0.5 to 1 g of a geological samples are sealed in polyethylene vials and pneumatically sent to the irradiation site at the research reactor. Irradiation is for a considerable time (hours to days) in the high thermal neutron-flux area. Reactions of reactor neutrons with the nuclei of atoms in the sample produce isotopes that generally decay by emission of gamma radiation.

After cooling the sample for about 1 week to allow for the decay of very short-lived isotopes, gamma-ray spectra are recorded by nuclear radiation detection systems (semiconductor radiation detector, multichannel analyser, read-out unit). Line intensity calculations from the complex gamma-ray spectra are performed by computer. A second radiation measurement of the sample is caried out after about 3 weeks of cooling.

While the first counting procedure leads to the determination of elements with relatively short half-lives (days to a few weeks), the second recording is for determination of long-lived isotopes, after decay of short-lived isotopes has progressed. About 30 elements in a sample may be analysed by this method. The techniques for analysing geological sample material are described in several detailed publications (e.g. Gordon et al., 1968; Plant et al., 1976). The method has been used extensively in the U.S. National Uranium Resource Evaluation (NURE) program (Price, 1978), where thousands of exploration samples were analysed by INAA.

Detection limits vary for different elements, generally depending on the yield (neutron cross section) with which neutrons initiate nuclear reactions in an element's nucleus. They are often at the ppb-level.

According to Plant et al. (1976) the elements determined by routine INAA are Na, Al, K, Sc, Ti, Cr, V, Mn, Fe, Co, Zn, Se, As, Br, Rb, Mo, Ag, Cd, Sb, Ba, Cs, La, Ce, Sm, Eu, Hf, Ta, W, Au, Th, U. Also this method is not free of both line and matrix interferences.

In the instrumental mode some economically important metals like Au may be determined rapidly at the ppb-level (e.g. Hoffman and Brooker, 1982). The same metal and the rare earth elements have also been determined in, e.g., metalliferous sediments from the East Pacific Rise (Kunzendorf et al., 1985).

Other analytical techniques. A number of analytical methods for special purposes have been used on board research vessels. Also in-situ measurements on the seafloor were developed during the past two decades which are briefly described in this section.

Because there is a general need for conducting chemical analysis directly on the ocean floor, several attempts were made to construct appropriate apparatus. The results of these efforts were recently reviewed by Noakes and Harding (1982). According to these authors the following techniques were studied in detail:

- Natural radiation measurements
- Neutron activation devices
- XRF probes

While the natural gamma radiation measurements of the ocean floor were conducted mainly in the USSR as a part of petroleum exploration, British efforts (Summerhayes et al., 1970; Miller and Symons, 1973) aimed at the detecting phosphorites due to their enrichment in uranium. Early investigations concentrated on the detection of total gamma radiation, but more recently, spectral instruments were developed that could assist in the geological mapping of the seafloor. U.S. seafloor natural gamma measurements were concentrated mainly on the phosphorite occurrences on the Georgia shelf (Noakes et al., 1974). Whereas radiation detection in the British system was housed in an "eel", the American construction consisted of a sled-mounted device.

In general, these systems may be important in outlining high-radioactivity seafloor EEZ areas. Research efforts are not yet finished so that more sophisticated constructions may be expected in the future. There are some major problems to be solved, e.g. keeping the instrumentation at a controlled distance above the seafloor to prevent the capture of a surface-going device in the sediments, and the absorption of radiation in the seafloor surface layer and in seawater.

Neutron activation techniques deployed by towed instruments were also developed. Important investigations were published by Senftle et al. (1969) and Wogman et al. (1972). Usually, a ^{252}Cf source is applied for neutron bombardment of the seafloor, and the gamma radiation is then detected by a cooled semiconductor detector. It is clear that such systems tended to be complicated and of massive construction. Prototypes of these instruments were lost due to especially bad weather conditions (Borcherding et al., 1977). At present, there is very little interest for further development of in-situ neutron activation devices.

XRF instrumentation for in-situ work has also been developed and tested (e.g. Wogman et al. (1975), but little is known on its application in marine mineral exploration.

There is little need for cable-mounted analytical equipment, but in the future remotely-operated vehicles may be employed for analytical purposes also.

GEOCHEMICAL EXPLORATION EXAMPLES

As already pointed out there exists no uniform geochemical exploration method for the different marine mineral commodities. Every type of marine mineral deposit (placers, phosphorites, massive sulfides at spreading ridges, ferromanganese concretions) has its own typical origin and therefore, distinct geochemical characteristics. Common strategies used in terrestrial exploration surveys like the investigation of primary and secondary dispersion halos of selected elements will often fail in marine geochemical exploration, since dispersion halos and trains are not or only weakly developed. Only in the search for massive sulfides at spreading ridges similarity with terrestrial geochemical exploration is observed because dispersion halos in the surrounding sediments and even in the water column may be developed and be the subject for exploration work.

Since a number of exploration examples already are included in Chapter 7, the examples presented in this section deal with placer exploration and with a reconnaissance survey conducted at the Mid-Indian Ocean Ridge.

Placer deposits off eastern Australia

Geochemical exploration for offshore placer deposits is considerably restricted. Dispersion halos do often not exist or are only weakly developed since placer minerals are especially stable. The only possibility to detect geochemical anomalies is a closely spaced sampling followed by the qauntitative determination of the placer minerals of interest in the recovered samples. However, the interpretation of the anomalies is precarious, because simple concentration gradients which may be traced back to the mineral source or deposit do normally not exist. This is the result of the complexity of the factors leading to the formation of placers, like currents, waves, tides, past and present sea level fluctuations and recent seafloor topography.

Usually, the direction of the sampling grids and the spacing between sampling are chosen on the basis of preceding geophysical investigations (bathymetric and seismic surveys). Sampling is carried out by use of grabs and vibrocorers. Recovered surface and core samples are then screened, dried and split and subsequently grinded, respectively treated with tetrabromoethane to obtain samples for chemical analysis and microscopic investigations.

Though detailed marine geological investigations were carried out in the 1960s and 70s by Australian mining companies and regional Geological Surveys, a combined Australian-German effort was conducted during a two-month cruise in 1980. Several areas between Newcastle and Fraser Island were investigated by geophysical and geochemical methods (von Stackelberg and Jones, 1982). Earlier work on offshore grid sampling off southern Queensland and northern New South Wales has been reported by Marshall (1980).

The results of more than 800 analyses of the fine fraction of surface sediments (32-315 μm) are decribed by Friedrich et al. (1982). These results were obtained by XRF analysis at the land-based laboratory, but ship-board XRF analyses already outlined a titaniferous mineralization near Cape Byron (Thijssen and Kunzendorf, 1980).

On board ship samples were wet sieved and the fraction 32-315 μm was used for all further investigations. For XRF purposes samples were ground in a disc mill employing agate components. Grinding the sample required about 7 min so that the analysis was not seriously delayed by sample preparation procedures. Fe and Zr were the main elements determined in the samples, but Ti was also estimated in samples with Ti > 0.4%. In many cases, Zr analyses were used to select those samples where heavy liquid separations for heavy minerals could be conducted directly on board ship.

Although the XRF equipment at that time installed on board RV "Sonne" was not able to detect Ti contents of less than 0.4% due to lack of vacuum equipment, the deposit could be traced by its high Zr content. Fig. 5.13 shows the ZrO_2 distribution off Cape Byron. Laboratory analyses later proved that there is a strong correlation between Ti and Zr in the samples, resulting in a similar distribution pattern for Ti.

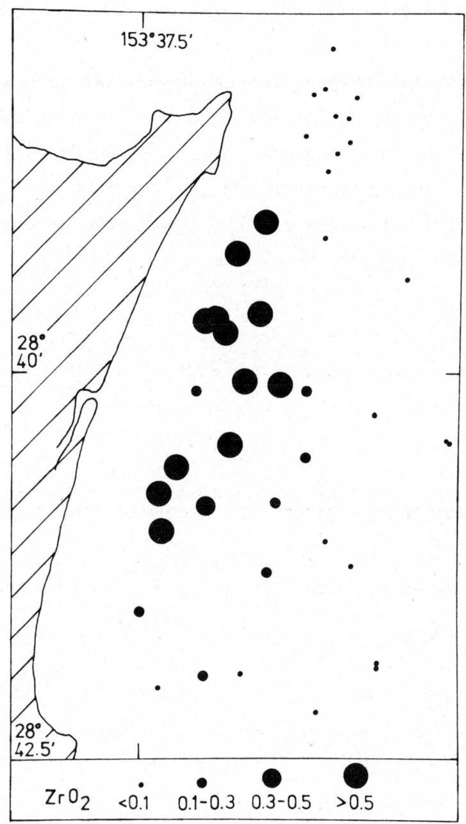

Fig. 5.13. ZrO_2 contents in grab samples from the Cape Byron area (off eastern Australia). Contents are in per cent.

During the exploration campaign, also other areas with Ti-enriched sediments were identified.

Geochemical methods applied at the Mid-Indian Ocean Ridge

Geochemical exploration of hydrothermal deposits on seafloor spreading ridges shows certain similarities with land-based surveys since often primary and secondary dispersion halos are well-developed. Hydrothermal deposits are related to areas with active seafloor spreading (i.e. divergent plate boundaries), to back-arc regions and to individual seamounts, preferentially in off-axis areas. Besides mineralization connected to momentary activ regions, fossil hydrothermal mineralization occurs in former active regions now covered by deep-sea sediments.

Exploration strategies to search for hydrothermal deposits have been described by Rona (1983), pointing out that the ideal exploration strategy has to follow a procedure of closing range on a prospective deposit by systematic progression from regional reconnaissance (range from 1000 to 1 km), to site-specific methods (from 100 to 0 m). Table 5.5 summarizes proposed geological, geochemical and geophysical methods.

TABLE 5.5.
Proposed exploration strategy of closing range to a mineral deposit (Rona, 1983).

Range	Method	Determination of
1000 - 10 km	Regional sediment sampling	Fe and Mn concentration gradients
	Regional water sampling (applies only to actively accumulating deposits)	Weak acid-soluble amorphous suspended particulate matter ^3He Total dissolvable Mn (TDM) Methane
10 - 1 km	Long-range side-looking sonar	
1000 - 10 m	Bathymetry, gravity, magnetics, side-scan sonar	
10 - 0 m	Bottom images, water sampling, sampling of mineralization and substratum	

A nearly complete listing of the known hydrothermal mineralization at seafloor spreading centers is presented by Rona (1984). The author summarizes 55 sites of hydrothermal mineral deposition, including data on location, water depth, spreading rate, structural setting, type of deposit and mineralogy. Besides these informations, a genetic interpretation of different types of hydrothermal mineral deposits is given.

During research cruise SO-28 with RV "Sonne" a combined geophysical-geochemical reconnaissance survey was carried out in four areas along the Mid-Indian Ocean Ridge (MIOR) (Plüger et al., 1984): Area A (6^{0} N), Area B ($1^{0}30'$ S), Area CD (6^{0} S) and Area EX (22^{0} S).

The four areas, about 50x50 nm in size, were mapped bathymetrically using a multiple-echosounder system (Seabeam). By means of subbottom-profiling with a 3.5 KHz echo-sounder the sediment-covered regions of Areas A to EX were distinguished from sediment-free regions in the vicinity of the axial valleys.

Subsequent sampling by water sampling chains, and box coring (3 m length) was carried out. The water samples were taken in the axial valley itself, whereas sediment samples were collected at discrete distances from it.

A modified gas-chromatograph with a flame-ionization detector (FID) was used onboard ship to determine dissolved methane contents of the collected seawater samples. Seawater is extracted by means of a gas-tight system directly coupled to the chromatograph. The instrument is a further development of that described by Swinnerton and Linnenbom (1967). The technique involves stripping the dissolved volatile organics by purging with high-purity, hydrocarbon-free helium. Adsorption of dispersed hydrocarbons occurs on cooled traps (-80^{0} C). Hydrocarbons up to hexane are quantitatively collected. Trapped gases are subsequently released by heating to 80^{0} C. At first, higher hexane to ethane are desorbed and finally methane, which usually is present in much higher concentrations. Desorbed gas mixtures are introduced on-line into the chromatograph by a stream of helium gas. Final detection is carried out by a FID, generating an electrical signal of each hydrocarbon component.

No methane anomalies were detected in the water samples of Areas A, B and CD. Only in the southernmost area, Area EX, an increase of methane in the bottom water at 3200 m depth was found. Compared with methane values from between 1000 and 3000 m depth (4 nl/l), anomalous bottom water showed 10 nl/l. The increase may be interpreted as a mantle-derived methane supply.

The deep-sea sediments of the four areas can be classified as calcareous oozes with carbonate contents of greater than 60%. The chemical analysis (major and trace elements) of the sediments by XRF and the subsequent data evaluation lead to the following conclusions:

- Factor analysis shows high scores of typical detrital, organic and diagenetic elements for samples from Areas A, B and CD

- Sediments from Area EX are characterized by a hydrothermal, a detrital and an organic factor.

Besides the already mentioned methane anomaly in Area EX, this is a further indication of probably enhanced hydrothermal activity. Higher heat-flow values in the sediments and the occurrence of Fe-Mn coatings on the recovered basalts from Area EX also support the idea, that the southernmost area investigated during SO-28 along the MIOR is a potential target for further research.

REFERENCES

Anonym, 1983. Pneumatic grab system. Preussag Data Sheet 165-311, Hannover, 1 pp.

Borcherding, K., Döbele, R., Eberle, H., Erbacher, I., Hauschild, J., Hübener, J., Lange, J., Müller, G., Rapp, W., Rathjen, E., Rusch, K.D., Schäf, A. and Tamm, U., 1977. Manganknollen-Analysensystem MANKA. KFK 2537, 194 pp.

Boumans, P.W.J.M., 1979. Inductively coupled plasma-atomic emission spectroscopy: its present position in analytical chemistry. ICP Inf. Newsl., 5: 181-209.

Brooks, R.R., Presley, B.J. and Kaplan, I.R., 1967. Determination of copper in saline waters by atomic absorption spectrophotometry combined with APDC-MIBK extraction. Talanta, 14: 809-814.

Cronan, D.S., 1980. Underwater minerals. Plenum, London, 356 pp.

Fite, L.E., Schweikert, E.A., Wainerdi, R.E. and Uken, E.A., 1971. Nuclear activation analysis. In: R.E. Wainerdi and E.A. Uken (Editors), Modern methods of geochemical analysis. Plenum, New York, pp. 319-350.

Fletcher, W.K., 1981. Analytical methods in geochemical prospecting. In: G.J.S. Govett (Editor), Handbook of exploration geochemistry. Elsevier, Amsterdam, Vol. 1, 255 pp.

Flanagan, F.J. and Gottfried, D., 1980. USGS rock standards, III: manganese nodule reference samples USGS-Nod-A-1 and USGS-Nod-P-1. U.S. Geological Survey Prof. Paper 1155, 39 pp.

Friedrich, G. and Plüger, W.L., 1974. Die Verteilung von Mangan, Eisen, Kobalt, Kupfer und Zink in Manganknollen verschiedener Felder. Meerestechnik MT, 5: 203-206.

Friedrich, G., Kunzendorf, H. and Plüger, W.L., 1974. Ship-borne geochemical investigations of deep-sea manganese nodule deposits in the Pacific using a radioisotope energy-dispersive X-ray system. J. Geochem. Explor., 3: 303-317.

Friedrich, G., Martin, E.J. and Kunzendorf, H., 1982. Geochemical studies on offshore sediments from the continental shelf of eastern Australia. Geol. Jb., D56: 165-177.

Gordon, G.E., Randle, K., Goles, G.G., Corliss, J.B., Beeson, M.H. and Oxley, S.S., 1968. Instrumental activation analysis of standard rocks with high-resolution γ-ray detectors. Geochim. Cosmochim. Acta, 32: 369-396.

Govindaraju, K., 1984. 1984 compilation of working values and sample description for international reference samples of mainly silicate rocks and minerals. Geostandards Newsletters, Vol. VIII, Spec. Iss.

Hawkes, H.E. and Webb, J.S., 1962. Geochemistry in mineral exploration. Harper and Row, New York, 415 pp.

Hawkes, H.E., 1982. Exploration geochemistry bibliography to January 1981. The Association of Exploration Geochemists, Vol. 11, 388 pp.

Hawkes, H.E., 1985. Exploration geochemistry bibliography January 1981 to October 1984. The Association of Exploration Geochemists, Vol. 11, Suppl. No. 1, 174 pp.

Hoffman, E.L. and Brooker, E.J., 1982. The determination of gold by neutron activation analysis. In: A.A. Levinson (Editor), Precious metals in the Northern Cordillera. The Association of Exploration Geochemists, pp. 69-77.

Jenkins, R. and De Vries, J.L., 1970. Practical X-ray spectrometry. Philips, Eindhoven, 190 pp.

Kunzendorf, H., 1972. Non-destructive determination of metals in rocks by radioisotope X-ray fluorescence instrumentation. In: M.J. Jones (Editor), Geochemical exploration 1972. The Institution of Mining and Metallurgy, London, pp. 401-414.

Kunzendorf, H., 1973. Die Isotop-Röntgenfluoreszenz Analyse und ihre Anwendung bei geochemischen Untersuchungen in Grönland. PhD diss., RWTH Aachen, Risø-M-1610, 125 pp.

Kunzendorf, H., Stoffers, P., Walter, P. and Gwozdz, R., 1985. Metal variations in divergent plate-boundary sediments from the Pacific. Chem. Geol., 47: 113-133.

Leake, B.E., Hendry, G.L., Kemp, A., Plant, A.G., Harry, P.K., Wilson, J.R., Coats, J.S., Aucott, J.W., Lunel, T. and Howarth, R.J., 1969. The chemical analysis of rock powders by automatic X-ray fluorescence. Chem. Geol., 5: 7-86.

Levinson, A.A., 1974. Introduction to exploration geochemistry. Applied Publishing Ltd., Maywood, Ill., 614 pp.

Lüschow, H.M. and Kraft, G., 1973. Non-dispersive X-ray spectrometry for manganese nodules. Meerestechnik MT, 4: 200-204.

Miller, J.M. and Symons, G.D., 1973. Radiometric traverse of the seabed off the Yorkshire coast. Nature, 242: 184-186.

Noakes, J.E., Harding, J.L. and Spalding, J.D., 1974. Locating offshore mineral deposits by natural radioactive measurements. MTS Journal, 8: 36-39.

Noakes, J.E. and Harding, J.L., 1982. Nuclear techniques for seafloor mineral exploration. Oceanology International 1982, 0182 1.3.

PE, 1976. Analytical methods for atomic absorption spectrophotometry. Perkin Elmer Co., Norwalk, Connect.

Plant, J., Goode, G.C. and Herrington, J., 1976. An instrumental neutron activation method for multi-element geochemical mapping. J. Geochem. Explor., 6: 299-319.

Plüger, W.L., 1975. Analyse von Manganknollen mit einem Isotop-Energiedispersivem Röntgenfluoreszenz-Spektrometer (System Tracor Northern). BMFT Forschungsbericht M-75-02, pp. 139-204.

Plüger, W.L., Friedrich, G., Herzig, P., Kunzendorf, H., Stoffers, P., Walter, P., Scholten, J., Michaelis, W. and Mycke, B., 1984. Prospektion hydrothermaler Mineralisationen am mittelozeanischen Rücken des Indischen Ozeans. BMFT Status Report, PLR-KFA, pp. 471-488.

Price, V., 1979. NURE geochemical investigations in the eastern United States. In: J.R. Watterson and P.K. Theobald (Editors), Geochemical Exploration 1978. The Association of Exploration Geochemists, pp. 161-172.

Price, W.J., 1972. Analytical atomic absorption spectrometry. Heyden and Son, London, 239 pp.

Rona, P.A., 1983. Exploration for hydrothermal mineral deposits at seafloor spreading centers. Mar. Mining, 4: 7-38.

Rona, P.A., 1984. Hydrothermal mineralization at seafloor spreading centers. Earth-Sci. Rev., 20: 1-104.

Rose, A.W., Hawkes, E.H. and Webb, J.S., 1979. Geochemistry in mineral exploration. Academic Press, London, 657 pp.

Senftle, F.E., Duffey, D. and Wiggins, P.F., 1969. Mineral exploration on the ocean floor by insitu neutron absorption using a Californium-252 source. MTS Journal, 3: 9-16.

Sipos, L., Ruetzel, H. and Thijssen, T.H.P., 1980. Performance of a new device for sampling seawater from the sea bottom. Thalassia Jugosl., 16: 89-94.

Smith, W. and McIntyrne, A.D., 1954. Spring-loaded bottom-sampler. J. Mar. Biol. Ass. U.K., 33: 257-264.

Summerhayes, C.P., Hazelhoff-Roelfsema, B.H., Tooms, T.S. and Smith, D.B., 1970. Phosphorite prospecting using a submersible scintillation counter. Econ. Geol., 65: 718-723.

Swinnerton, J.W. and Linnenbom, V.J., 1967. Determination of the Cl to C4 hydrocarbons in seawater by gas chromatography. J. Gas Chromat., 5: 570-573.

Thijssen, T. and Kunzendorf, H., 1981. Röntgenfluoreszenzanalyse von Ti, Fe und Zr. Unpubl. cruise report SO-15, Bundesanstalt fur Geowissenschaften und Rohstoffe, Hannover, pp. 28-31.

Turekian, K.V., 1985. Die Ozeane. Enke, Stuttgart, 202 pp.

van Veen, J., 1936. "Onderzoekingen in de Hoofden". Landsdrukkerij, The Haag, 252 pp.

von Stackelberg, U. and Jones, H.A., 1982. Outline of Sonne cruise SO-15 on the east Australian shelf between Newcastle and Fraser Island. Geol. Jb., D56: 5-23.

Wainerdi, R.E. and Uken, E.A., 1971. Modern methods in geochemical analysis. Plenum, New York, 397 pp.

Wogman, N.A., Rieck, H.G., Kosorok, J.R. and Perkins, R.W., 1972. In-situ activation analysis of marine sediments with Cf-252. BNWL-SA-4434.

Wogman, N.A., Rieck, H.G. and Kosorok, J.R., 1975. In situ analysis of sedimentary pollutants by X-ray fluorescence. Nucl. Instr. Meth., 128: 561-568.

CHAPTER 6

DATA EVALUATION OF EXPLORATION SURVEYS

J. P. LENOBLE

INTRODUCTION

Marine mineral exploration accumulates large quantities of data of various kinds. After a survey of data storage facilities, general methods of data verification and validation have to be considered. Analysis of data is achieved by several mathematical methods including geostatistics. The undersea prospector has to build an image of the scenery of the sea bottom by a careful assemblage of individual indirect measurements, the way the on-shore geologist does for sub-surface geology. Moreover, the off-shore geologist must also do the same for sub-bottom geology.

During this blind search for knowledge, accumulation of data is needed not only to secure the truth but to achieve its understanding. Consequently, the explorer handles a considerable amount of data for which he has to require the use of computers (Schimmelbusch et al., 1975).

The evaluation of ore and metal content needs a careful approach with multiple iterations. Data representation using graphs, profiles, sections, maps, etc., now drawn by computers, are essential for improved understanding of complicated data structures.

In this chapter we will consider successively:

- the collection and storage of data in computers and the validation of the data after their introduction into the base,
- the analysis of the structure of the data by classic and more sophisticated statistical methods,
- the representation of the data by computerized graphical methods,
- the interpretation of the data in terms of evaluation of the physiographic environment and assessment of mineral resources.

Many of the examples given in this chapter deal with French exercises in marine mineral exploration and experience from the AFERNOD (Association Francaise pour l'étude et la recherche des Nodules) program is especially called on. Other examples can be found in the respective literature.

DATA COLLECTION AND STORAGE

The tool-box: the data bases

Increasing development of computerized data storage facilities makes the choice of suitable software a nightmare. However, most software is machine dependent, and as the choice of which computer to purchase is generally determined by other considerations, the choice is less intricate.

With access to large computers of the last generation, everything is permitted, including the possibility of asking the system to solve your problem, and to build the most suitable data base (Lesk, 1984). The size of the available memory is likely to authorize the storage of all the data in the same base, with few problems of access time. Packages of software are provided by the machine manufacturer for most data-treatment operations, as statistics, graphic representation, even contour mapping. Data extraction for more sophisticated treatment needs the writing of some interface software in a commonly used advanced language, e.g. FORTRAN, PL1, BASIC, etc.

Whatever data management system one is using, the construction of the base structure is a decisive step. In an electronic data storage system, each item is stored in a field, several related fields are grouped in a record, a bunch of records constitutes a file. With a network base, several files can be connected through key fields. Key indexes can also be prepared for hierarchical bases to facilitate the search of special values in key fields. However, their efficiency is rather poor, as it is necessary to rebuild the index each time a new set of data is introduced in the base. In the most sophisticated relational data bases, all fields can be considered as keys to information retrieval.

The internal organisation of the data management system is rather indifferent, as for most purposed the less complicated, i.e. the hierarchical one, is adequate (Monget and Roux, 1976). However, some comfort is not forbidden, and a network or even a relational model is helpful for searching scattered values of some parameters. For instance, if one wants to know where sea cucumbers have been encountered on sea bottom photographs, because he suspects them to play a role in the in-situ shear strength of the sediment, he will be satisfied, if by chance the occurrences of the holoturies were entered in its base, and by using a relational model, it will be a child's play to extract both data for correlation.

As marine exploration is carried out by exploration cruises, it would be convenient to affect a file or a set of files for each cruise. However, most data are space related, and it is better to show a preference for that variable. Some bases especially developed for geosciences are structured in data tables organised around the coordinates of the point of observation.

Since 1970 AFERNOD has used the data management system INFOL (INFormation Oriented Language), which was developed by the University of Geneva. The base is structured in records of 280 items (fields), themselves divided into several sub-items. Each file is devoted to one station, corresponding to a series of measurements and/or sampling made in the close vicinity of a point defined by its geographic coordinates. This base is being converted to a new data management system named APPEL IV, which is better adapted to the present generation of computers.

Nature and type of data

Space relationship. In the previous chapters various techniques of exploration were introduced. The data resulting from the use of these techniques are diverse (e.g. Siapno, 1976, Cronan, 1980). They are all referred to a portion of space (locality) where the measurement or the sampling were made. However, the definition of this portion of space is somewhat confused.

Firstly, the absolute position of the geometric center of this portion of space is known with an uncertainty that can introduce errors in the relative position of two or more successively close measurements. This inaccuracy makes it also difficult to return to the same point of measurement. Any operation involving a space relationship below the range of accuracy is questionable. This range varies from some meters for a sea-bottom acoustical positioning to some kilometers for obtaining an estimated position between two satellite fixes.

Secondly, the portion of space to which the measurement refers, can be misleading. For instance:

- The depth measurement made from a surface ship with a normal acoustic sounder, is unrelated to the point in a direction vertical to the ship, but relates instead to the integrated signal of all the echos returning from an area of the sea bottom which, for at a 4000 m depth e.g., is close to a circle 3 km in diameter (Bastien-Thiry, 1979). However, this diameter is reduced to only 200 m with a narrow beam precision depth recorder.
- It is commonly reported that the resolution of a mud penetrator or seismic device is only a few meters or even decimeters; these figures refer only to the vertical resolution, and the horizontal size of the responding area is about 1000 times larger.

Thirdly, it is not wise to consider the measurements made on a cubic decimeter to represent the average value of the ocean floor surrounding the point of measurement to the mid distance to the next point, especially when these two points are separated by several tens of miles (David, 1977).

Accordingly, it is convenient to complete the set of data with their range of accuracy expressed as two times the standard error of measurements, if known.

Table 6.1
Nature and type of data in oceanographic surveys.

Techniques	Device	Original data	Data to record in the base	Type	Final desired data
NAVIGATION	Satellite, loch (speed), compass Radionavigation	Latitude, longitude	Time, coordinates in degrees, accuracy Idem or idem in meter	N	Navigation map
BATHYMETRY	Normal echo-sounder	Bottom profile	Time, depth, (or X-Y-Z),accuracy	N	Bathymetric map
	Narrow-beam E.S.	Bottom profile, tape	Time, depth, (or X-Y-Z),accuracy	N	
	Multibeam E.S.	Bottom profile, lane Contour map, tape	Time, set of traverse distances and depths, or grid of X-Y-Z	N	
SONAR	Side scan sonar	Bottom acoustic imagery, tape	To be defined		Bottom surface structural map
SEISMIC	Mud penetrator,seismic reflection, refraction	Sub-bottom analog profile of tape	Time,set of depths of sub-bottom structures, or their X-Y-Z	N	Isopachs & isobaths map
MAGNETISM	Magnetometer	Graph or tape of mag. field vs time	Time, mini-maxi field values	N	Map of anomalies after corrections
GRAVITY	Gravitometer gravity vs time	Graph or tape of gravity field	Time, mini-maxi field values	N	Map of anomalies after corrections
SEA SURFACE	On board equipment Surface buoy	Sea state,wind,air temperature,moisture	Time,X-Y,wave/heave hight/period direction,wind force/direction air/sea temperature,air moisture	AN	Weather forecasts
WATER COLUMN	Bathysonde or other device	Temperature, salinity vs depth	X-Y-Z, temperature, salinity time	N	Anomalies maps and profiles
	Sound velocity	S.V. vs depth	X-Y-Z, sound velocity, time	N	
	Current meters	Current speed and direction vs time	X-Y-Z, set of time + 3D speeds	N	Current maps and profiles vs time
HEAT FLOW	Thermistances along core barrel	Temperature,depths on tape	X-Y-Z, heat-flow value	N	Map of anomalies
BOTTOM PICTURES	Free-fall camera Towed camera Camera on submersible Television	Picture frame Film Film Video-tape	Picture description: see text	AN	Map of bottom features
SOLID SAMPLES	Corers, dredges, grabs core-drill, etc.	Operating procedure	Sampler type, X-Y-Z,penetration depth, recovery, etc.	AN	
	Mineralogical laborat.	Micro description	See text	AN	
		Mineral composition	Abundance of each mineral	AN	
	Chemical laboratory	Chemical composition	Abundance of each element	AN	
	Physical laboratory	Macro description, grain size, water content, soil & rock mechanics, etc.	See text	AN	
LIQUID SAMPLES	Bottles, pumps	Chemical composition,	Abundance of each element	AN	
		dispersed particles	Nature & abundance of particles	AN	

N = Numeric, AN = Alphanumeric.

Type of data. They can be numbers (numeric), string of characters (alphanumeric), or logical operators (YES, NO, etc.). All are accepted by most data bases. For numbers, it is necessary to define the range, the mode (real, decimal, exponential, integer), eventually the number of decimal places, etc. Care should be taken in assigning the measurement units.

Nature of data. Table 6.1 summarizes data corresponding to the different techniques of exploration commonly used. This list was established for deep sea nodule exploration, with some additions referring to sulfide deposits. It will be easy to adapt the list to any other mineral exploration.

Reference is made to the devices or facilities that are used to obtain the information, to its original form when acquired, to the desirable format for introduction in a data base, to the type of data, and to some of the final desired information.

Original navigation, bathymetric, seismic, magnetic and sonar data are now often recorded on tape which are already data bases. However, it is convenient to extract significant parts of this information to be introduced in a more general descriptive data base.

Description of samples or pictures of the bottom is not easily recorded properly due to the rigidity of most data bases (Defossez et al., 1980). Flowing description is feasible only for the most sophisticated information storage systems, with the disadvantage of long access times.

AFERNOD developed two different bases to treat this problem, these data bases are now in the process of being merged into a new single data base. In the first data base, managed by INFOL, the information was structured in specialised descriptive items as shown in Table 6.2, and used for stations where only a single sample and photograph are described. The second data base was defined to collect description of photographic surveys made by RAIE (a still camera with throbes towed by a steel cable) or EPAULARD (an unmanned free vehicle equipped with still camera and monitored by acoustic). Each record refers to a series of pictures alike, as far as their descriptive characteristics are concerned.

Verification of data

Data storage is not an achievement in itself. Computers are utilized to help the search and retrieval of data for further treatment. Consequently, data must be:

- selected and arranged to permit such treatment and avoid overstocking of the base, and
- verified after entry into the base to locate any error that could lead to disastrous results.

TABLE 6.2

Structure of the data base used by AFERNOD for description of samples and sea-bottom photographs.

Items	Descriptive characteristics
	Description of grab samples
66	Weight of waste
67	Weight of nodules
	Morphology. For each morphologic group series of sub-items in each following items:
120	- Roundness and flatness
121	- Percentage in weight of the nodules in the group
122	Name of the morphologic group
	Granulometry. For each size class (8 sub-items in the following items):
123	- Size range (10 mm intervals)
124	- Percentage in weight of nodules in the class
130	- Number of nodules in the class
125	Lengths of the longest axis of the smallest and largest nodules
126	Minimum and maximum dimensions of flagstones or slabs
127	Roughness of nodule surface
129	Other observations about ore
141	Dimensions of waste: average, minimum, maximum
142	Shape (roundness/flatness) and colour of waste
143	Nature of waste
144	Roughness of waste surface
145	Thickness of crust
147	Other observations about waste
	Description of sediment
151	Colour (Munsell code), name
	Microscopic observation of smear slides. For each class, series of sub-items in each following item
152	- Grain size class (>63 μm, 63 to 2 μm, <2 μm)
153	- Abundance of each class %
154	- Nature of phase (terrigenous, clay, calcareous/siliceous bioclastic, authigeneous)
155	- Abundance of each phase %
156	$CaCO_3$ content: from analysis, estimated from smear slides
157	Cohesion
158	Water content
162	Other observations about sediment
	Description of photographs
101	Quality of the picture
105	Population (percentage of the picture surface occupied by nodules)
	Granulometry. For each class of diameter, series of sub-items in each following items:
106	- Class limits
107	- Population of each class
	Morphology. For each morphologic class, series of sub-items in each following items:
109	- Class description (shape, roundness/flatness)
110	- Population of each class
111	Heterogeneity of the distribution in the picture
112	Living organisms
113	Other observations (tracks, identified or unidentified objects etc.)

Standardization of the measurements. Consistency of the data is essential. Data issued from different sources or acquired by different equipments cannot be merged without verifying their homogeneity. Furthermore, some apparatus can deliver different results depending how they are used (Frazer, 1979). To avoid such misadventure, instructions for use have to be prepared carefully.

Some original information is too abundant and must be shortened. However, this should be done without important loss of information. For instance, the grain size of a basket of nodules, defined originally by the weight percentages of ten size grades, can be reduced to the average, the lower and upper quartiles sizes, if and only if the sorting is regular.

Visual descriptions of samples or phenomena are very difficult to handle and no further treatment will be possible if the semantics are careless. It wastes time to introduce such data in the data base. Roughly, one has to call a spade a spade and keep the same name all the years to come. Lists of standard names or adjectives, specifying the gradation of the properties, are helpful, as they are associated with reference examples, for the training and later the self-calibration of the observers. Colour charts for marine samples are available (e.g. Munsell color chart for sediments), and other charts can be prepared for parameters like shape, size, surface appearance, percentage of surface occupied by round objects, etc. (Lenoble, 1980a).

Indirect measurements refer to physical parameters, sometimes with some chemical significance, expressed as a chemical value (Greenslate et al., 1972). Two main types of physical phenomena are considered: the time transit of wave signals (acoustic or seismic) and the measure of a field of force, e.g. gravity, magnetism, electromagnetism. A third type which seldom is used in situ, is radiation, e.g. X-ray, natural or induced radioactivity. Most of these physical properties are recorded on continuous analog graphic recorders and magnetic tapes (Richter, 1973). Actually, the continuity of the recording is only apparent and comes from the juxtaposition of a number of discrete measures.

The measurements are made by ways that differ according to the manufacturer of the equipment, and it is advised that the methods used by the equipment be examined with care. It could be necessary to get additional information from the manufacturer. Main factors are the frequency of the sampling, the time interval of the measure, the threshold and the kind of measurement (threshold, peak, integration, etc.); for signals, the frequency ranges, noise filtration, etc. are required; for fields of force, it will be the direction of the measurements, etc.

After such audit of each techniques, it is possible to decide what kind of sampling can be made in order to limit the recording to significant values in the data base. The method commonly used with analog recording is the digitalization of the main events. The method commonly used with analog recording is the digitalization of the main events. The last word refers to the original recording versus time; it has to be transformed into data versus space by using a special file of coordinates versus time. This operation is now fully automatised for bathymetry, magnetism, gravimetry in most laboratories and on board major oceanographic vessels.

Consistency of the data. Errors could be brought in anywhere, anytime, and for any reason. Of both kinds of errors, random and systematic, the latter is likely to be the more pernicious, but it is also the easiest to find.

There are several ways to prevent random errors.

Some safeguards can be introduced in the base structure, causing the rejection of inappropriate data entries. Numbers are accepted in definite fields only if they fall between a minimum and maximum. Their specific nature: integer, real, exponential, negative or positive, can be specified. A reference glossary can be defined for each field or a group of fields so that any unlisted word will be refused.

Immediately after data entry, a careful investigation must be undertaken to search for any possible errors. A visual check of the data on a video terminal or printed listing is often insufficient. For numbers, a statistical analysis associated with a plot of data (if possible two sets of correlatable data) is useful to localise ambiguous data (Fig. 6.1). For words, an alphabetical list with the numbers of occurrence serves to track down spelling mistakes, the use of synonyms, or those that are undefined.

The search for systematic errors is more delicate (Frazer, 1979). In general, this kind of error is associated with some drift in the calibration of an apparatus or a method of measurement or of an observation. If the drift is instantaneous and important, the consecutive sudden change in a series of values can be detected. But if the change approximately equals the usual variation (standard deviation), or if the drift is progressive, its discovery will not be so easy to make.

If there is only one piece of equipment and one operator and if the measurements or the observations can be repeated on the same object, repetition tests must be made systematically on already treated specimens. The rate of these tests depends on the known stability of the operator-apparatus combination. Statistical treatment of the results, particularly regression analysis and correlation diagrams between the two sets of measurements, will disclose a possible systematic drift from the normal instrumental error.

If there are several pieces of equipment and/or several operators and if the measurements can be repeated, the tests can be organized with exchange of operators and apparatus. Such intercalibrations are commonly done successfully between chemical laboratories (Fig. 6.2).

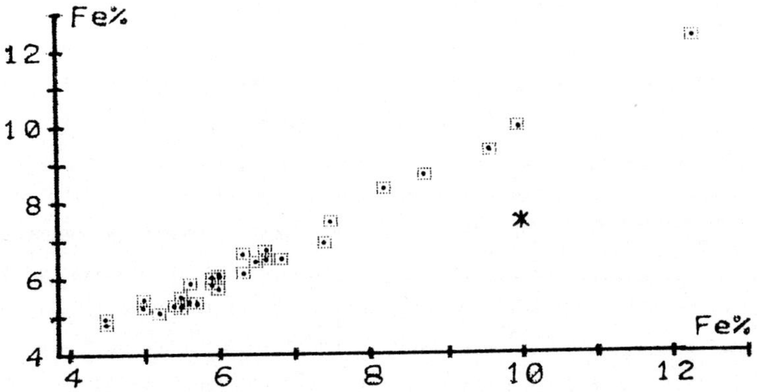

Fig. 6.1. Correlation diagram of two sets of iron analysis of the same samples (intercalibration of laboratories). One erroneous data point (*) is out of range of the normal distribution.

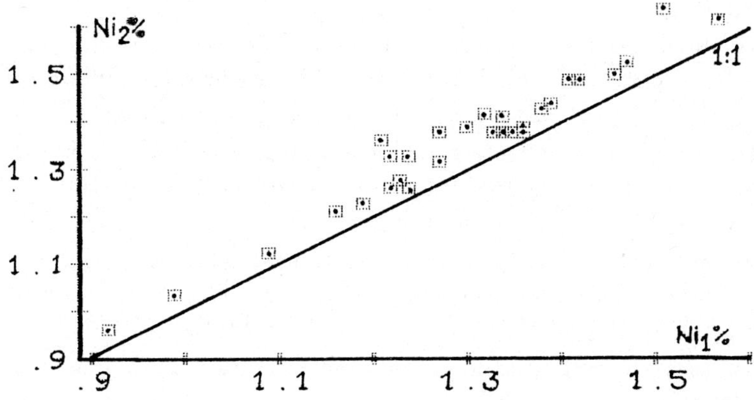

Fig. 6.2. Correlation diagram of two sets of nickel analysis done by two different laboratories on the same sample. The systematic error shown by the shift of X and Y scales was due to differences in drying procedures.

If repetition of the measurements or observations cannot be done, the only solution is to simulate comparative tests by sampling the population of data over time, apparatus and operators. The number of tested data must be large enough to give a significant result, and sometimes the entire population is considered to be the sample. The sampled data are then submitted to statistical treatment. The standard statistical parameters computed on each set of data (collected through time, apparatus, operator, etc.) are compared. Similarity tests are run between the different series., However, separation between geographical variations and systematic errors is often dubious.

Correction of errors is often difficult. Some errors are simply a mistaken reading or typing, and data are then immediately restored. Sometimes, although the error is obvious, its cause cannot be determined, nor the true value reconstituted. A few random errors that appear scattered in a large amount of information can be simply discarded because it is better to allow a gap of information than retain an incorrect value. Systematic errors can be rectified when a strong correlation exists between the sampled set of mistaken data and the calibration test (Fig. 6.2). Certification of errors occasionally needs complementary tests. When a systematic error is suspected and cannot be corrected, it is necessary to go back to the field and make new measurements.

DATA STRUCTURE ANALYSIS

Facing a large amount of information could be disturbing. To obtain a clear understanding of the reality hidden behind thousands of figures or words, many methods have been proposed (Laffitte et. al., 1972, Agterberg, 1974). Only a few of them are presented below.

Valuable minerals exist everywhere, but recoverable one are seldom found. To constitute potential resources, they must be concentrated in identified structures, which are called ore deposits when it is proved that they can be mined economically. Identification of these structures is the purpose of mineral exloration. Similarly identification of structures of the data leads to discovery of possible geological structures.

Typical data structure analysis involves standard statistics, correlation analyses, likeness coefficients, multiple regression, discriminant and cluster analyses and geostatistical techniques. These methods are shortly reviewed below. For detailed discussion, the readers referred to the standard statistical literature (see references).

Standard statistics

Standard statistical methods include the determination of minimum and maximum values, the calculation of arithmetic average, standard deviation or variance, as defined hereafter:

$$\text{Average:} \quad m = \frac{1}{N} \sum_{1}^{N} (v_i)$$

where v_i are the values when i varies from 1 to N, N being the total number of available values.

$$\text{Variance:}\quad V = \frac{1}{N}\sum_{1}^{N}(m - v_i)^2$$

$$\text{Standard deviation:}\quad S = \sqrt{V}$$

Most computers are equipped with software for these calculations.

A first step is the comparison of the average with minimum and maximum. If m is very different from the mid-value, i.e.:

$$\frac{\text{Mini. + Maxi.}}{2}$$

then the distribution is likely to be asymmetric. If the standard deviation is very large relative to the average, the population of the values is dispersed or multiple.

In these cases it is necessary to proceed with a more detailed investigation. A plot of the frequency of the values is useful (Healing et al., 1975;1979). Values are clustered in groups, generally ten to fifteen. The bar graph or histogram is well adapted. (Fig. 6.3). Several trials are often necessary to optimize the division of groups by adapting the range of some of them. Software is available to compute and trace such plots. The mode is determined as the most frequently observed value, or more easily as the mid-value of the group showing the highest frequency.

A multimodal distribution, as presented in Fig. 6.3, comes generally from the mixing of several groups of objects or phenomena. The study of their spatial distribution and their relationships to other parameters can lead to some important clue. For instance, in addition to their South Pacific affinity, the group of low-grade nodules of Fig. 6.3 that occurs in the Clarion-Clipperton zone, was mainly found near the cliffs which skirt along the N-S hills.

However, when no explanation can be found for multimodal or largely dispersed data, it is wise to re-examine the data collection, as described previously, before proceeding.

For alphanumeric data, statistical functions are meaningless, but occurrence frequency of descriptive parameters is often a valuable piece of information and can be obtained by similar methods.

Correlation analysis

Natural phenomena are defined by the assemblage of characteristics, some being obvious, others concealed by the abundance of information. Correlational analysis allows us to find the relationships among different parameters.

Data are organised in tables (matrices) and refer to objects (e.g. nodule) or phenomena (e.g. current), and characteristics (or variables: e.g. metal

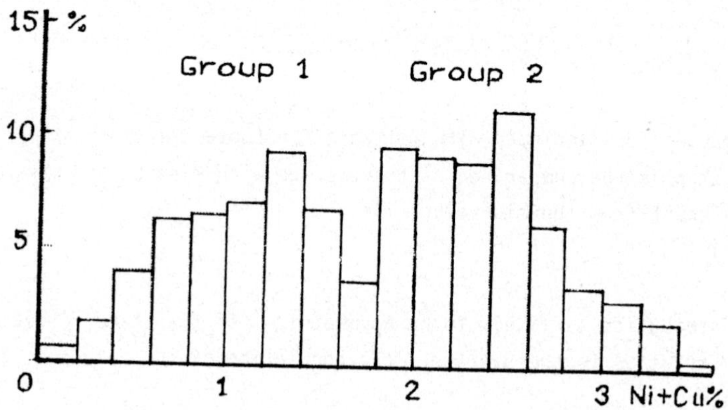

Fig. 6.3. Histogram of nickel+copper content of Pacific nodules. The two modes represents two groups of nodules of different geochemistry and mineralogy. The richest group was found mainly in the Clarion-Clipperton zone (Lenoble, 1980a).

content, shape, speed or direction). Objects or phenomena can be grouped into categories defined by certain common variables. The purpose of correlational analysis is to determine the categories of objects, their common characteristics or the objects belonging to different categories.

Table 6.3 summarizes the different multivariate statistical methods that could be used. Three cases are envisaged, subject to the partition of the data matrix by variables or objects, or without partition.

Likeness coefficients. The determination of the similarity or dissimilarity of objects or phenomena is done by computing numerous coefficients. The most common are listed below.

- Correlation coefficient of Pearson:

$$R_{xy} = \frac{\Sigma n(X_i - X)(Y_i - Y)}{\sqrt{\Sigma n(X_i - X)^2 \; \Sigma n(Y_i - Y)^2}} \qquad -1 < R_{xy} < +1$$

where Σ n is the sum of the values of n individuals

X_i, Y_i are the values of variables X and Y (i=1, 2, ..., n)

X, Y the arithmetic means of the values of variables X, Y.

TABLE 6.3

Multivariate statistical methods.

	MATRIX PARTITION MODE		
	BY VARIABLES	BY OBJECTS	NONE
NATURE OF PROBLEMS	Search of links between variables	Sorting of objects	Classification Numerical taxonomy
METHODS	Multiple regression Canonical correlation	Discriminatory analysis	Factor analysis Correspondence analysis Cluster analysis
LIKELINESS COEFFICIENTS	Correlation coefficients		Association, correlation and distance coefficients

- Rank coefficient of Spearman:

$$R_s = 1 - \frac{\sigma \Sigma n(x_i - y_i)^2}{n(n^2 - 1)} \qquad -1 < R_s < +1$$

which is used with non-numerical (qualitative) variables that are arranged and aritificially numbered from 1 to n.

For both methods, the larger are the correlation coefficients, the greater is the similarity between two series of variables.

- Distance coefficient:

If two specimens A and B are described by n characteristics:

$X_{A1}, X_{A2}, \ldots, X_{An}$ and $X_{B1}, X_{B2}, \ldots, X_{Bn}$

their "distance" is expressed by:

$$d_{AB} = \sqrt{\Sigma n(X_{Ai} - X_{Bi})^2} \quad (i = 1, 2, \ldots, n).$$

This "distance" is calculated by analogy with the true distance between two points A and B owing to the coordinated (X_A, Y_A), (X_B, Y_B), which is: $d_{AB} = \sqrt{(X_A - X_B)^2 + (Y_A - Y_B)^2}$.

- Association coefficients:

Several parity functions are in use to express the number of observed concordances and discordances when several objects, defined by different variables, are compared. Whatever nature, number or quality, the variables are, a certain value (or realisation) will be marked as 1, whereas any other value (or absence) is marked as 0. When two specimens are compared, the concordant variables will appear as 1,1 (positive) and 0,0 (negative), the discordant ones as 1,0 and 0,1. If n_{11} is the number of concordances and n the total number of compared variables, the quantity:

$$S_{SM} = \frac{n_{11}}{n}$$ (Simple matching coefficient of Sokal & Michener)

where n_{11} is the number of positive concordances, is a measure of their degree of similarity $(0 < S_{SM} < 1)$.

A similar formula gives more weight to the positive concordances:

$$S_D = \frac{2n_{11}}{2n_{11}+u}$$ (Dice's coefficient) $0 < S_D < 1$

where n_{11} is the number of positive concordances (1,1), and
u the number of discordances (0,1) and (1,0).

The efficiency of these coefficients depends on the careful choice of significant variables.

All previous coefficients, except that associated with distance, vary in absolute value between 0 and 1. When their value is 0, one can say that no correlation exists between the two variables. An absolute value approaching 1 is likely to mean a strong correlation. However, the interpretation must be made carefully as coincidence is not a proof of a link from cause to effect.

Table 6.4 presents a matrix of the Pearson's coefficients of correlation of the geochemistry of the North Pacific manganese nodules (Hein, 1977).

Multiple regression. The multiple regression method searches linear equations that could link a variable Y with p other variables X_j, such as:

$y_i^* = b_1x_{i1} + b_2x_{i2} + \ldots + b_px_{ip}$

where y_i are the observed values of Y (i = 1,2,3, ..., n),
y_i^* its estimated values,
x_{ij} the observed values of X (j = 1,2,3, ..., p) and
b_j the regression coefficients to be computed.

The validity of the equation is tested by a correlation coefficient:

$$R^2 = \frac{(\Sigma(y_i - Y)(y_i^* - Y^*))^2}{\Sigma(y_i - Y)^2 \ \Sigma(y_i^* - Y^*)^2}$$

where Y and Y* are, respectively, the means of all the observed and calculated variables y_i and y_i^*.

TABLE 6.4
Correlation coefficients for selected elements on North Pacific manganese nodules (LI = loss of ignition).

	LI	Mn	Fe	Ni	Co	Cu	Na	Mg	K	Ca	Si	Al	Ti	Zn
Mn	.59													
Fe	-.12	-.64												
Ni	.43	.80	-.73											
Co	.09	-.29	.78	-.40										
Cu	.34	.80	-.82	.82	-.49									
Na	.05	.36	-.50	.32	-.33	.33								
Mg	.28	.18	-.21	.28	-.08	.19	.28							
K	-.65	-.66	.10	-.49	-.12	-.48	.08	.00						
Ca	-.10	-.08	.03	-.10	-.03	.00	-.05	.31	-.22					
Si	-.76	-.74	.08	-.50	-.19	-.46	-.03	-.05	.86	-.10				
Al	-.59	-.75	.16	-.47	-.12	-.51	-.20	.00	-.80	-.12	.87			
Ti	-.03	-.58	.92	-.67	.68	-.79	-.56	-.35	.01	.11	.02	.17		
Zn	.45	.84	-.74	.76	-.47	.48	.31	-.52	-.01	-.55	-.59	-.70		
Pb	.16	-.42	.88	.50	.84	-.67	-.38	-.06	-.11	.04	-.17	-.03	.83	-.52

The computing of the regression coefficients is done so as to maximize the correlation coefficient. The quantity F:

$$F = \frac{R^2 \qquad (n - p - 1)}{(1 - R^2) \qquad p}$$

follows Snedecor's law with p and (n-p-1) degrees of freedom. If the computed F is smaller then the threshold in Snedecor's tables, the regression is insignificant (e.g. Rao, 1965).

Simple regression considers only two variables. It is possible to calculate linear equations of second (or more) order, and even exponential equations. For instance, Hein (1977) computed the following equations for metal contents of

North Pacific polymetallic nodules:

- copper versus nickel:

$Cu = - 1.23\ (Ni)^4 + 3.47\ (Ni)^3 - 2.70\ (Ni)^2 + 1.18\ (Ni) + .06$

which expresses the saturation in Ni for high values, although copper still increases.

- copper and nickel versus manganese:

$Cu = .0009\ (Mn)^2 + .016\ (Mn) + .053$

$Ni = -.0006\ (Mn)^2 + .073\ (Mn) + .241$

where nickel varies quite linearly, copper, however, shows a slow increase for high manganese values. This high-value copper is likely to be adsorbed by manganese hydroxides rather than be included into their lattices.

Canonical correlation. When the same or two different but likely related objects are described by two different sets of variables, their relationships can be studied by the canonical regression. For instance, the geochemistry of polymetallic nodules and associated sediments were studied by this method.

Studied observations, (e.g. nodules and associated sediment at the same location), are defined by their chemical elements contents. The two matrices of analysis:

$Y_K(y_{K1}, \ldots, y_{Kq})$ and $X_K\ (x_{K1}, \ldots, x_{Kr})$.

corresponding to the q variables (analysis) of k Y-observations (nodules) and to the r X-observations (sediments), are combined to find the coefficients l_i, and m_j, so that the two sets of linear equations:

$(l_1 y_{K1} + \ldots + l_q y_{Kq})\ (m_1 x_{K1} + \ldots + m_r x_{Kr})$

are correlated to the maximum for all observations.

As previously, a test of validity can be made on the canonical correlation coefficients.

It is sometimes difficult to give a naturalist explanation of the equations, even when the correlation test is significant. Most statisticians prefer to use the correlation methods comparing objects, such as the following ones.

Discriminatory analysis. When groups of objects or phenomena are defined, as for instance the two geochemical families of manganese nodules mentioned above, discriminatory analysis can sort the observations into these groups.

If two (or k) groups are defined by p significant characteristics, each specimen can be represented by a virtual "point" in p-dimensional space. All specimens of one group are inside a virtual "solid" which can be identified by the "coordinates" of its virtual "center" and its average "dimensions". Those coordinates are the arithmetic means of the p observed characteristics, and the dimensions can be defined by their variances and covariances. Each group is determined by:

$G_i\ (M_i, C_i)\ (i = 1, 2, \ldots, k)$

where M_i is the matrix of the p means and C_i is the square matrix of the variances and covariances.

The characteristics of a new specimen X:

$(x_1, x_2, \ldots, x_p)$

are compared with the k groups two by two, and a linkage probability is computed for each group, known as a discriminant score:

$L_i = f_i p_i(X)$

where f_i is the a-priori probability of existence of the group i (frequency of previously observed occurrences), and

$p_i(X)$ the density of probability of the distribution of the X_i in the group i (supposed to follow a normal law).

Several computing methods have been proposed for the discriminant score. However, their results are rather uncertain when the number of groups exceeds two.

Cluster analysis. This method was established by biologists to classify units in accordance with the similarity of their characteristics. The starting point is the matrix of correlation (or distance) coefficients of the p variables, as defined above. The most correlated variables are associated in groups (taxons). During successive cycles, isolated variables are aggregated to previous taxons, themselves associated in clusters.

The clustering is done in accordance with a linkage criterion that is reduced by a constant number (e.g. .01) at each cycle. The original criterion is the similarity coefficient of the first group of clustering. Clustering is decided for units that have a similarity coefficient equal or greater than the linkage criterion. The calculation of the similarity coefficients of the successive taxons differs from one method to another. In one of the most sophisticated methods, the coefficient between two new taxons P and Q, grouping respectively p and q units, P(I = A, B, D, ...) and Q (J = L, M, N, ...) is computed by:

$$R_{PQ} = \frac{\Sigma_{pq} R_{IJ}}{\sqrt{(p + 2\,\Sigma_{p-1} R_{II})(q + 2\,\Sigma_{q-1} R_{JJ}}} \qquad \text{(Spearman's formula)}$$

where $\Sigma_{pq} R_{IJ} = R_{AL} + R_{AM} + \ldots + R_{BL} + R_{BM} + \ldots$
is the sum of the intragroups similarity coefficients, and

$\Sigma_{p-1} R_{II} = R_{AB} + R_{BD} + \ldots$, and

$\Sigma_{q-1} R_{JJ} = R_{LM} + R_{MN} + \ldots$

are the sums of the introgroups similarity coefficients.

The average similarity coefficient is also used:

$$R_{PQ} = \frac{\Sigma_{pq} R_{IJ}}{pq} .$$

The result of a cluster analysis is often presented as a dendrogram showing on the Y-axis the linkage coefficient used for each clustering (Fig. 6.4).

Factor and correspondence analysis. The factor analysis was introduced during the first half of the century and principally used by psychologists to analyse the factors of behaviour. The method has been considerably improved by the works of Benzecri and his team of the Sorbonne statistical laboratory during the seventies (Benzecri, 1973). They developed a new method known as correspondence analysis, which is based on a less number of hypothesis and therfore is more commonly applicable.

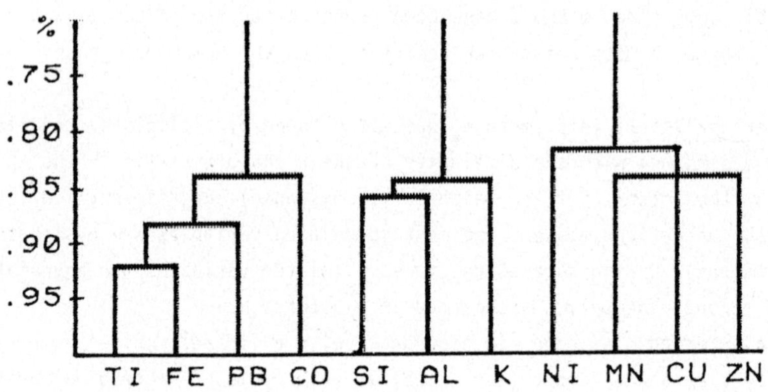

Fig. 6.4. Dendrogram of clustering of the geochemistry of North Pacific manganese nodules.

If x_{ij} is the value of the characteristic j for the specimen i, (i = 1, 2, ..., p; j = 1,2 ..., q), each specimen i can be represented by a "point" in p-dimensional space with coordinates:

$(x_{i1}, x_{i2}, \ldots, x_{ip})$.

The similarity between two specimens k and l can be measured by the "distance" d(k, l) as:

$$d^2 \ (k, l) = \Sigma_q \left(\frac{1}{\Sigma_p x_{ij}} \left(\frac{x_{Kj}}{\Sigma_q x_{kj}} - \frac{x_{lj}}{\Sigma_q x_{lj}}\right)^2\right),$$

where $\frac{x_{kj}}{\Sigma_q x_{kj}}$ are the values of the characteristics j of the specimen k weighted by the sum of the q values of the q characteristics of the specimen k, and

$\Sigma_p x_{ij}$ is a weighting factor equal to the sum of the p values of the characteristic j of the p specimens i.

Symmetrically, each characteristic j can be represented by a "point" of coordinated (x_{1j}, x_{2j}, ..., x_{qj}) in a p-dimensional space named as the space of the variable. The correlation between two characteristics m and n is measured by:

$$d^2(m,n) = \Sigma_p\left(\frac{1}{\Sigma_q x_{ij}}\left(\frac{x_{im}}{\Sigma_p x_{im}} - \frac{x_{in}}{\Sigma_p x_{in}}\right)^2\right),$$

where $\frac{x_{im}}{\Sigma_p x_{im}}$ are the values of the characteristic m of the specimens i weighted by the sum of the p values taken by the characteristic m for p specimens, etc.

The method determines the main "axes" of the "clouds" including all the "points" in both the spaces of the variables and of the specimens. Those axes are used to display in 2-dimentinal diagrams the projections of the representative "points" or "clouds" (Fig. 6.5).

As the dimensions are unified by the weighting, both the characteristics and the specimens can be displayed on the same diagram (Monget and Roux, 1976). This representation allows us to visualize the more significant characteristics of each group of specimens.

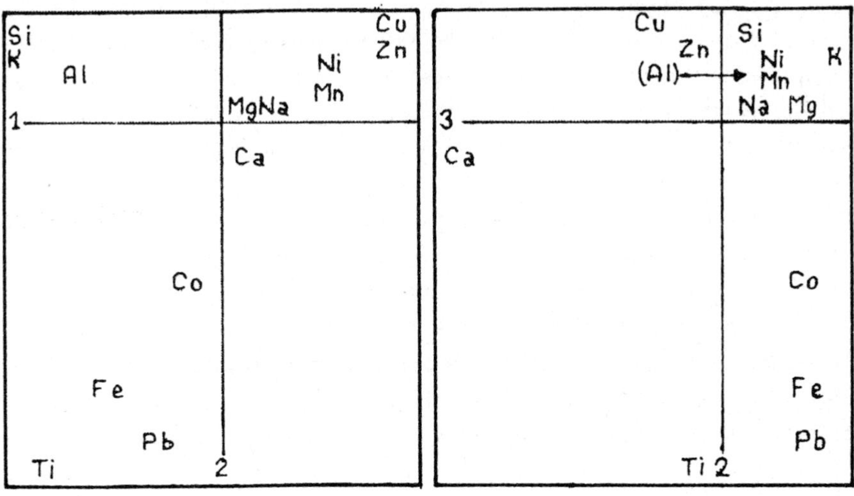

Fig. 6.5. Correspondence analysis of the geochemistry of North Pacific manganese nodules (from Hein, 1977). The left diagram (1st-2nd axes) and the right one (2nd-3rd axes) show a more detailed correlation between Ni-Mn, Cu-Zn, Si-K, etc. than the dendogram of figure 6.4.

Methodology. The use of the above-mentioned methods is sometimes confusing. The natural process should be:

- computing of standard statistics and correlation coefficients;
- plotting of the distribution;
- use of correspondence analysis to determine the relationships between variables and between specimens;
- use of cluster analysis or other classification methods to separate the different classes or species;
- search of algorithm of correlation between strongly correlated variables.

At each step an interpretation of the results should be made. A satisfactory explanation must be found to each apparent correlation or other statistical feature before proceeding to the next step. The geoscientist must take full control over the mathematical process to avoid inconsistency.

Geostatistics

The previous methods related only to the properties or qualities of objects or phenomena with none or few references to their location. Actually, a natural phenomenon is characterized by the distribution in space of one or more variables that can be called regionalized variables (Journel and Huijbregts, 1978).

The geostatistic, developed by Matheron (1962), considers each observed value as a particular realization of a random variable at the point of observation. The set of variables relative to all the points of the considered space constitutes a random function. The values taken by the random variable at two neighbouring points are not, in general, independent but related by a correlation which is itself space related. The problem is to define the spatial correlation between the various random variables at the different points of the space being considered.

To do so geostatisticians start by a structural analysis of the observed variables using the concept of variogram. Let $z(x)$ and $z(x+h)$ be the values of the random variable at the points x and $x+h$ separated by the vector h. To simplify the notation, x represents the point of coordinates (x, y, z) of the 3-dimensional space, $x+h$ a neighbouring point (with coordinates $x+h$, $y+k$, $z+l$), and h a vector of projections (h, k, l) on the three coordinate axes. The variability between these two quantities can be characterized by:

$$2G(h) = \frac{1}{N} \Sigma_h (z(x_i) - z(x_i+h))^2$$

where N is the number of experimental pairs $(z(x_i), z(x_i+h))$ of data separated by the vector h.

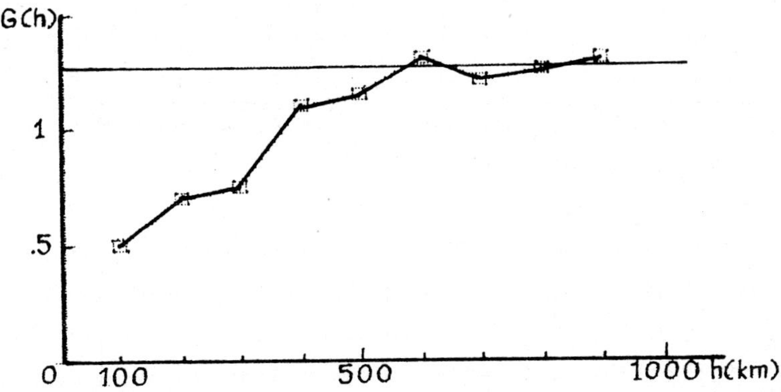

Fig. 6.6 . Experimental variogram of nickel+copper content of Pacific manganese nodules. A sill can be defined above 500 km, which is the average size of the geochemical provinces. This property was used to adjust the distances between observations (Lenoble, 1980).

This quantity is an acceptable estimate of the variogram, provided the variogram depends only on the vector h and not of the location x (intrinsic hypothesis). This hypothesis, which applied only to the differences between values, is normally true if the phenomenon is homogeneous within the considered space.

The variogram is represented on a graph where the h-modulus is reported on the X-axis and G(h), the semi-variogram, on the Y-axis (Fig. 6.6).

Experimental variograms can be fitted by a limited number of theoretical models which are characterized by their behaviour at the origin and the presence or absence of a sill. Among the models with a sill, two have a linear behaviour at the origin, the spherical and the exponential models, and one a parabolic behaviour, the Gaussian model. Two others are without a sill, the logarithmic and the h^5 $(0<S<2)$ models (Fig. 6.7).

Furthermore, a nugget effect can be found, which corresponds to a discontinuity at the origin; its name comes from the analogy with gold placers where two very close samples may have very different grades when only one of them contains a gold nugget.

The fitting and the interpretation of variograms are delicate but also worthy operations. Sometimes it is necessary to combine several models to describe the experimental variogram (nested models). When a sill exists, the corresponding value of G(h) is the a-priori variance of the variable and it may be said that above a certain distance (range) the values are no more correlated (Fig. 6.6 and 6.7).

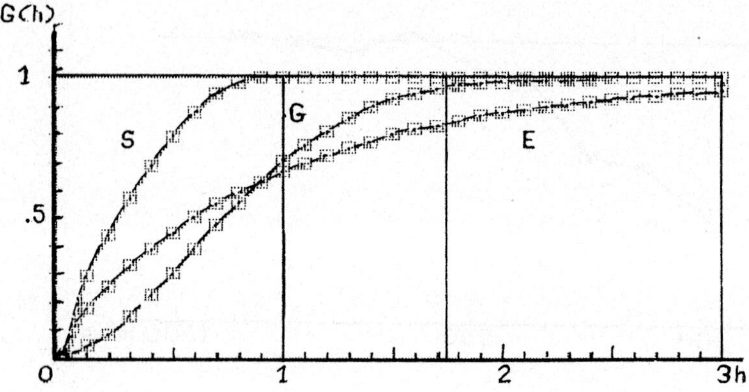

Fig. 6.7. The three models of variogram with a sill: S = spherical, E = exponential, G = Gaussian. Their range are respectively 1, 3, and $\sqrt{3}$. Table 6.5 gives the main characteristics and formulas of these models.

TABLE 6.5

Characteristics of common mode models of the variogram.

Model	Formula	Behaviour at origin	Range before sill
SPHERICAL	$G(h) = \frac{3r}{2a} - \frac{r^3}{2a^3} \quad r<a$	linear	a
	$= 1 \quad r>a$		
EXPONENTIAL	$G(h) = 1 - \exp(- r/a)$	linear	3a
GAUSSIAN	$G(h) = 1 - \exp(- r^2/a^2)$	parabolic	$a\sqrt{3}$
LOGARITHMIC	$G(h) = \log r$	logarithmic	no sill
POWER	$G(h) = r^3 \quad 0 < S < 2$	various	no sill

The most important application of geostatistics is the estimate of appropriate values for a given spatial domain from the observed values of another domain, with the calculation of the range of error (estimation variance). This possibility is described below in the paragraph dealing with resources assessment, but it can be used for all kind of estimates. Another utilisation of geostatistics is the estimate of local values by "krigeage" which is useful not only in ore deposit evaluation, but in mapping techniques as well, as explained in the next paragraph.

DATA REPRESENTATION

Lists of data are sometimes disturbing and it is difficult to get any ideas from them on the structure of phenomena. We already described some graphical methods to show the statistical structure of the data:

- histograms (Fig. 6.3),
- correlograms (Fig. 6.1 and 6.2),
- dendrograms (Fig. 6.4),
- correspondence analysis diagrams (Fig. 6.5),
- variograms (Fig. 6.6).

However the geographical distribution is not shown by any of these graphs. Mapping is the convenient way to display their spatial repartition. The development of mapping techniques using computers has been a tremendous challenge during the last ten years. The different methods are not detailed here, but some remarks will be made about the main problems, particularly:

- the projections and scales of the maps, and
- different ways of data representation, especially contouring.

Projections and scales of the maps

Maps are representations of a part of the earth's surface on a sheet of paper. It is obvious that as the earth is a somewhat pear-like geoid, and as a sheet of paper is a plane, some difficulties will be found in representing the surface of the earth on paper. Projections are the methods used to solve these problems.

Projections. Seamen and oceanographers use nautical charts in the Mercator projection. Suppose the geoid to be projected on a cylinder of axis coincident with the polar axis, and which is tangent to the earth at its equator. When the cylindar is produced, meridians and parallels are represented on the resulting plane by straight orthogonal lines. The Mercator projection is an alteration of the precedent, where the distances between parallels have been adjusted to allow the representation of loxodromic (route connecting two points on the geoid at a constant heading) by a straight line. Actually this projection is not isometric: distances and surfaces are altered on large maps, the polar regions are completely exaggerated in dimension. However, Mercator projection can be used on small areas where the distance and surface distortion are insignificant.

The Universal Transverse Mercator (U.T.M.) is a "cylindrical" projection with its axis on the equator plane, and a tangent "circle" along a meridian. Only the areas close to the tangent meridian are represented with minor surface deformation. The whole earth is mapped by 60 sectors, 6 degrees wide in longitude.

The Lambert's projection utilizes a "cone" with polar axis, tangent to the medium parallel circle of the area to be mapped. As a cylinder the cone can be produced as a plane, on which meridians appear as straight convergent lines and parallels as arcs of circle. It is the most common projection used for mapping on land.

The stereographic projection is used for polar areas. Every point of the geoid is projected along a radius on a plane tangent to the pole.

All the previous projections are called "conformal" as they preserve angles. However, they cannot be used for very large area representation as they do not keep the surface nor the distances. Some "equivalent" projections are able to do so. One of the most useful is the projection of Goode where the earth take the shape of a jelly fish.

Coordinates. Geographical coordinates are determined by the meridian (longitude) and parallel (latitude) passing through the observed point. These lines are the intersections with the geoid of planes respectively passing through the polar axis and parallel to the equator. The longitude is measured by the arc intersected on the equator between the meridian plane and a standard meridian (Greenwich). The latitude is measured along a meridian by the arc intersected between the equator and the parallel plane. Consequently, the distance corresponding to one minute of latitude decreases from one nautical mile at the equator to zero at the pole.

Lambert's and U.T.M. projections developed their own system of coordinates based on rectangular grids on the plane of projection. However, each sector, related to a parallel (Lambert) or a meridian (U.T.M.), has its own grid; near the borders of two adjacent sectors both grids do not coincide. The coordinates are measured in kilometers, and consequently are more suitable for the computing of spatial variables (e.g. geostatistics).

Scales. The scale of a map is the ratio of the distance measured on the map between two representative points to the actual distance in the field between the real points. For most systems of projection the scale is true only along the reference geodetic line: parallel for Lambert's and meridian for U.T.M. The scale of a Mercator map is computed for a given reference parallel. The scale of detailed maps are larger, the area mapped smaller; small-scale maps are convenient in order to have an overall view of large areas. At large scale

(1 to some ten thousand), whatever the projection used, the surface deformations are negligible compared to those introduced by further duplication of the document. Consequently, "home-made" rectangular grid and projections can be used for most purposes to simplify the making of the map.

Plotting the data.

Direct plotting. The simplest display of the data is the direct plotting of their values near the representative point of observation. However, when the density of data to display is very high, the result is a jumble and even unreadable. Some computer softwares manage the available space for printing depending on the density of the data. The major interest of direct plotting is the possibility, whenever the scale versus the density of the data permit, of printing together the values of several characteristics on the same sheet. However, the spatial variations will scarcely be seen.

The use of symbols with size or tone gradation is recommended. The choice of the symbols and of their representative values need some experience and could be guided by the statistical analysis as well as an art feeling. Threshold or cut-off values must be emphasized by special designs.

Contouring. For many years, people have been accustomed to "read" contour maps as they do with on-land topographic maps, although this process is not natural. If the observed characteristic can be defined as a continuous function of space, as is obvious for altitude on land, the contour line that links adjacent points of similar value, is also a continuous function of space and can be projected on the reference "plane" as an isopleth. By extension, some discrete variables, as the metal content of nodules, can also be represented by "isovalues", provided that a spatial structure of the data exist in the geostatistics meaning.

The choice of the representative values of the isopleths and their intervals depends on the range and the statistical distribution of the data, their accuracy, and the scale of the map. A high density of data is essential to obtain an accurate contouring. Until recently, most bathymetric maps were established with direct printing of the depth values, because of the lack of data density.

Isopleths can be drawn by hand, but a lot of computer facilities have been developed in the last ten years that work faster and more efficiently. The most sophisticated methods proceed by computing a high density grid of estimated values from the network of observed data. Their estimate can be made by simple linear interpolation between observed values or better by kriging. This latter method uses an estimator that weights the contribution of n neighbouring values $Z(x_i)$, as:

$$Z_K(x) = \Sigma_n k_i Z(x_i)$$

where the n weights k_i are so that $\Sigma_n k_i = 1$.

According to a general feeling of a so-called "area of influence" the weighting factor k will decrease with the distance. A rigorous and adequate weighting is given by geostatistics that calculate an estimator to ensure that the estimation variance is minimal. The n weighting factors must then verify the set of n+1 Lagrangian equations:

$\Sigma_n K_i G(x_i - x_j) + M = G(x_i - x)$

$\Sigma_n k_i = 1$, with i=1 to n and j=1 to n

where $G(x_i - x_j)$ designated the mean value of the variogram function G(h) when x_i and x_j take all the positions two by two of the points of observation, x being the point to evaluate, and m is the Lagrange parameter, which is the (n+1)th variable.

The kriging variance is then:

$V_k = \Sigma_n k_j G(x_j - x) + m - C_0$, j=1 to n

where C_0 is the nugget effect if it exists.

As geostatistical kriging needs a large computer and time; it is costly and should be used only when strong and unquestionable structures are proved.

The grid of estimated values serves as a basis for contouring by interpolation inside each cell of the grid. Smoothing techniques are provided by mapping softwares to avoid unaesthetic broken lines.

Other methods. Coloured maps are now obtained by computers. They permit the display of several characteristics either by symbols of different colours, superposition of coloured contours, and/or coloured areas.

Cross-sections are intersections of representative "surfaces", either real (e.g. sea-bottom) or abstract (e.g. metal content), with planes perpendicular to the main projection plane. They can be used to display internal structure: sub-bottom geology, isochrones of acoustic or seismic surveys, or the variation in a given direction of the space of several correlated phenomena. Associations of cross-sections are used to show the evolution of such phenomena in two spatial directions.

Block-diagrams are perspective representations of "surfaces". The feeling of relief is obtained either by contour lines shifted from lateral cross-sections, by a series of contiguous and parallel cross-sections, or by two orthogonal series of parallel cross-sections figuring a net. By colouring the different segments of the isopleths or cross-sections, a second characteristic can be displayed in the form of coloured splashes.

DATA INTERPRETATION

This whole stock of data processing facilities must be mastered. No calculation no matter how sophisticated will replace the interpretation of data in terms of geological, environmental, or economical significance. Computer data processing is only a means not an end.

Mineral exploration aims to define:

- identification of mineral occurrences,
- localisation of geological settings suitable for accumulation of these minerals, and
- delineation of the physiographic environment of such accumulation in terms of 1) guidance for further prospection of similar mineral deposits and 2) characteristics of found mineral deposits to evaluate their qualification as ore deposits.

Ore deposit is considered here as any mineral deposit which could be mined at once or in a definite future. Such an ore deposit enters in the category of reserves or subeconomic resources as defined by the U. S. Bureau of Mines (Fig. 6.8). Mineral deposit can enter in the other categories of resources.

Characterization of mineral deposit

The interpretation of data must delineate the main features that define the deposit. Among these are:

- nature of the mineral deposit, i.e. what features characterizes this kind of deposit. For instance:
  - massive sulfides of iron, zinc and copper in small mounds or chimneys;
  - concretions of manganese and iron hydroxides, enriched in nickel, copper and cobalt, laying on top of sediments.
- nature of the geological environment, i.e. to which geological features does this kind of deposit seem to be attached. For instance:
  - base metal sulfides are presently known to be found in areas of submarine volcanism with hydrothermal activity and until now, mostly at mid-oceanic ridges;
  - abyssal hills with low sedimentation rate for rich polymetallic nodules and particularly where radiolarian argilaceous mud occurs.
- hypothesis that could explain the genesis of the mineral deposit and consequently could aim further prospection, or that could localise efficiently the most promising areas to locate other deposits at lower cost. These two prospects are complementary and not mutually exclusive. For instance, the knowledge of the mechanism of manganese concentration in interstitial water of the sediment is hopeless for prospection: the recovery of interstitial water for analysis is so delicate and expensive an operation that it cannot be considered as an efficient exploration method. On the other hand, the coexistence of radioalarians in sediment and of rich and abundant nodules did not receive any definitive explanation until now.

All these factors or hypotheses can be discovered or statistically "proved" by using the different methods previously exposed. Correlation or correspondence analysis on numeric or alphanumeric data may lead to the discovery or the verification of association between several variables. Mapping of several variables may display geographic conjunction. A statistical analysis may separate several families of phenomena. The use of correspondence analysis at the start may enlighten one on a complex correlation of numerous parameters.

Resources and reserves assessment

The characteristics used for qualification of an ore deposit include the following:

- location and condition of access,
- descriptive morphology, dimensions and nature of the ore body, and
- mineral constitution, nature and abundance of useful minerals and waste, amount of potential ore by quality (tonnage versus metal grades).

Many other variables will likely be necessary to evaluate the economic interest of a possible ore deposit. Most are dependent on the available technology. Inversely, the technologies to use depends on the characteristics of the ore body. Iterative computations will be needed to define the usefulness of all variables and the efficiency of the technologies envisaged.

Accurate estimates of the variables are necessary to achieve any economic evaluation. Geostatistics provide an efficient means for such an estimate.

Global estimation of in-situ parameters. At the early stage of exploration, to determine if a mineral deposit can be ranked as an ore deposit to be mined economically estimates of the different parameters may be made by the arithmetic mean of the n observed data $z(x_i)$:

$$Z = \frac{1}{n} \Sigma_n z(x_i) \quad (n = 1, 2, \ldots, n)$$

The reliability of the result can be quantified by the estimation variance given by:

$$V_E = 2G(d,D) - G(D,D) - G(d,d)$$

where d is the domain of space on which the single observations have been made (support),

D is the domain in which the variable Z is being estimated,

G(d,D) is the mean value of G(h) when one extremity of the vector h described the domain d and the other extremity independently described the domain D, etc.

These G(...) quantities, which depend only on the respective geometry of the domains and of the variogram, are easily computed from the fitted variogram owing to tables and charts that have been established for various geometric configurations (Journel and Huijbregts, 1978).

For instance, to estimate the manganese nodule abundance of a 220 000 km^2 area in the North Pacific, AFERNOD used a roughly 45 km x 30 km rectangular grid. At each intersection, a locality of 5 stations was made on a 2.5 km segment. The total manganese nodule tonnage was computed from the actual 163 localities by:

$$T = \Sigma_j s_j a_j \quad (j=1, 2, \ldots, 163)$$

where s_j is the surface of the rectangle surrounding the locality j, and a_j the average abundance calculated in j by:

$$a_j = \frac{1}{5} \Sigma_i a_i \quad i = 1, 2, \ldots, 5$$

where a_i is the abundance of manganese nodules at each station i.

The estimation variance was then determined by combining two terms:

- the estimation variance V_1 of the localities by all the stations i, computed from the estimation variance of the central point of each locality from the data observed at the 5 stations, with reference to the spherical variogram (Fig. 6.8):

$$G(h) = 5 + 41.1(h - \frac{h^3}{6.75}) \quad \text{in } (kg/m^2)^2 \text{ for } h<1.5 \text{ km, and}$$

$$G(h) = 46.1 \text{ for } h>1.5 \text{ km}$$

- the estimation variance V_2 of the 220 000 km^2 area by the 163 localities, computed from the extension variance of each locality to the surrounding rectangular area.

The result gives an approximate relative error, at a 95% confidence level, of $\pm$20% of the global tonnage of manganese nodules in situ.

Kriging can also be used for estimating smaller areas and will give the best estimate when a great density of data is available. In this case, the mean will be established using an estimator as:

$$Z_K = \Sigma_n k_i Z_i \quad i = 1, 2, \ldots, n$$

where Z_i are the observed values, and k_i the solutions of the set of Lagrangian equations (i = 1 to n):

$$\Sigma_i k_i G(d_i, d_j) + m = G(d_j, D) \quad j = 1 \text{ to } n$$

$$\Sigma_i k_i = 1$$

where m is the Lagrange parameter,

d_i the successive domains of the observations i (e.g. photography of the bottom),

D the domain to estimate, and

G(d,D) the mean value of the variogram G(h) when one extrimity of the vector h described the domain d and the other extremity the domain D.

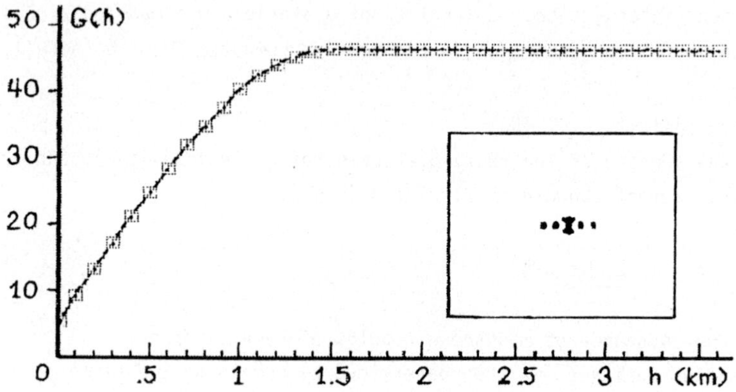

Fig. 6.8. Variogram of the abundance of manganese nodules in 220 000 km^2 surveyed by AFERNOD. The nugget effect may correspond to structures hidden by the too large spacing between observations. The right corner sketch shows the configuration of stations, locality and its surrounding area.

The kriging variance is then:

$V_K = \Sigma_i k_i G(d_i, D) + m - G(D,D)$.

Estimation of recoverable reserves. The estimate of the recoverable tonnage has to take in consideration several constraints as:

- minimum abundance of nodules by area unit of mining (e.g. area mined by hour, day, week, ...) depending on the mining method,
- minimum recovered metal content (e.g. Ni, Cu, Co, ...),
- maximum impurities content (e.g. As, Mg, ...),
- topographic features (e.g. maximum slope grade or step height, ...),
- bearing strength of the sediment, etc.

At early stages of exploration, the above-mentioned density of observations will be insufficient to analyse the effect of the geographic distribution of all combined parameters on the recoverability of the nodule tonnage. A global approach can be made based only on the statistical distribution (histogram) of observed values.

A more sophisticated method is provided by the geostatistical simulation of one or several deposits. A conditional kriging of a dense grid of simulated data is achieved from the observed values. The simulated data will have the same mean, variance and variogram than the observed ones and must fit the observed data values at the actual data locations. On these simulated deposits, several simulations of mining methods can be made to evaluate their relative efficiency.

Those methods are out of the scope of this book and the reader should refer to the specialized bibliography.

CONCLUSION

Data management is a master key for successful research activities. In marine mineral exploration, the quantity and diversity of the data are particularly important and require a considerable management effort. Fortunately, the constant improvement of computer hard- and softwares provides better facilities year by year. The time is coming where most data will be acquired by remote controlled sensors and transferred automatically to data bases. Sophisticated user-friendly softwares will be available not only for data retrieval, but also for their verification, statistical treatment, etc. Expert systems will help the decision maker by referring to previous experiences in exploration management.

Will the daily work be more intellectually confortable? Is the amount of knowledge needed for mastering the computers smaller than the exploration expert one? The answers are still to come.

REFERENCES

Addy, S. K. and Lindsay, J. F., 1977. Cluster analysis of manganese nodules and micronodules. CNRS Inst. Geol., 1, pp. 1247-1258.

Agterberg, F. P., 1974. Geomathematics - Mathematical background and geoscience applications, Elsevier, Amsterdam.

Bastien-Thiry, H., Lenoble, J. P. and Rogel P., 1977. French exploration seeks to define mineable nodules tonnages on Pacific floor. Engineering and Mining Journal, July 77, pp. 86-87.

Bastien-Thiry, H., 1979. Sampling and surveying techniques. In: The United Nations Ocean Economics and Technology Office, Manganese nodules: dimensions and perspectives. D. Reidel Publishing Co. Drodrecht, pp. 7-20.

Benzecri, J. P., 1973. L'analyse des données. V. 2, Dunod, Paris 617 p.

Craig, J. D. and Andrews, J. E., 1978. A factor analysis study of deep sea ferromanganese deposits in the equatorial Pacific ocean. Marine Mining, 1, 4, pp. 301-322.

Cronan, D. S., 1980. Underwater minerals. Academic Press, London, 362 pp.

Cruickshank, M.J., 1965. The compilation of data on marine mineral resources from sea floor photographs and bottom samples. Ocean Science and Ocean Engineering, 1, pp. 24-53.

David, M., 1977. Geostatistical ore reserve estimation. Developments in geomathematics. Elsevier, Amsterdam, 364 pp.

Defossez, M., Monget, J.M., and Roux P., 1980. Contribution of statistical methods to the study of some South Pacific sediments in relation with manganese nodule deposits. In: Varentsov, I. M. and Grasselly, G. Y. (Editors), Geology and geochemistry of manganese. Akademiai Kiado, Budapest, pp. 413-442.

Frazer, J. Z., 1977. Manganese nodule reserves: an updated estimate. Marine Mining, 1, 1/2, pp. 103-123.

Frazer, J. Z., 1979. The reliability of available data on element concentrations in sea-floor manganese nodules. In: The United Nations Ocean Economics and Technology Office. Manganese nodules: dimensions and perspectives. D. Reidel Publishing Co. Dordrecht, pp. 21-36.

Frazer, J. Z., 1980. Resources in seafloor manganese nodules. In: Kildow, J. T., (Editor), Deep sea mining. The MIT Press, Cambridge, Massachusetts, pp. 41-83.

Glasby, G. P., 1982. Manganese nodules from the South Pacific ocean - an evaluation. Marine Mining, 3, 3/4, pp. 231-270.

Greenslate, J. and al., 1972. Chemical mapping of the ocean floor: a rapid analytical system integrated with an operational computerized data bank. In: Horn, D. R. (Editor), Conference on ferromanganese deposits on the ocean floor, pp. 213.

Healing, R. A. and Archer, A. A., 1975. Manganese nodules - an example of a Lasky distribution? Resources Policy, 1, 6, pp. 306-312.

Healing, R. A., Frazer, J. Z. and Archer, A. A., 1979. The frequency distribution of nickel, copper and cobalt grades in sea-floor manganese nodules. In: The United Nations Ocean Economics and Technology Office, Manganese nodules: dimensions and perspectives. D. Reidel Publishing Co. Dordrecht, pp. 37-58.

Hein, P., 1977. Géochimie des nodules de Pacifique nord-est, étude statistique. Rapports scient. & techn., Centre National pour l'Exploitation des Océans (CNEXO/IFREMER), 35, 74 pp.

Howarth, R. J., Cronan, D. S. and Glasby, G. P., 1977. Non-linear mapping of regional variability of manganese nodules in the Pacific ocean. Institution of Mining and Metallurgy Transactions, B, 86, pp. B4-B8.

Jounel, A. G., and Huijbregts, Ch. J., 1978. Mining Geostatistics. Academic Press, London, 600 pp.

Laffitte, P. (Editor), 1972. Traité d'informatique géologique. Masson, Paris, 624 p.

Lemaire, J., 1982. Les travaux d'AFERNOD sur la faisabilité d'une exploitation des nodules polymetalliques GEOEXPO 82. Association des Geologues du Sud-Ouest Les Ressources du sous-sol, colloque 3 "Les Minerais", Pau, France.

Lenoble, J. P., 1980a. Technical problems in ocean mining evaluation. Paper presented at the symposium on economic geology of the sea floor in Sidney, August 1976. In: Varentsov, I. M. and Grasselly, G. Y. (Editors), Geology and geochemistry of manganese. Akademiai Kiado, Budapest, 3, pp. 327-342.

Lenoble, J. P., 1980b. Exploration for polymetallic nodules. In: Haley, K. B. and Stone, L. D. (Editors), Search theory and applications. Plenum Press, New-York, pp. 173-182.

Lenoble, J. P., 1981, Polymetallic nodules resources and reserves in the north Pacific from the data collected by AFERNOD. Ocean Management, 7, 1/4, pp. 9-24.

Le Suavé, R. and Morel, Y., 1985. L'environnement géologique des gisements de nodules. Société géologique de France, in press.

Lesk, M., 1984. Computer software for information management. Scientific American, 251, 3, pp. 115-124.

Magnuson, A. H., 1983. Manganese nodule abundance and size from bottom reflectivity measurements. Marine Mining, 4, 2/3, pp. 265-296.

Matheron, P., 1962, Traité de géostatistique appliquée. Vol. 1 and 2, Technip, Paris.

Monget J. M. and Roux, P., 1976. Management and statistical analysis of a data file for under sea mining of manganese nodules. Computers and geosciences, 2, pp. 321-324.

Rao, C. R., 1965. Linear statistical inference and its applications. John Wiley and Sons, New York, London, 522 p.

Richards, A. F. and Park, J. M., 1977. Geotechnical predictor equations for East Central Pacific nodule mining area sediments. OTC, 2773, pp. 377-384.

Richter, H., 1973. Results of digitally recorded seismic for manganese nodules exploration. Meerestechnik-Marine Technology, 4, 5, pp. 151-152.

Schimmelbusch, H. and Danzer, P., 1975. The use of computers in evaluating the results of manganese nodule exploration. Metallgesellschaft. Mitt. Arb. (Berlin), 18, pp. 50-53.

Siapno, W. D., 1976. Exploration technology and ocean mining parameters. Mining Congress Journal, 62, 5, pp. 16-22.

Sundkvist, K. E., 1983a. Spacing statistics of manganese nodules on the ocean floor. Marine Mining, 4, 2/3, pp. 255-264.

Sundkvist, K. E., 1983b. Size distribution of manganese nodules. Marine Mining, 4, 2/3, pp. 305-316.

CHAPTER 7

MARINE MINERAL EXPLORATION EXAMPLES

H. BÄCKER and R. FELLERER

INTRODUCTION

The first discoveries of the main types of marine mineral resources were made during general oceanographic research operations using conventional oceanographic equipment. For example, manganese nodules were among the products recovered from the ocean floor during the famous H.M.S. "Challenger" expedition (Murray and Irvine, 1895). Dredges of the type still used today served as the main sampling gear to collect hard-rock and crusts from the sea floor, mainly as a by-product of benthos research. The first sampling of the Red Sea metalliferous sediments dates back also to the last century, when, during the Austrian Pola expeditions in 1897/98, iron-rich sediments were retrieved from below a 2000-m depth in the central Red Sea (Natterer and Luksch, 1901).

Good bathymetric maps are necessary for all kind of mineral research, though much better resolution is required for hydrothermal deposits than for, e.g., uniform placer deposits or manganese nodule fields. Methods to predict, locate, and sample the various types of minerals on the ocean floor differ for the different mineral commodities, and part of the necessary gear have been developed for the special purpose. It therefore appears convenient to describe separately marine mineral exploration procedures by typical exploration examples or exploration case histories. Examples selected in the following sections include Red Sea metalliferous muds, massive ridge sulfides, manganese nodules, Co-rich ferromanganese crusts, and phosphorites.

METALLIFEROUS SEDIMENTS: THE RED SEA

There are several types of metalliferous sediments which ultimately could serve as resources for various metals (Meylan et al., 1981). Most occurrences can be related to submarine volcanism and hydrothermal circulation. While some metalliferous sediments are the result of bulk precipitation (mixing of ocean water with unsolvable hydrothermal fluids) others are the result of differentiation processes leading to the removal or concentration of certain metals.

The understanding of the genesis of hydrothermal-sedimentary deposits and hence the development of successful exploration strategies require considerable multi-disciplinary efforts involving geophysics, structural geology, volcanology, geochemistry, sedimentology, mineralogy and oceanography.

Despite of the many well-equipped research vessels operating in the Red Sea since 1969, only one deposit of metalliferous muds (Atlantis II Deep) has been investigated in any detail. The Atlantis II metalliferous sediments can now be considered a measured resource.

Initial considerations

In the Red Sea the metal-bearing fluids usually have extremely high NaCl content which originates in Miocene evaporites adjacent to the volcanic axial zone of the Red Sea mid-ocean ridge. Such brines may attain exit temperatures of up to 400°C. They are retained in morphological depressions and constitue semi-closed systems with only limited interchange with the deep sea water body (see Figs. 7.2 and 7.5).

Figure 7.1 shows well-stratified but poorly crystallised Red Sea metalliferous muds. The sediments precipitated within the brine pools tend to be enriched in metals like Zn, Cu and Ag. They occur in the form of sulfides that could be readily oxidised in oxigenated deep water. Brine contents in the muds are rather high, between 80 and 95%, the youngest sediments being unconsolidated to semi-liquid. This poses certain technical problems for the recovery of representative samples.

Though metalliferous sediments are known to occur several tens of km away from their source (transport of fine particles by ocean currents) high-grade occurrences are linked to the hydrothermal vent areas which in the case of the Red Sea is the brine-covered sea floor. The sources are preferentially found within or near the neo-volcanic zone with a pelagic sediment cover of less than 10 m. In the southern Red Sea, the neo-volcanic zone is produced by sea floor spreading producing long, parallel graben/horst structures as oceanic plates move away. In the central and northern Red Sea, where most brine pools occur, only an incipient form of sea floor spreading can be recognised by the presence of isolated deep troughs floored by volcanic rocks. Very often an axial high is developed (Fig. 7.3).

Fig. 7.1. Sediment core from the Suakin Deep, Red Sea, recovered by a kasten corer. Light coloured normal Red Sea sediments (marls) and dark layers influenced by hydrothermal precipitation. 15 cm boxes.

The axial zone is rugged, characterised by well-developed small-scale morphology. Subbottom profilers show no penetration. If smooth surfaces and acoustic bottom penetration indicate local basin fillings this might be due to metalliferous sediments which have sedimentation rates of generally two orders of magnitude higher than normal pelagic sediments. However, salt flows and turbidities could have caused similar features.

Red Sea exploration history (1963-1984)

Early indications of the presence of brines and metalliferous sediments in the Red Sea were not recognised as important findings (Natterer and Luksch, 1901; Bruneau et al., 1953). Only during the International Indian Ocean expedition (1963-65) did the combined effort of several ships from different nations demonstrate the importance of present submarine hydrothermal activity and connected metallogenesis not only in terms of marine resources but also regarding the understanding of the formation of many fossil deposits. A first cruise entirely devoted to the newly discovered brine pools Atlantis II, Discovery, and Chain was carried out in the fall of 1966 by Woods Hole Oceanographic Institute research vessel "Chain" (Degens and Ross, 1969).

During the early cruises, conventional oceanographic research equipment was used but this proved to be entirely inadequate. For instance, reversal thermometers calibrated for temperatures exceeding 50°C were unavailable at that time. Also, CTD sondes did not allow the measurement of the high temperatures and salinities of brines. Furthermore, acoustic devices such as pingers stopped working after some minutes of exposure to hot brines and release weights of piston corers penetrated several meters into the unconsolidated metalliferous mud before triggering.

Economically directed exploration started in 1969 with the "Wando River" cruise to the Atlantis II Deep and continued in 1971/72 with two cruises by R/V "Valdivia" (Bäcker, 1975). During the "Valdivia" operations which were carried out under exploration licences issued by various Red Sea coastal States, work was focussed on the search for active hydrothermal sites represented by the presence of brine pools.

The most efficient tool to detect brines from the sea surface at normal cruising speeds (6 - 12 knots) was a stabilised narrow-beam high-frequency echosounder (beam: 1.4°, frequency: 30 kHz). Free brine surfaces appeared as typical horizontal mid-water reflectors (Fig. 7.2). Since this tool also provides well-defined bottom images without side echos, even at steep slopes (Fig. 7.4), bathymetric mapping of the axial graben structures of the Red Sea appeared to be the most convenient way to record major brine pools, recognising at the same time important structural elements. A total of 27 000 line km was run during the cruises (Bäcker et al., 1975). Line spacing varied between 5 and 100 m depending on detail requirement. High navigational accuracy was obtained by Decca Hi-fix stations operated along the African shoreline. The bathymetric maps obtained were later used during more localised operations with different navigation systems. To secure identical coordinates, some bathymetric profiles were run across known topographical structures at the beginning of the follow-up operations.

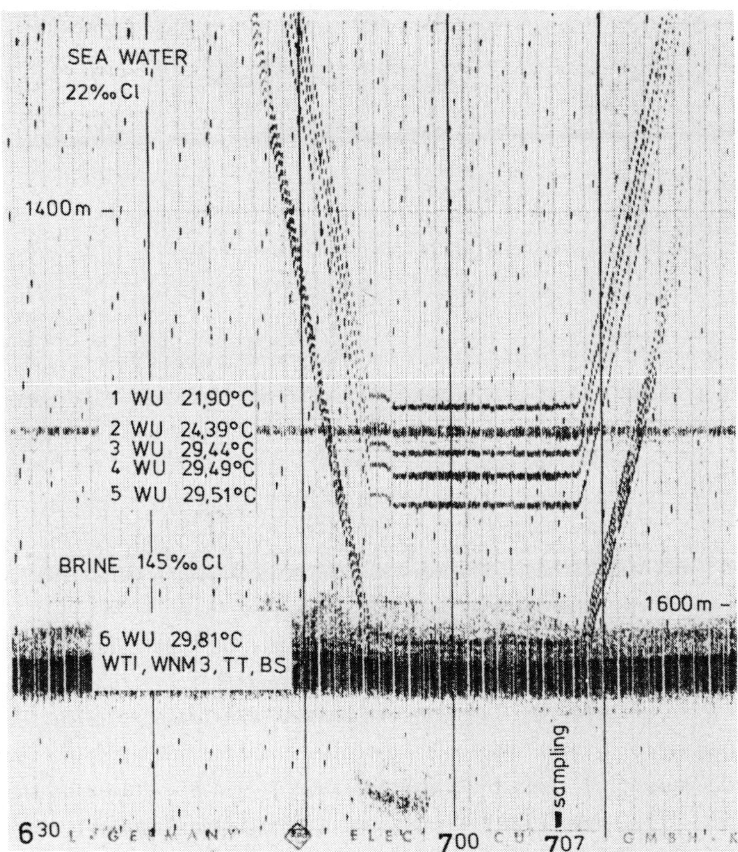

Fig. 7.2. Echo sounder record of an oceanographic station in the Valdivia Deep, Red Sea. Elac stabilized narrow-beam echo sounder 30 kHz. The sea floor, the brine/sea water interface, and the position of instruments (water samplers and CTD sonde) are displayed.

Some of the bathymetric surveys were accompanied by gravimetric and magnetic measurements. These allowed the recognition of igneous rocks and their attribution to sea floor spreading processes (Roeser, 1975).

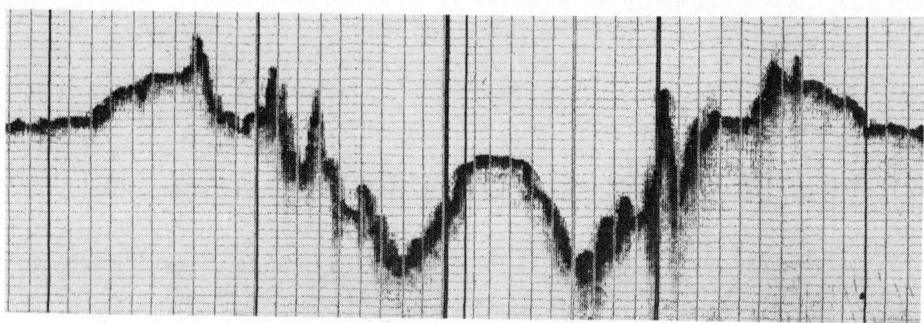

Fig. 7.3. Typical cross-section of a Red Sea deep (Nereus Deep). Elac narrow-beam echo sounder 30 kHz. Relative depth of the through is 1,200 m, width 14 km.

With the help of structural maps oceanographic measurements and sampling could be limited to the most promising areas which at first were those hosting brine pools. The pools were investigated by an especially developed CTD sonde operated with a water sampler chain. The latter were furnished with reversal thermometers which were released by messenger weights at right positions with respect to the horizontal and vertical extension of the brines. The narrow-beam echo sounders were used for exact positioning of the water samplers (see Fig. 7.2).

The sound velocity within brines could be calculated by simply measuring the known distances between the single instruments on the echo sounder record outside and inside the brines (Hartmann, 1972). Following the oceanographic survey, bottom sampling was carried out using a variety of equipment, depending on the suspected nature of the bottom and the sample requirements. In unknown areas, piston corers with heat flow measurement apparatus were preferred (Fig. 7.4). Areas covered by brines were sampled by box corers (15x15 cm^2 area, 3-12 m long) because better conservation of the stratification and of upper sediments were obtained (see Fig. 7.1).

Large amounts of ore sludge was required from the Atlantis II Deep to provide sufficient material for technical testing and developing preconcentration processes. For this purpose, large Van Veen-type grabs with capacities of 1 to 2 m^3 were used (Fig. 7.5).

The launching and recovery of long sediment corers is carried out with the help of a ship corer support frame (see Fig. 7.4) which avoids bending of the barrels during rough sea states. For bottom approch control an acoustic pinger attached to the wire 20 to 50 m above coring device is used. At good navigational conditions the path of the sampling gear into the sediment can be visualised on the narrow-beam echo sounder record (Figs. 7.2, 7.5, and 7.6). Even the correct function of a sampler can be controlled in this way (Fig. 7.5).

Fig. 7.4. Piston corer and corer support frame for handling of long cores.

Brines and metalliferous sediments are usually collected in morphological depressions. Therefore, bottom sampling was preferentially carried out in graben structures produced by tensional forces along the Red Sea axis, and also in some marginal depressions that possibly are formed as collapse structures on Miocene salt formations. Lacking detailed morphological maps, the floor of such depressions can be reached by adequate maneuvering of the ship with the help of the echo sounder record (Fig. 7.6).

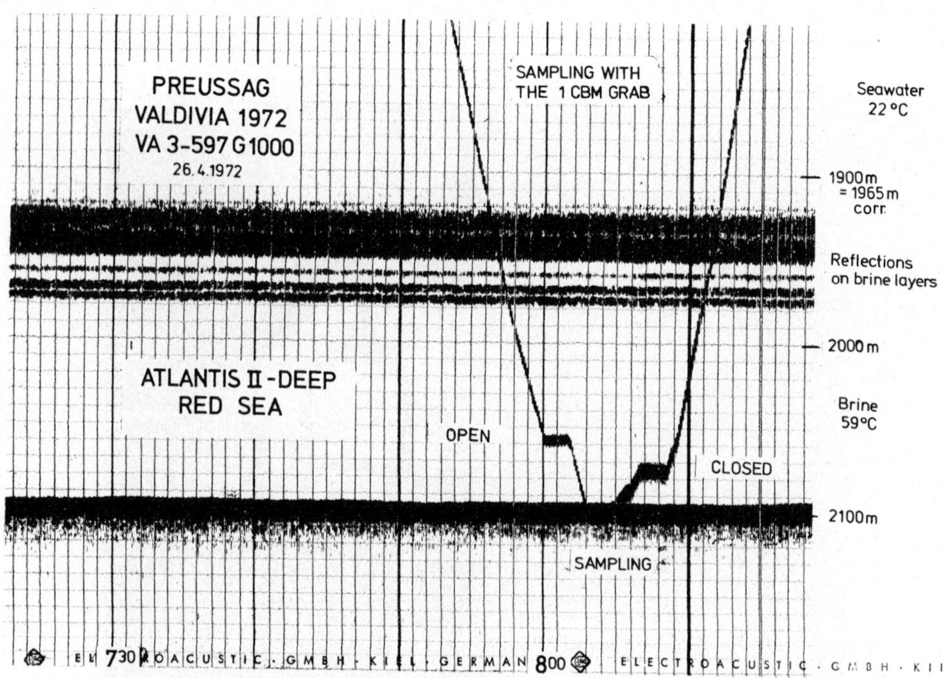

Fig. 7.5. Sampling of metalliferous sediments by a large (1 m^3) Van Veen grab. The narrow-beam echo sounder (30 kHz) display shows the seafloor in the Atlantis II Deep, the brine reflectors and the grab before and after sampling. The different reflections of the grab indicate proper functioning.

Laboratory work on economically oriented exploration cruises is essentially restricted to (a) representative subsampling of cores by collecting average split samples, (b) sample description including photography, and (c) to chemical analyses of only those samples which might have influence on ongoing exploration. Detailed analytical work was usually performed in the land-based laboratories. Metalliferous sediments from the Red Sea are a relatively difficult material to handle, especially if representative mineralogical results are required. This is caused by the high brine content, the heterogeneity, the extremely fine grain size, and the poor crystallinity. A chemical analysis is usually carried out by atomic

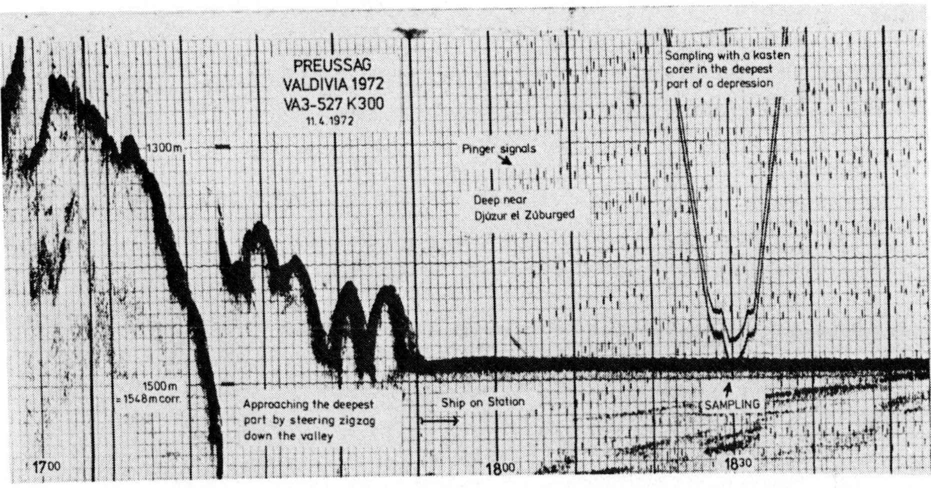

Fig. 7.6. Bathymetric control of coring in the deepest part of a morphological depression in the Red Sea. Elac narrow-beam echo sounder 30 kHz.

absorption spectrophotometry (AAS). Prior to this samples are reduced to a salt-free dry basis to avoid clogging of analytical instruments and to account for the highly varying interstitial brine content.

Following the systematic exploration steps in the early seventies, further work on the metallic potential of the Red Sea proceeded along two lines:

1) Several scientific research cruises organised by various countries considerably increased the understanding of the geological and geophysical background of the metallogenesis in the Red Sea. They did not, however, lead to the discovery of other hydrothermal sites. This work included the first submarine operations in 1979/80 by a Russian team (Monin et al., 1982) and detailed investigations in some of the lesser known deeps, such as the Nereus Deep (Bignell et al., 1974).

2) More economically directed investigations were initiated in 1975 by the establishment of the Saudi Sudanese Red Sea Commission. Work focussed on the Atlantis II Deep as the major known occurrence of metalliferous sediments. To obtain a reliable deposit evaluation, more than 600 long sediment cores had to be produced from the 60 km^2 large deposit. Specially designed heavy box corers (30x30 cm^2 and 25x25 cm^2 in area, up to 18 m in length) were used for this purpose. Chemical analysis as well as detailed

bathymetric measurements were fed into a geostatistical evaluation system using modern geostatistical methods (e.g. Kriging).

Acoustical techniques to measure sediment thickness continuously along survey lines were employed as well (see Fig. 7.7). However, the resolution of normal seismic systems (10 to 30 m) is not sufficiently high;

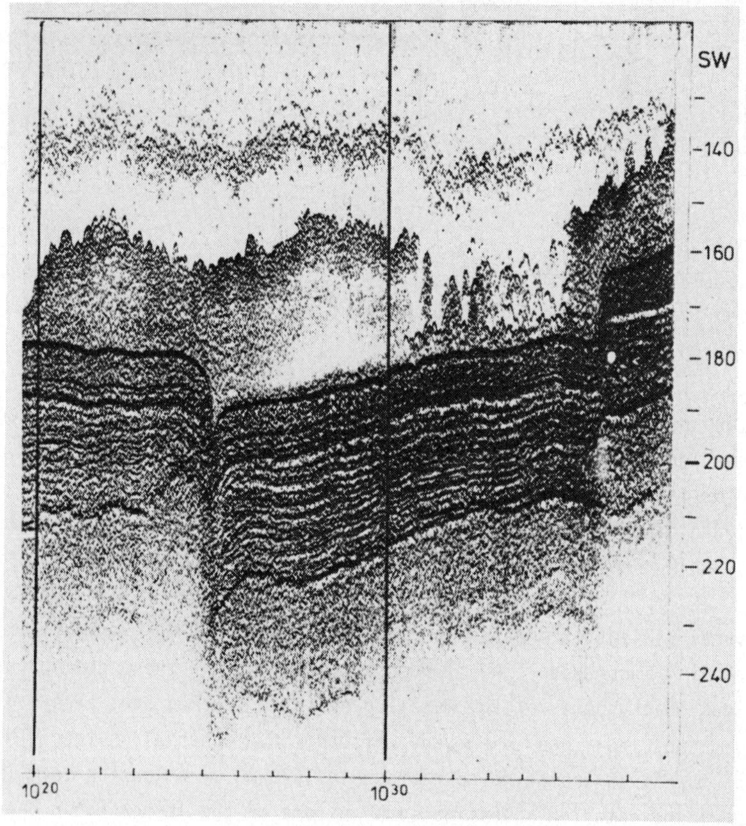

Fig. 7.7. Uranium enriched sediments in the Ghoubbet et Kharab, Gulf of Aden. Sediment echo sounder 18 kHz, 13^0 beam width. Depth in meters (1500 m/s).

they take data from subbottom profiler systems (3.5 to 18 kHz) operated from the sea surface at water depths exceeding 400 m usually obscured by side echo signals typical for rough sea spreading topography. The operation of remotely operated vehicles (ROV) in a Red Sea deep poses several navigational and safety problems.

Fig. 7.8. Box or spade corer used to sample undisturbed surface sediments.

During recent years, Red Sea research has concentrated on the northern portion of the basin, where deep graben structures and volcanic intrusions are rare. The morphological character is defined by salt diapirism and sedimentation rather than by faulting and volcanism. In this environment, a systematic combined geothermal and geochemical survey appeared to be more efficient. A number of anomalies could be detected including one major brine pool, occurrences of massive sulfides, and also some not-brine pool related hydrothermal sources and metalliferous sediments.

MASSIVE SULFIDES AT THE EAST PACIFIC RISE

Until 1978 only certain metalliferous sediments were regarded as marine hydrothermal products. The discovery of massive sulfides of Fe, Cu and Zn during a "Cyana" dive near the axis of the East Pacific Rise at 21° N (CYAMEX, 1978) was a major event in ocean exploration.

These occurrences escaped discovery by previous research efforts due to their relatively small size combined with an extremely rough sea floor topography to which they are related. Furthermore, hardly anyone would have suspected the existence and preservation of sulfides in an oxigenated deep-sea environment. Extremely high deposition rates, i.e. 8 cm/day sulfide chimney growth (Hekinian et al., 1983) suggest that one may classify the deposits as a renewable mineral resource.

For the exploration of these deposits, highly sophisticated equipment is required. In the following section an account is given to the presently discovered sulfide sites at the East Pacific Rise (EPR).

Massive sulfides and hydrothermal products

The known occurrences of recent marine massive sulfides form within volcanic terrain at sea floor spreading centers by precipitation at and very near to the exits of high-temperature hydrothermal sources (see Chapter 2). Cooling of the hot fluids (200 to 400° C) and mixing with oxigenated cold sea water occurs very rapidly, within centimeters to a few meters from the source. Outside this range temperature anomalies rarely exceed one degree C.

The sulfides form complicated chimney structures that after reaching a height of a few to about 20 m, eventually break down and form sulfide talus mounds through which fluids continue to percolate filling interstices. The internal structure of the sulfides is characterised by zoning, recrystallisation and by remnants of voids which are frequently produced by polychaete vent worms (alvinella pompeiana, Desbruyere and Laubier, 1980). The presence of these voids makes the determination of the geotectonical engineering properties of the sulfides rather difficult (Crawford et al., 1984).

Though the specific gravity of crushed samples exceeds 3.5 g/cm^3, the bulk dry densities generally are below 2, dependent on porosity and the occurrence of large vesicles (Fig. 7.9).

Fig. 7.9. Massive sulfides of iron, copper and zinc from the East Pacific Rise at 21,5^{o}S.

The bulk composition of the ore is relatively constant but the relation between the various components is highly variable. Frequently high-temperature phases co-exist with low-temperature mineral assemblages. The dominant minerals are pyrite, marcasite, cubic cubanite, wurtzite and sphalerite (Oudin, 1982; Tufar et al., 1984).

It is evident that only small portions of the metal load introduced by the hydrothermal fluids into the marine environment is deposited as sulfides around the vents. Significant components of the fluids such as silica and manganese, but also iron, leave the chimneys as "black smoke", a cloud of suspended, extremely fine-grained precipitates which drift for tens to hundreds of kilometers in the residual current direction before eventually settling. In fact, hydrothermal components in the deep sea sediments along the EPR have long been recognised (e.g. Skornyakova, 1965; Boström et al., 1969).

The dispersed metals may be used as exploration guides to identify areas of increased hydrothermal input and possibly also outline the direction towards the source if prevailing transport directions are known. Typical components of sediments related to hydrothermalism are iron, manganese, zinc and arsenic (Marchig et al., 1982; Kunzendorf et al., 1985). Discrimination between the origin from high- and low-temperature hydrothermal sources, from hydrogeneous, biogeneous and detrital input is especially important (Dymond, 1981). A high-silica content in sediments may point to local hydrothermal activity (e.g. Galapagos hydrothermal mounds), but could also have a biogenic background (e.g. Gubbet et Kharab, Fig. 7.7).

In areas with rock outcrops ferromanganoan crusts constitute an additional exploration guide. These crusts which normally grow over long periods of time, up to millions of years, are predominantly composed of hydrogeneous components, but in the vicinity of hydrothermal vents they may incorporate significant amounts of material originating from the sources. Typical hydrothermal crusts show irregular Fe/Mn relations, either high Fe or Mn and usually low amounts of minor and trace elements such as Co and Ni. On the other hand, basaltophile elements like Zn show evident enrichment towards active areas when compared with Co. Dissolved gases, such as helium and methane, recently became important indicators of ongoing hydrothermal activity but they are not useful in the search for inactive sulfide sites. For example, $^{3}He/^{4}He$ anomalies could be related to the hydrothermal fields at the Galapagos Rift at 86^{0} (Lupton et al., 1977), East Pacific Rise at 21^{0} N (Lupton et al., 1980), Juan de Fuca Ridge (Lupton et al., 1981) and above Guayamas Basin (Lupton, 1979). CH_n is a slightly less sensitive tracer, but has the advantage of being analysed on board ship.

In addition to tracers that are directly transferred from the hydrothermal fluids into sea water and from there eventually to the ocean floor, biological indicators are used as well to locate the submarine sources to which they are related. The primary production within these specialised communities is at least in part based upon sulfide-oxidising bacteria (Felbeck and Somero, 1982). "Black smoker" bacteria grow at temperatures exceeding 250°C (Baross and Deming, 1983). The food chain starting with these bacteria consists of a great variety of species which in part are sessile and they include vent worms Riftia pachyptila (Jones, 1981) and Alvinella pompeiana (Desbruyeres and Laubier 1980). The worms are therefore good proximity indicators. Predator, such as crabs and fishes are more widely distributed but they point at least to an increased primary production. A high biological activity in general also increases the turbidity of the water which can be measured by light attenuation devices.

Sulfide occurrences on the East Pacific Rise

The important massive sulfide findings made so far are all located in the eastern part of the Pacific Ocean. This does not necessarily reflect the world distribution pattern of these deposits. In fact, indications of submarine hydrothermal activity as well as sulfide mineralisation were reported at numerous locations around the world, at divergent plate boundaries as well as at island arc environments and even at top areas of intra-plate volcanoes (e.g. Lonsdale et al., 1982). Batiza (1983) estimates that seamounts hold about 10% of the sulfides generated at the mid-ocean ridges. More than 12 000 seamounts are known from the Pacific.

Though little is known about the distribution of sulfide occurrences along divergent plate boundaries, the assumption that active hydrothermal systems are more likely to be found along medium- to fast-spreading ridges played a major role in the choice of the exploration targets and hence on the distribution of discovered sulfide occurrences. The first discovery was made in February/March 1978 at the EPR at 21° N and 109° W (CYAMEX, 1978) resulting in numerous scientific investigations and publications. Occurrences were discovered at 2620 m depth about 240 km south of Tamayo and about 90 km north of Rivera transform faults.

After the collection of additional information along the EPR system, a theory was presented which linked the sulfide formation to magmatic processes taking place at axial morphological highs, about midway between transform faults (Ballard and Francheteau, 1982; Francheteau and Ballard, 1983). Another hydrothermal field which has been studied by several research cruises fitting the above mentioned theory is situated at the EPR, at 12°50' N (Hekinian et al., 1983). The sulfide mounds there are distributed within a 15-km long section of the neo-volcanic zone at about 2630 m depth. Sulfides were also discovered on a seamount about 5 km away from the axis. An additional sulfide occurrence was located at the EPR, at 6°42' N in 2740 m deep water (Boulegue et al., 1984). The axis of the southern branch of the EPR between 10 and 22° S shows on the eastern flank metalliferous sediments (Boström, 1973; Kunzendorf et al., 1985) and a large

^{3}He anomaly (Lupton and Craig, 1981). The largest sulfide occurrences located so far within this EPR section were discovered in the axial graben structures at 18.5° S and 21.5° S (Bäcker et al., 1985). Within the Juan de Fuca and Explorer Ridge system, numerous research cruises were carried out by American and Canadian teams in recent years. Most of these areas are now covered by detailed Seabeam and Seamark maps. As of 1985, four localities with sulfides were discovered (Koski et al., 1982; Normark et al., 1982; Canadian American Expedition, 1985; Kingson et al., 1983).

At the Galapagos Rift important sulfide occurrences were found during "Alvin" dives near 85°50' W (Malahoff et al., 1983), and during "Sonne" cruises further west.

All of the known sulfide deposits occur in diverse structural settings:

- Axial seamounts
- Deep axial grabens
- Lava lakes
- Faults accompanying shallow graben structures
- Deep rift valleys
- Off-ridge seamounts.

In general, visible sulfides are restricted to the neo-volcanic zone, within a hundred meters to a few kilometers from the spreading axis. They align along tensional faults which also dominate the general direction of lava lake walls (see Fig. 7.10). Some chimney structures even grow at the upper edge of deep collapse structures. Individual chimneys and long chimney groups (up to several 100 m long) are present. They usually root in relatively fresh lava. In talus where sulfide mounds are cut by young faults, stockwork mineralisation may become exposed. Included are highly deformed lava and lava breccias which are more or less mineralised. Such breccias have been found in the EPR grabens at 18.5° and 21.5° S (see Fig. 7.11) and at the Galapagos Rift occurrences.

Massive sulfides grow rapidly and will be oxidised afterwards if no lava burial occurs. Assuming a sedimentation rate of 1 cm/1000 y and a chimney height of 2 m, sulfides or their decomposition products will probably not remain for more than about 200 000 years. After this time the deposit will have migrated between 4 and 12 km away from the zero-age axis. Therefore, the exploration of massive sulfides can be limited to an approximately 30-km wide strip, with the exception of those areas where ongoing volcanism and hydrothermal activity is indicated.

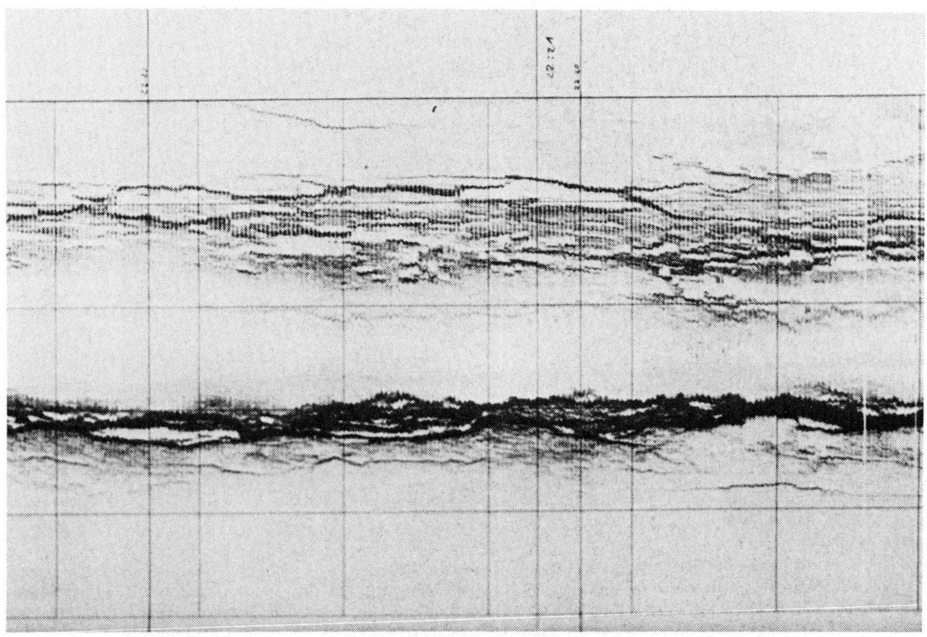

Fig. 7.10. Side scan sonar track along the EPR rift axis at 13°N. AMR deeptow, 100 kHz, no tow speed correction; double swath width 400 m, track about 600 m long. The predominance of faults and collapse features is evident.

Proposed exploration strategy for massive sulfides

The methods used for the discovery of the known occurrences of massive sulfides were largely dependent on the exploration techniques that were available at that time. Some of the modern equipment such as Seabeam, Seamark, deep-towed side-scan sonar and manned submersibles were not available for all exploration cruises. Furthermore, site-specific methods were employed at first instead of systematic regional surveys. At present, only methods for reconnaissance are worked out and tested. The detailed exploration of massive sulfide deposits is only in a primordial state of development. Important necessary methods and equipment such as deep-sea bare-rock drilling equipment and refined geophysical methods still have to be developed. Detailed work is also still hampered by unsufficient deep-sea positioning methods.

Fig. 7.11. Sulfidic basalt breccia from the Galapagos Rift, 86°W, indicating the existence of stockwork mineralization below the sulfide chimney structures.

A schematic collection of exploration methods for massive sulfide exploration is shown in Fig. 7.12. The proposed steps in the exploration for these deposits are geotectonic considerations, operational planning, geophysical and oceanographic survey, reconnaissance sediment sampling, deep-tow operations, reconnaissance rock sampling, manned submarine operations, and analysis and evaluation steps.

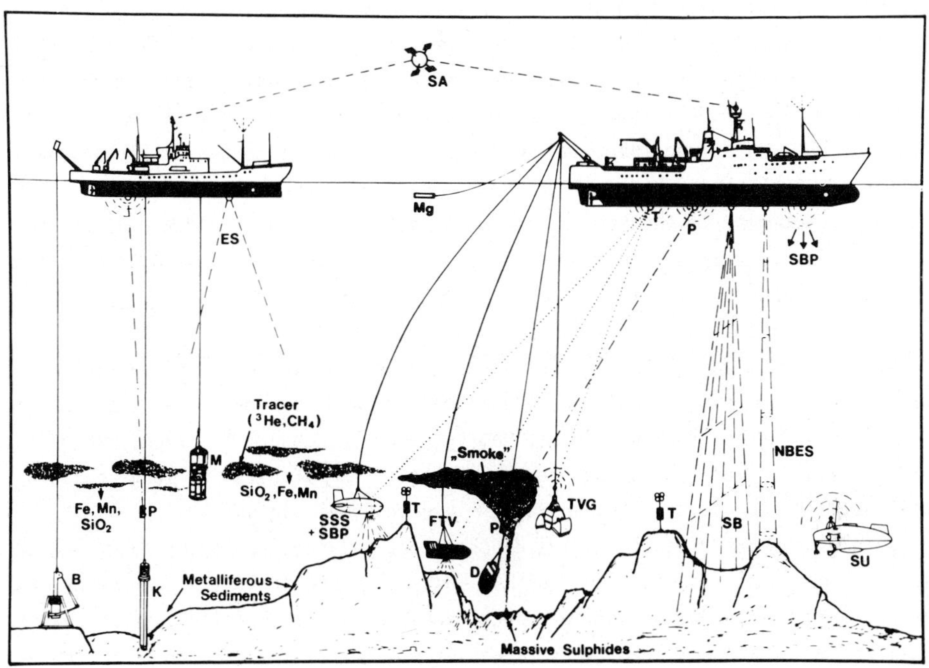

Fig. 7.12. Diagram showing equipment used for massive sulfide exploration. Abbreviations: SA satellite navigation, ES echo sounder, Mg magnetometer, T transponder, P pinger, SB seabeam, NBES narrow-beam echo sounder, SBP subbottom profiler, B box corer, K kasten corer, M multisonde (CTD sonde), SSS side scan sonar, FTV foto-TV frame, D dredge, TVG television grab, SU submersible.

Geotectonic considerations. In order to choose exploration areas that are favourable for massive sulfide deposits, emphasis at present is laid on the active divergent plate boundaries because they are the site of ongoing submarine volcanism. Most of these plate boundaries are, however, situated outside Exclusive Economic Zones and thus for the time being the possibilities for future exploitation remain uncertain. In any event scientifically oriented exploration may focus efforts on these open ocean structures.

Other targets for massive sulfide exploration are volcanically active portions of convergent-plate boundaries. They include back-arc basins and collision zones. We know from the stratigraphic record that many important fossil sulfide deposits were formed in such environments, often in relatively shallow water. Our knowledge of such recent deposits is still scarce but there are a number of encouraging indications of active hydrothermalism. For instance, barite/opal rocks were dredged from the Lau Basin ten years ago (Bertine and Keene, 1975) and a number of cruises were performed in the basin by American, New Zealand and German research vessels. At present, activity is focussed on the Marianna Basin where highly scattered heat flow values (Anderson, 1975) and geochemical sediment anomalies (Leinen and Anderson, 1981) indicate hydrothermal activity.

An important target for sulfide exploration is also isolated seamounts and seamount groups which can be detected easily by wide-angle bathymetric or side-scan surveys. Sulfide metallogenesis was observed in the Pacific (Lonsdale et al., 1982) and in the Mediterranean (Honnorez, 1969; Minnitti and Bonavia, 1984). It appears that the ring faults of large calderas are well-suited to the transport of hydrothermal fluids.

Some important fossil sulfide deposits are related to major faults and they lack any indication of volcanism. Lonsdale (1979) describes a hydrothermal site with barite deposition on a strike-slip fault off California. Transform faults as targets for sulfide formation are regarded as controversial at present. The theory of topographic highs (Ballard and Francheteau, 1982; Francheteau and Ballard, 1983) excludes major sulfide occurrences at and near transform faults. On the other hand, some disseminated Fe, Cu and Zn sulfides are frequently found in volcanic rocks dredged from major fracture zones.

Operational planning. Planning of a massive sulfide exploration cruise will depend to some degree on available equipment. Since most targets are suspected to occur in open sea areas with water depths exceeding 1500 m, an ocean-going vessel with excellent winch facilities and handling gear is required. Expedition of exploration phases depends greatly on the selected ship. Most of the exploration activities and of equipment handling needs experienced and specialised personnel in numbers often exceeding the available space on the ship.

A separation of geophysical and oceanographic surveys from other exploration steps is highly desirable. The reason for this is that heavy equipment is not required in the following exploration steps and results from the geophysical survey are needed for site-specific exploration.

Geophysical and oceanographic survey. A regional geophysical survey includes multiple-beam bathymetry (e.g. Seabeam), narrow-beam echosounding (for slope angle determination), wide-angle side-scanning (e.g. systems Gloria or Seamark 2), subbottom profiling (3.5 kHz), magnetometry and gravimetry. Satellite navigation is usually applied.

Results of the geophysical survey are maps. Large exploration targets are usually not entirely covered by the geophysical profiling. Spacing of geophysical lines depends on the required structural detail. In most cases, a spacing of 5 nautical miles is sufficient for recognising major structures. Automatic charting systems (Seabeam, side-scan sonar) allow on-line observation of important lineaments such as spreading axes and hence will reduce the ship time drastically.

An oceanographic survey may be combined with the geophysical survey, but it can also be conducted by a smaller, less well-equipped vessel. A hydrographic winch and satellite natigation is required however. The main equipment used is a CTD sonde and a number of water samplers. A combination of both (Fig. 7.13) permits water to be samples at temperature, oxygen, salinity, or light attenuation anomalies. Such anomalies are often found downstream from the source above 200 m from the exit depth . In order to reduces sampling efforts deep-sea current measurements via tracers are included. Tracer analysis can be performed onboard ship.

Reconnaissance sediment sampling. Regional sediment sampling is often neglected in massive sulfide exploration. However, more general knowledge about the size of hydrothermal activity in an area can be obtained by sediment sampling. Generally, hydrothermal precipitates are not deposited very close to the source but drift for several miles before settling is observed.

Detection of most recent activities is by surface sediments which only are sampled best by a box corer (see Fig. 7.8) or by free fall equipment. For quantification purposes, metal accumulation rates have also to be determined. This necessitates recovery of long sediment cores using long box or piston corers (see Fig. 7.4). Sediment sampling can be combined successfully with heat flow measurements. Heat flow values mainly from scattering may indicate ongoing hydrothermal activity. Heat flow equipment is generally attached to coring devices for in-situ temperature measurements. The thermal conductivity of the core sections is determined after recovery. A different approach is transfer of in-situ parameters by acoustic telemetry (Wright and Fang, 1984).

Deep-tow survey. Deep-tow surveying is an effective tool in more site-specific operations. The most sophisticated deep-tow systems such as the Scripps Oceanographic Institutes deep-tow system can accommodate a large number of geophysical tools (e.g. side-scanning, echo sounding, sediment echo sounding, magnetometry, CTD measurements, TV and still cameras). The deployment and recovery of such an instrument package requires specialised handling gear and a 8000-m long coaxial cable on board the research vessel.

Fig. 7.13. CTD sonde, in combination with a water sampler array. The equipment allows the measurement of salinity, temperature, oxygen, velocity of sound and turbidity relative to the depth, and the subsequent release of water samplers on command.

Some institutions have split the measurement functions of the deep-tow system into three to four operations. By doing so investment costs and risks are reduced, but the operating of the ship time is then increased. Deep-towed platforms carrying side-scan sonars and subbottom profilers are used to provide a detailed picture of the seafloor morphology, especially of tectonic features and lava flows (see Fig. 7.10). They also indicate sediment cover or pockets.

Horizontal measurement of temperature, salinity, light attenuation and oxygen content from vehicles towed about 100 m above the seafloor provide direct evidence of hydrothermal activity. Television and photographing are done separately or in combination, depending in part on the availability of a suitable cable winch. Vehicles without television cameras can be controlled only by pingers pictures being released continuously or through bottomtouching messenger weights. Television allows better control, continuous visual coverage and can be used for selecting photo targets.

A visual deep-tow survey along profiles gives information cn the type of lava terrain, on small-scale tectonic features, on collapse structures, relative age of lavas via glass alteration and sediment cover, on benthos population and on the hydrothermal sources and their products. During a visual survey several ten thousand color slides are processed and reviewed onboard the ship. A fairly good coverage allows the construction of tectonic and volcanological maps (Ballard et al., 1981).

Instrument packages are generally towed about 1 km behind the ship. Positioning of the towed fish is therefore important for follow-up operations. Satellite navigation is inadequate for deep-tow and site-specific operations. A transponder network is therefore used for ship and towed vehicle positioning yielding navigation accuracies of 1 to 2% of water depth.

Reconnaissance rock sampling. Until recently, blind dredging using chain or pipe dredges (Fig. 7.14) was usually carried out without geophysical surveying. Recovery of such samples with the help of tensiometer readings and visual inspection of dredge tracks is rather erratic. Sampling uncertainties may be reduced considerably by lowering the dredge vertically into the target area and then moving the ship horizontally to a distance of about 1 km controlled by a transponder attached to the wire above the dreg.

Selective sampling is possible only by using grabs guided by a TV camera (Fig. 7.15). Grabs should open and closed on command from the ship (Burkhardt, 1984). Reconnaissance rock sampling leads to large quantities of rock samples which are carefully examined for indications of metallogenesis (hydrothermal alterations, presence of Mn/Fe crusts, crack precipitations). Occasionally, sulfides may be sampled.

Visual sampling and sampling lead to discrimination between small mineral shows and more important mineralisations. In the latter case, an intense sampling and survey operation is envisaged.

Fig. 7.14. Pipe and chain dredge, the most common instruments to recover rocks from the ocean floor.

Manned submarine operations. Specific survey operations are best facilitated by manned submarine operations. For non-military operations, there are only a few submersibles available operating at water depths between 1000 and 3000 m. These submersibles are "Pisces", "Alvin" and "Cyana". Recently, a French submersible for operating depths of up to 6000 m has been launched.

Submersibles carry 3 persons within a 2-m sphere. They are furnished with manipulators, still cameras, TV and various other devices. Operating time is usually 6 hours at vehicle speeds of 2 knots. Employment is strictly limited by weather conditions during launching and recovery. Submersibles are unsuited to large-scale exploration purposes but may be employed for inspecting sulfide deposits already discovered. Typical applications are vent temperature measurements, hydrothermal fluid sampling, surficial rock sampling, biological observations and display and recovery of instruments.

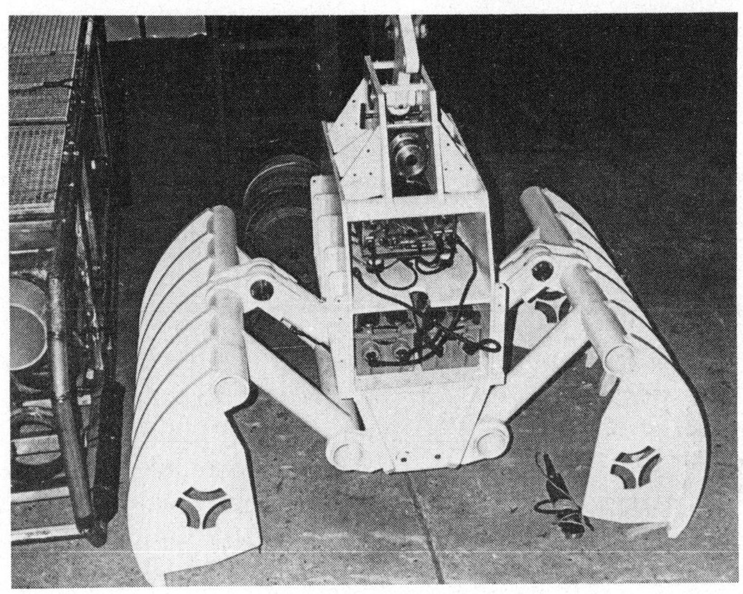

Fig. 7.15. Electro-hydraulic TV grab for the recovery of rocks from the deep-sea bottom. Weight 3.7 t in air, volume 1 m^3. Left side a deeptowed TV-photo package.

Manned submersible in the future may be replaced by remote-operated vehicles (ROV). The French "Epaulard" has already been used in sulfide research (Galerne, 1983). This ROV is an untethered 5-m long vehicle that is able to take 5000 photographs at a speed of 1 m/s.

Sampling and analysis. Bulk sampling is necessary to provide sufficient ore material for mineral processing and geotechnical tests. Up to several tens of tons of material are needed depending on the variability of the ore. Instruments suited to this type of sampling are large TV-guided hydraulic grabs (see Fig. 7.15).

Chemical analyses and the evaluation of material and data produced by massive sulfide exploration cruises requires about one year of work. Applied work will concentrate on chemical analyses of representative samples and on those ore characteristics that influence ore values, mineral processing operations, and on detailed exploration and mining methods.

DEEPSEA MANGANESE NODULES IN THE EQUATORIAL PACIFIC

History and project development

Manganese nodules have been known since 1873, when they were dredged from the seafloor of the Pacific Ocean by the British research vessel H.M.S. "Challenger". One hundred and eleven years later, first exploration licences were issued to companies that have been investigating the nodules intensively since the late sixties. These investigations included exploration, pilot mining tests and metallurgical process evaluation. In the following sections, German exploration efforts in the evaluation of these deposits are described.

Table 7.1 displays the history and development of the nodules project from a German point of view. However, from the table one can also deduce the world-wide political importance which has been attached to manganese nodules in the past.

According to the ideology of the United Nations organisations, nodule exploitation shall be ruled by a new Law of the Sea (see Chapter 8). Since this U.N. convention, especially the part which regulates deepsea mining beyond national jurisdiction was unaccepted by some of the members for momentous reasons, so that most western industrialised States have passed national laws. Exploration licences have already been granted being mutually recognised under the umbrella of a Reciprocating States Agreement which in 1984 became effective between the United States, the United Kingdom, Japan, France, Italy, The Netherlands, Belgium and the Federal Republic of Germany. Thus, two legal regimes exist that are competitive and incompatible as long as no moderation of certain U.N. regulations has a chance to be negotiable again.

The manganese nodules of the deepsea basins contain larger metal quantities than are known from exploitable onshore deposits (see also Chapter 2). The economically most interesting occurrences concentrate in the North Pacific between 10° and 20° N, between the Hawaiian Islands and Mexico. They comprise the so-called nodule belt bordered by the Clarion Fracture Zone in the north and the Clippertan Fracture Zone in the south.

Larger coherent nodule fields of probable economic value were also delineated in the South Pacific and in the Indian Ocean (Fig. 7.16), whereas environmental conditions of the Atlantic Ocean seem to be less favourable for forming valuable deposits.

One single "nodule mine" with an annual production of 4.3 millon tons ore could meet, for example, between 3 and 100% the demand of copper, nickel, manganese and cobalt of the F.R. of Germany, depending on the metal considered. An investment of about 4 billion DM is however necessary. This is only an order of investment magnitude and the economic impact of this

Table 7.1
History of the manganese nodule project.

Year	Event
1873/74	The British research vessel H.M.S. "Challenger" recovers the first manganese nodules from the Pacific Ocean by a dredge haul.
1898	The German research vessel "Valdivia (I) is dredging the first nodules from the Atlantic Ocean. For ninety years the nodules were considered an exceptional mineralogical curiosity.
1957/58	Investigations during the International Geophysical Year and on a world-wide scale lead to the discovery that nodules are wide-spread on the ocean floor.
1965	First comprehensive publication on the economic importance of marine mineral resources by J. Mero.
1968	German mining companies consider manganese nodule project.
1970	First joint cruises of German scientific institutions and companies with Deep Sea Ventures Inc., USA, on board R/V "Prospector".
1971	Commissioning of FS "Valdivia (II)" as an exploration vessel.
1972	Start of systematic exploration with FS "Valdivia (II)" in the Pacific and Indian Oceans continuing until 1981. Numerous joint missions with third parties.
1973	Formation of "Arbeitsgemeinschaft meerestechnisch gewinnbare Rohstoffe" (AMR) with Metallgesellschaft AG, Preussag AG and Salzgitter AG.
1975	Formation of the international joint venture "Ocean Management Inc." (OMI), with AMR, Deep Ocean Mining Corp. (DOMCO, lead by Sumitomo, Japan), INCO (Canada) and SEDCO (USA) with 25% shares each. Seven other consortia were established at the same time.
1976	Commisioning of the second German vessel for marine exploration (MS "Sonne").
1978	World´s first successful deepsea mining test by OMI with substantial contribution by AMR. Continuation of UNCLOS III.
1980	Establishment of national laws for deepsea mining beyond the nationalised Exclusive Economic Zone in the USA and the F.R. Germany (other States, among them the USSR, follow). First application for exploration licences in the USA, F.R. Germany and France.
1981	Commencement of the negotiations for a Reciprocating States Agreement (RSA) between eight larger western industrialised countries (USA, UK, France, Japan, Belgium, The Netherlands, Italy, F.R. Germany). Interconsortial negotiations for conflict solution of overlapping licence applications as a prerequisite for the governmental RSA. Successful final seatrials of the HSES exploration system of the OMI group.
1982	Passing of the U.N. Law of the Sea Convention (legislation by signature limited to December 9, 1984; valid on the basis of ratification by at least sixty nations).
1983	Conflict settlement among six private consortia and signature of various binding and suable settlement agreements.
1984	Amended licence applications in the USA, France, Japan and F.R. Germany. Accession of AMR as an independent consortium to the inter-consortial agreements in addition to participation as a member of OMI. Signature of the governmental Reciprocating States´ Agreement. Rejection of the U.N. Convention by the German government because of unacceptable regulations and restrictions in the field of marine mining beyond the EEZ. Accession of the European Community to the U.N. Convention. Granting of exploration licences on the basis of national laws in the USA and the UK.
1985	Granting of exploration licences in the F.R. Germany.

new mineral resource can be judged from Fig. 7.17.

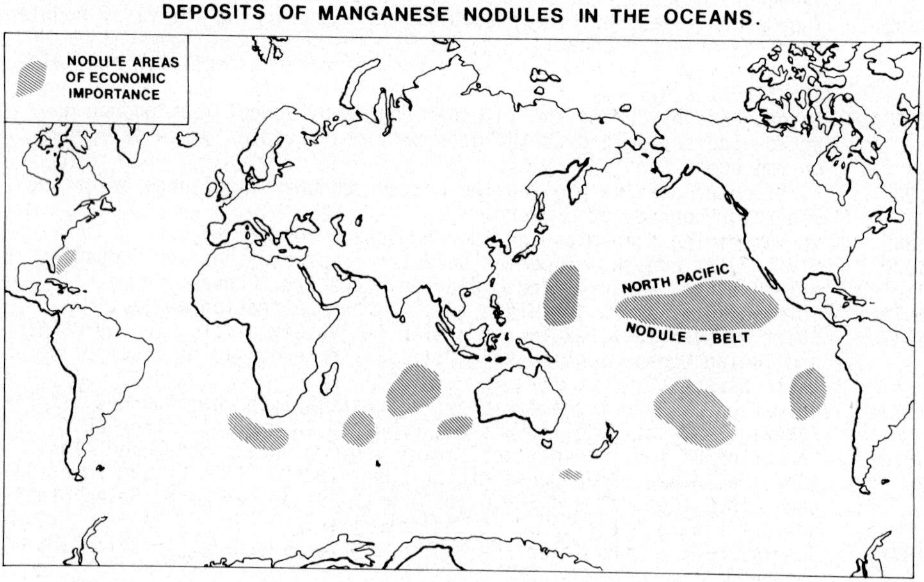

Fig. 7.16. Economically valuable deposits of manganese nodules in the oceans (1985).

Although the list of publications on the various aspects of manganese nodules had become almost endless in the past (see Fellerer, 1975; Meylan et al., 1981), most of the exploration information and the exploration data concerning the deposits are however embodied in confidential reports at private corporations or even confined in their safes.

Characteristic features of manganese nodule deposits

Deepsea manganese nodules, and just such one attract an economic interest at the moment, belong to the low-grade bulk ores for manganese with content of 25 to 30% Mn but also, and this makes their true value, with non-ferrous metals content of Ni, Cu and Co of altogether 2 to 3%. Contrary to onshore deposits where such metallic combination is scarcely found, manganese nodule occurrences are two-dimensional formations. This means that the ore content of a deposit of economic size is

Mio. t	Reserves onshore (1983)	NODULE POTENTIAL	
		North Pacific belt	Ocean Floor 3-6000m depth
Mn	1835,0	330,0	24 650
Ni	49,0	15,6	850
Cu	511,0	14,0	765
Co	2,7	2,4	170

Fig. 7.17. Total metal reserves onshore versus nodule resources.

scattered over a large area of the seafloor. The scattering is irregular but greatly varying coverage is observed, with a scattering of areas that are unexploitable for technical or economic reasons (see Fig. 7.18 a+b). In reality, an average coverage of 8000 tons per square kilometer within an area of 100 km^2, for example, means that the seafloor is covered by an ore layer only 4 mm thick. In some places this layer could be slightly thicker but may be thinner or missing in others. Also, the chemical composition as well as the size and the shape of the nodules may change and vary along relatively short distances.

A deposit of economic order of size is supposed to yield a minimum total of about 110 Mt of ore. One has to consider that nodule coverage (nodule density) is realistically not greater than 10 kg/m^2 and that there exist areas where nodules are absent or where they are buried, where mountains or other obstacles prevent any exploitation or where areas contain nodules that are below a certain cut-off grade of economics. Finally, when taking into account the technical restrictions of mining efficiency, it is easily understood why nodule fields to be explored have extensions between 130 000 and 200 000 km^2 which is nearly 80% of the area of the Federal Republic of Germany. Only areas of comparable sizes are expected to yield the required tonnage of exploitable ore and meet the technical and economic demands at a sufficient level of confidence. These areas have to be surveyed and sampled without gaps. Staking out of such an exploration licence area is preceded by initial exploration phases which still had to cover considerable portions of the Pacific Ocean.

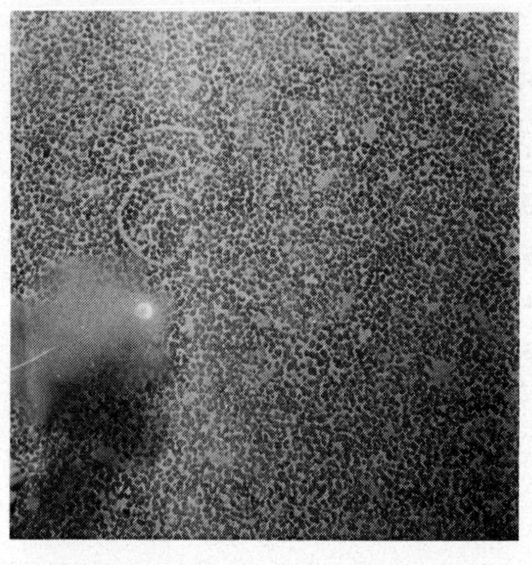

(a) (b)

Fig. 7.18 (a). Typical nodule field in the North Pacific.
Fig. 7.18 (b). Various types of nodules.

Exploration of deepsea manganese nodules has not only to cope with the huge extension of their deposits but also with their unique environmental conditions, both demanding new methods and tools. Among others these conditions are:

- Overburden of 5000 m sea water with ambient pressures of more than 500 kp/m^2 at the ore site.
- Considerable hydrodynamic resistance which becomes critical and almost unsurmountable for deep-tow equipment at ship speeds greater than 12 km/h.
- Absolutely darkness on the seafloor and a very limited range of light.
- Impermeability of sea water for radio waves and therefore the necessity to use umbilical cables or sonar techniques for data transmission.
- Remoteness from land, adverse or even prohibitive sea state and weather conditions.
- Corrosion problems.

German manganese nodule exploration efforts

Initial activities. Since 1969, the Federal Republic of Germany has invested a total of 210 mill. DM from governmental and private funds into the manganese nodule project, an essential part of which was spent on basic research and exploration. In the F.R. of Germany, which regards the manganese nodules as a new resource for manganese, nickel, copper and cobalt, the "Arbeitsgemeinschaft meerestechnisch gewinnbare Rohstoffe" (AMR), a national consortium consisting of three private mining companies (Metallgesellschaft AG, Preussag AG and Salzgitter AG) cooperates closely with scientific institutions (e.g. Geological Survey of Germany (BGR), universities of Aachen and Clausthal) is the main unit engaged in manganese nodule exploration and research. AMR is also a member of the international consortium "Ocean Management Inc." (OMI) which also members a Japanese group lead by Sumitomo (DOMCO), the Canadian International Nickel Corp. (INCO) and American South-Eastern Drilling Company (SEDCO). World-wide, four international and five governmental groups are exploring and developing manganese nodules (Table 7.2).

The field work was carried out with two German vessels: FS "Valdivia" (since 1972) and MS "Sonne" (since 1978). Both ships conducted basic and environmental research, economically oriented studies and technical tests of surveying and mining equipment partially supported by the German Ministry of Research and Technology. These ships were reputed to belong to the leading ones of an international fleet of exploration vessels.

Exploration strategy. The licensed areas held by AMR and OMI are the result of systematic field work which at the time of writing (1985) has been lasting for 13 years.

Since nodule deposits represent large dimensions and they are distributed over large areas from which exploitable best portions have to be selected, it soon became clear that the most suitable strategy to manage the connected exploration problems is a systematic sieving of exploration efforts with progressively decreasing mesh size. In the case of manganese nodules (better: polymetallic nodules) which are scattered on the seafloor like black potatoes on a huge field but are exploitable only from a few portions of the field, it is important at an early stage of exploration to know where not to go, either because nodules are either absent or poor or because mountains and obstacles are prohibitive for mining. The essential questions of deposit evaluation are the presence or absence of ore or underwater landscape favourable/unfavourable for exploitation. They can already be answered by a coarse exploration grid.

Contrary to this, areal mapping of nodule coverage (density), type of ore, and metallic content which has to include all variations, and a quantitative surveying of mining obstacles requires investment of time, costs and techniques which is larger by orders of magnitude. Therefore, these investigations have to be limited and applied to those areas which have proven prospects in preceding exploration steps.

TABLE 7.2
List of consortia and groups being engaged in the development of manganese nodules as a new metal resource.

Consortium	Countries involved
KCON Kennecott	USA, Canada, Japan, UK
OMA Ocean Mining Assoc.	USA, Belgium, Italy
OMCO Ocean Minerals Comp.	USA, The Netherlands
OMI Ocean Management Inc.	Canada, F.R. of Germany, Japan, USA
GEMONOD	France former AFERNOD
DORDO Deep Ocean Research and Development Organ.	Japan
AMR	F. R. of Germany
YUZHMORGEOLOGIYA	USSR
DOD/NIO	India

Figure 7.19 shows the targets and locations where nodule exploration has to deal with. It also outlines the tasks and the principal aims of evaluation which have to cover a considerably wider field of investigations than the survey of most of the traditional metal deposits onshore, especially when considering the survey for environmental impact and protection.

According to the principle of sieving with increasingly smaller mesh sizes, which permits us to eliminate unfavourable areas and focus step by step on the more expensive and time-consuming methods on increasingly smaller portions, one could subdivide the exploration activities on deepsea manganese

PLACE	TASKS	AIMS
SEA FLOOR		
NODULES, CRUSTS, PHOSPHORITES*	Distribution, variability, chemical composition; mineralogy; biology	Ore reserves; economic and technical assessment; conception of mining method; protection of environment
MORPHOLOGY*	Mapping of the submarine landscape; observation of natural and man-made obstacles	Recoverable ore reserves; technical assessment of mining operations
SUBSTRATUM		
SHALLOW	Soil mechanics; hard layers, slabs, rocks, outcrops	Exploitability of seafloor; design of mining systems; planning of mining operations; mining risks
DEEP	Sediment thickness	Prospectivity of areas; prediction of hazardeous areas and mining risks
SEAWATER OVERBURDEN	Currents; sound velocities, temperature, salinity, turbidity, chemistry, biology	Planning of mining operation; calibration of sonar methods; protection of environment
ATMOSPHERE – SEA SURFACE	Weather and sea conditions; waves, cyclones	Design of transport vessels and mining unit; mining risks

* top priority

Fig. 7.19. Targets, exploration tasks and evaluation aims of nodule exploration (simplified).

nodules into five phases. The sampling and surveying grids within distinct phases are a matter of practice. The availability of new tools and methods such as deep-towed wide-angle systems (e.g. HSES, Deeptow) or multiple-beam echo sounders (e.g. Seabeam) may influence and change the grids but they cannot replace the basic principles. When passing through the single phases, their tasks and their needs, it is advisable to recall the area size we are talking about (see Fig. 7.20).

Exploration phase I. During this phase the selection of the target areas for exploration by means of the available data and guided by knowledge on the genesis of nodule deposits has to be conducted. The initial phase was purely desk work in the interval 1968 to 1970.

Exploration phase II. After a period of first experiencing in practice manganese nodule occurrences, and sampling and surveying equipment at use in those early years gathered aboard the R/V "Prospector", real field work of

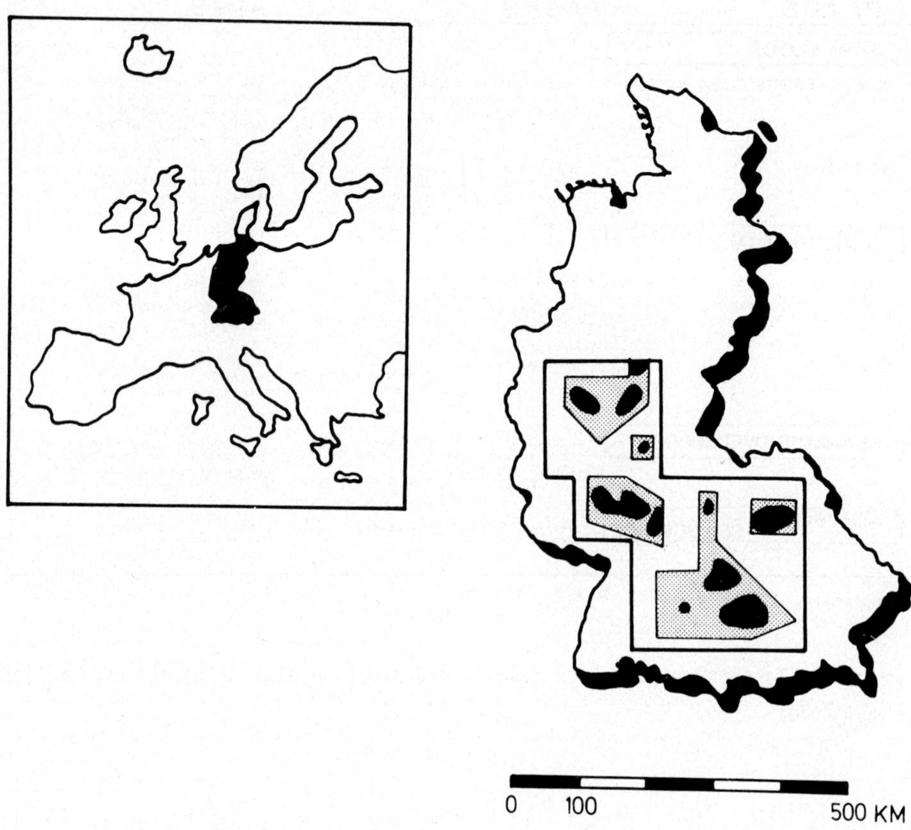

Fig. 7.20. Area size of nodule fields and exploration licences, compared with the size of the Federal Republic of Germany.

AMR started in 1972. It was carried at separations of considerable length 150 to 185 km. Geological sampling with wire-bound spade and box corers (see Figs. 7.8 and 7.12) and a free-fall grab sampler combined with seafloor photography took place at distances between 55 and 90 km.

Sampling was supported by geophysical profiling using narrow-beam echosounders (12 and 30 kHz), 3.5 kHz subbottom profiler and airgun seismic. These investigations were carried out to answer the following questions:

- Do indications exist of large areas along the survey lines which seem to be prospective and exploitable by evidence of seafloor morphology?
- Do these favourable sections of the survey lines show sediment cover which has a sufficient thickness to veil the basaltic basement?
- Are there nodule occurrences with sufficient grade and abundance within the morphologically favourable areas?

Exploration phase III. In principle, the phase III survey aimed at the same targets as the preceding phase II, but it took place as an area-covering operation. As an outcome of this, the regional distribution of morphologically exploitable and unworkable areas was outlined as well as those that were nodule-bearing and barren (or buried areas). Even the variations of the metal content and coverage were recorded to a certain degree.

The distance between geophysical lines (single-beam echo-sounding, subbottom profiling and optionally air-gun seismic) was 18 km. The distance between the sampling localities which were composed of 5 to 8 free-fall grab stations (bottom photographs included) and a wire-bound sediment sampler or an oceanographic station was 33 to 90 km, depending on the properties of the nodule occurrence and the sediment. Individual sampling stations were 900 m apart, either along a line or as a cluster with the wire-bound sample taken at the center.

Main sampling tools were free-fall grabs with mounted single-shot cameras and free-fall sediment corers. A certain priority in this exploration phase had wire-bound sampling of undisturbed sediment columns collected mainly for the purpose of soil-mechanical investigations (bearing capacity of the seafloor) and for sedimentological, geochemical and mineralogical analyses to determine nodule genesis. For that purpose, spade corer, box corer and piston corer in this order of frequency were also applied. Dredge hauls using large and heavy kasten and chain dredges yielded bulk material for metallurgical trials (see Figs. 7.14 and 7.12). Oceanographic probes combined with sea-water sampling supplied data on salinity, temperature, oxygen content, suspended matter and were indispensable for the calibration of sonar instruments, and for the determination of vertical distribution of sound velocities.

The profiles and sampling grids of the following exploration phases were placed between the open spaces of the previous survey phases depending on the ability to prospect these areas. The decisions that had to be made after exploring phase III were (a) where to continue with the next and more sumptuous step, (b) where to postpone or stop further investigations.

Exploration phase IV. When entering this phase of exploration, increasingly higher demands on the positioning and navigating accuracy for the survey vessel as well as for underwater equipment have to be made, justifying the term marine surveying for this phase.

The grid width of parallel geophysical lines was reduced in two steps to 9 km and less. Since a multi-beam echo-sounder became available (1981) some smaller nodule areas were bathymetrically mapped without gaps, i.e. with line distances of 1.5 km at 5000 m water depth. The distance between nodule sampling stations dwindled to less than 1000 m.

Besides various echo-sounding systems and subbottom profilers with penetration depth into subsoil of 50 to 100 m, main sampling tools again were free-fall grabs with mounted still cameras, spade corer, long box corers and dredges. Important additional instruments at this stage were deepsea TV, oceanographic probes, current meters for epi- and bathy-currents. At a later stage the application of integrated multipurpose systems ("underwater surveying satellites") would be advisable.

Investigations were focussed on detailed and shallow seismic mapping, analysis of the submarine landscape and structures in terms of roughness and accessibility (i.e. slope angles, frequency and amplitude of elevations, dislevels), interrelation between sediment and basement (intrusions, extrusions, faults), sediment facies, soil mechanics, average nodule coverage and its variation, metal content and variations, physical and chemical parameters of the nodules (e.g. size distribution, shape, tape, admixtures, friability, water content, slag-forming components), trend of nodule properties and environment, oceanographic parameters and meteorological and sea-state observations.

Within this exploration phase there is a domain on computerised data handling, data processing and evaluation as on application of various mathematical and statistical methods as well. At the end of this phase areas under survey were mapped showing zones of good, medium and low nodule quality in terms of both technical and economic exploitability.

Since the ultimate working step of exploration phase V hardly jeopardises the deposit as a whole, but will yield data with higher resolution and level of confidence, a feasibility study on the particular nodule field under consideration would be possible at phase IV. However, the level of confidence would be insufficient for profitability analysis and mine layout.

Exploration phase V. This exploration phase is dedicated to the filling of gaps within the survey grids of both geophysical detailed mapping and sampling. It yields, if required, additional data in order to support statistical methods. Furthermore, ultimate field work aims at area-covering data as precise and detailed as possible to geologically recording small mining obstacles and hazardous areas which may be areas with rocks, crusts and slabs, or zones with sediment gliding, dykes, faults and escarpments. In general these are smaller than detectable from the sea surface. Even zones with nodules of an anomalous type (e.g. oversize) and composition (phosphorite nuclei) have to be recorded carefully.

Another target is the small-scale variation of the nodule coverage and its possible influence on mining operation and production output. At this stage, long-term current readings from the sea surface to the bottom become very important.

These aims can be achieved only by employing advanced exploration systems. The efficiency of special navigation systems including underwater reference subsystems is also indispensable.

Exploration phase V serves to provide visible (proved) ore reserves for a minimum mining period of 5 to 6 years and probable ore for a further 6 to 8 years. It will then accompany the mining operations in order to convert these probable ore reserves into proved ones and to provide new probable and possible reserves. Before mining eventually can start, investigations on the environment and possible impacts with special tasks have to be conducted.

Exploration tools and methods

Of the large numbers of instruments that have been proposed and developed in the past, only relatively few have survived the hard qualifying contests during 15 years of practice, and these have gained a firm standing in the field for routine operations. The demand for some new, highly sophisticated multi-purpose systems exist. Their availability is expected within short time. Fig. 7.21 represents the quintessence of practical exploration at sea. Three pieces of information can be gathered: on the top is the frequency of application of the main tools and methods, in the vertical direction the universatility of a tool visualised by the number of targets being covered, and horizontally the specialisation of the instruments.

Basically, the exploration equipment can be subdivided into five main groups:

1) Geophysical methods
2) Sampling tools
3) Optical and high-frequency methods
4) Auxiliary devices
5) Evaluation tools and methods.

Their function or mode of operation has to be taken for granted.

A few remarks should be made with respect to the integrated, tethered, deep-tow multi-purpose systems which belong to group (3), since they represent the most advanced technology and will be indispensable in the future. Such systems are based upon sonar methods and they will solve tasks with

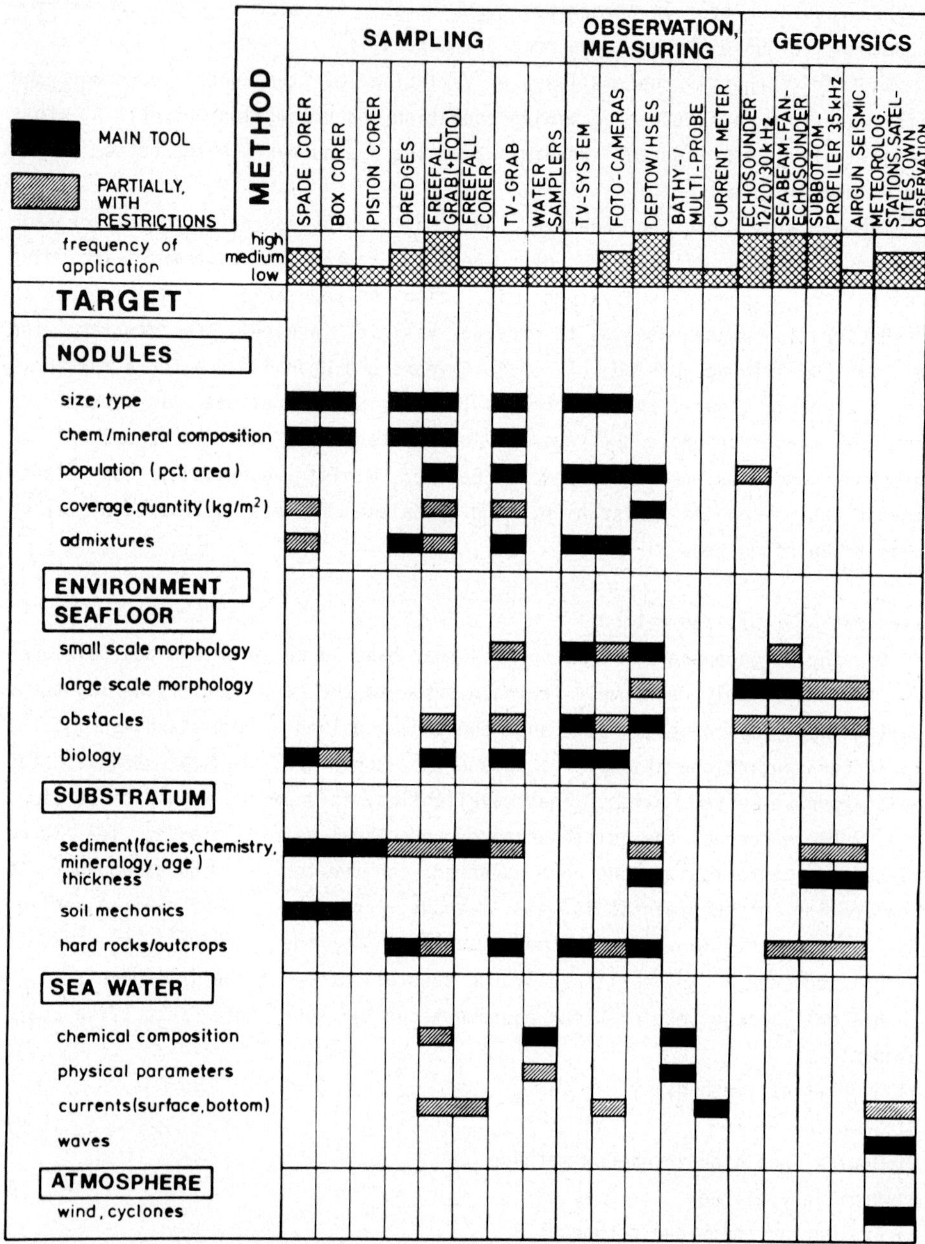

Fig. 7.21. Exploration tools and their employment to specific tasks of nodule exploration.

high accuracy and resolving power, e.g. quantitative recording of nodule coverage, sensing and mapping of the micromorphology of the seafloor, measuring the sediment thickness and subsoil structures. OMI/AMR have developed such a system (working since 1973) and have successfully tested a prototype in the Pacific in 1981. This deep-towed system is able to map nodules quantitatively and continuously on a strip 5000 km width, similar to areal photogrammetry. Its effectiveness is about 500 times higher than a deepsea TV system calculated on a purely geometrical basis. In reality, however, such systems as the High Speed Exploration System (HSES) are of much higher efficiency because they allow every square meter to be visualised without gaps and overlapping, if required. This cannot be achieved by conventional and optical methods.

Status of the manganese nodule project

Regarding the status of manganese nodule exploration it can be stated that, in the licenced areas (granted to German companies or OMI), exploration phase III has been conducted to almost 80% (of the 1985 level). Concerning exploration phases IV, approximately 20% of the total area has been surveyed, but systematic exploration was not conducted over more than 1%. In the course of these activities metal quantities on the seafloor of 1800 Mt of Mn, 79 Mt tons Ni, 58 Mt Cu and 8 Mt of Co can be regarded as probable and proved manganese nodule reserves. A considerable portion thereof could be mined if resources become profitable and competitive with traditional reserves onshore.

An example of a manganese nodule survey that has advanced to the limits of a level of confidence of the seventies reached by defundable investments using conventional methods is partly presented in Fig. 7.22. This area had to be prepared for OMI's successful pilot mining test in March 1978. A multiple-beam echo-sounder and HSES were unavailable at this time. A total of 90 days of intensive surveying at sea had been spent in this area with extension of 40 by 60 km. Figure 7.22 shows about 25% of the total area. In the evaluation program, expenditures for this task were at 6.5 million DM (1985 values). Nevertheless, the data were insufficient in all aspects of a profitability study, and also the level of confidence for some parameters was lower than required.

Figure 7.23 presents a pattern of the exploration survey grid the distances of which were achieved applying a multiple-beam echosounder system and HSES for nodule mapping. The grid of Fig. 7.22 on the other hand, is based on conventional methods leaving large gaps without any information. The distance of multiple-beam systems and HSES lines produce even overlapping survey strips of 3000 to 4000 m width, leaving very little

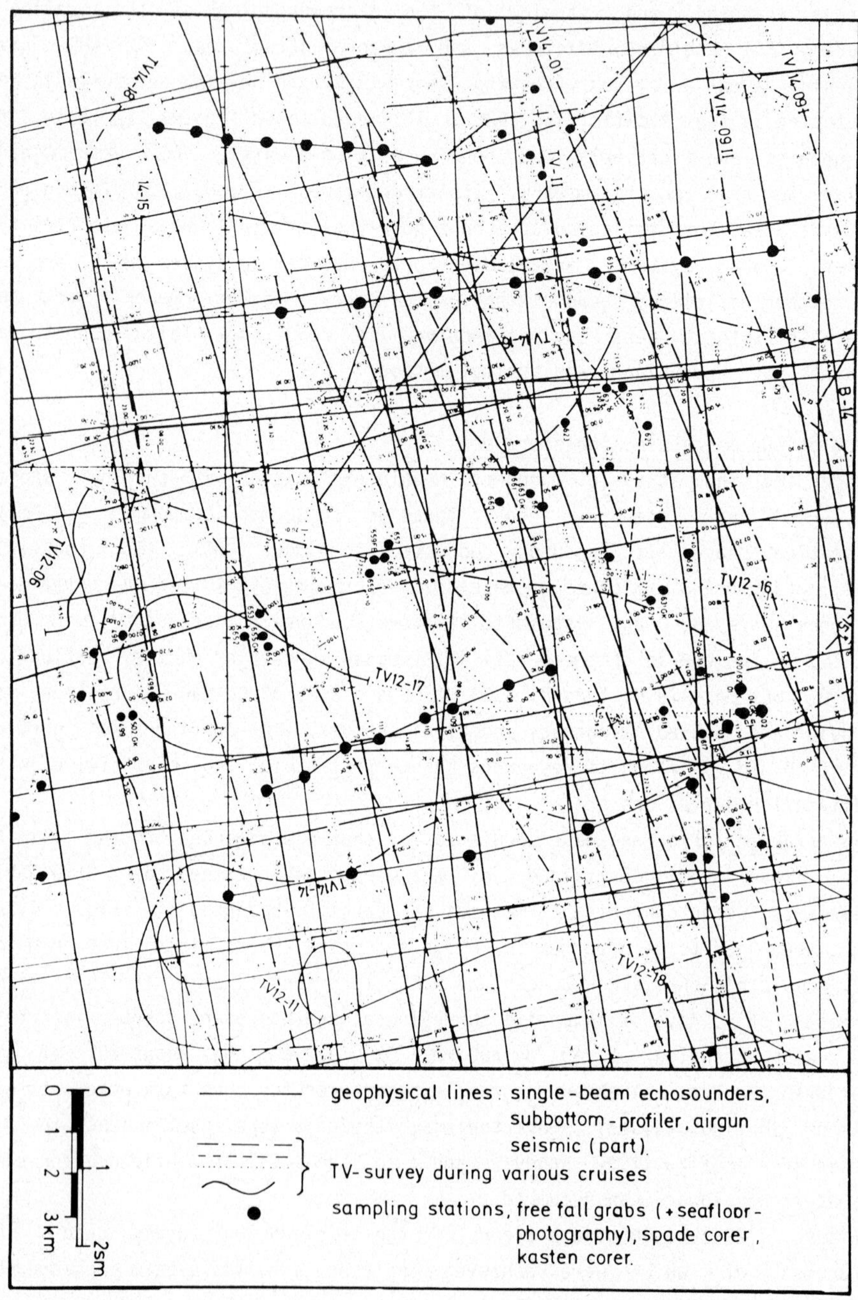

Fig. 7.22. Survey and sampling grid for preparation of a mining test area in the North Pacific (1974 - 1976).

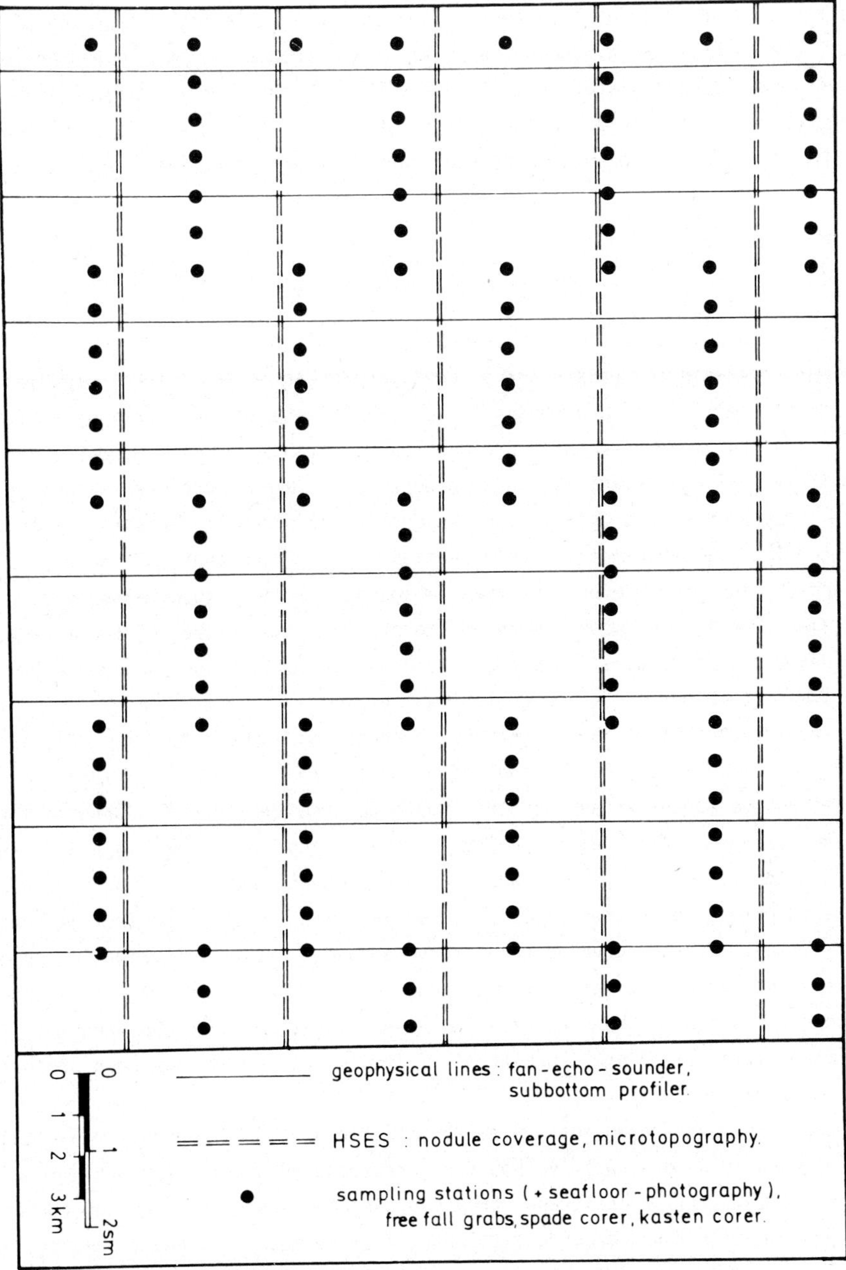

Fig. 7.23. Survey and sampling grid of nodule Exploration Phase III with advanced techniques of the year 1985.

of the area unsurveyed. Nevertheless, expenditures were only one third of those for the survey area shown in Fig. 7.23 despite the inclusion of 60 additional sampling stations. The results of such a survey would meet all requirements of nodule exploration with respect to a profitability study and a mining operation plan whereas the survey exemplified in Fig. 7.22 was sufficient only to prepare an area for sea trials of mining equipment although being three times more costly.

COBALT-RICH CRUSTS OF PACIFIC SEAMOUNTS

Marine manganese crusts have been known at least since companies and international mining consortia have started their exploration activities directed to manganese nodule occurrences (Table 7.3). However, at the beginning of the seventies, crusts were considered more as annoying and they represented a detoriation in quality of a nodule deposit, since their content of valuable metals was essentially lower than that of nodules. This changed abruptly when in the following periods such manganese crusts became known from seamounts more and more, from the slopes of the Mid-Pacific ridges and archipelagos, and when their almost ubiquitous occurrence as well as their anomalously high Co content was proved.

The systematic investigations started on the Hawaiian Archipelago (Morgenstein, 1974; Glasby and Andrews, 1977; Craig et al., 1982) and they moved to the southwest and to the south to the region of Johnston Island (Line Island Archipelago), Palmyra Island and Christmas Island (Fanning Archipelago), Baker Island and Titov Seamount and to other prospective structures of the Central Pacific region (Halbach, 1982). The fact is that in nearly all of the surveyed areas ferromanganese crusts with high Co content were discovered. It was confirmed that crusts indeed deserve an economic interest as a potential resource (Halbach, 1982; Halbach and Manheim, 1982; Commeaus et al., 1984; Clark et al., 1984), as has already been proposed by Mero in 1965.

Crusts from the deep sea, i.e. from water depths of greater than 4000 m, have grown much more rapidly than the corresponding nodules (probably due to hydrothermal influences) and they contain only one-tenth of the content of valuable metals found in nodules. Crusts from seamount plateaus, however, have grown extremely slowly, about 2 mm per million year or only a single layer of ions per week. They contain often 1% Co and more. The Co content increases with decreasing water depth. Nickel and especially Cu content are considerably below those of deepsea manganese nodules. Additionally, 0.06% V and 0.14 to 0.8 ppm Pt were reported (Halbach and Puteanus, 1984). A byproduct of crusts may be phosphorus, mainly occurring in the deeper parts of the crusts, partially as phosphorite encrustations (see Fig. 7.24).

TABLE 7.3
History and project development of cobalt-rich crusts.

Year	Event
1965	Comprehensive description of manganese crusts by J. Mero
1970/80	Increasing evidence of the semi-continuous distribution and economic value of shallow-water crusts, proceeding from the Hawaiian archipelago
1981	Exploration cruise SO-18 (Midpac '81) with MS "Sonne" to the Mid-Pacific Mountains
1983	Survey of the U.S. Geological Survey with R/V "Lee" in the Mid-Pacific Mountain area. Exploration by the Geological Survey of Japan
1984	Exploration cruise SO-33 with MS "Sonne" in the Central Pacific
1985	Detailed exploration cruise SO-37 with MS "Sonne" around Line Island, Line Island Ridge and Kelly Ridge of Johnston Island Establishment of a "Hard Mineral Consortium" by Brown & Root International (USA), Nippon Kokan (Japan) and Preussag, (F. R. Germany)
1986	Appraisal study (planned)

In favourable areas, crusts reach a thickness of 2 and up to 10 cm growing mainly on volcanic rocks and thereby disposing long, complicated and eventful growth histories. For this reason and since the seafloor morphology is very irregular and rugged, the same mining methods as those for seafloor-scattered manganese nodules cannot be applied in the case of crusts. This also implies that exploration surveys are more difficult to conduct.

Area-covering multiple-beam echosounding in combination with a highly accurate navigation system is indispensable as is subpositioned deepsea TV and seafloor still-camera surveying. Side-scan sonar survey by means of

Fig. 7.24. Cobalt-rich manganese crusts from the Kelly Ridge, S. Johnston Island, North Pacific, approx. 1,500 m water depth, size of specimen: 35 x 20 cm; crust thickness: 7 cm; growth on altered volcanic rock with phosphoritic impregnations.

deep-tow systems are regarded as very useful. Up to now, a crust thickness has to be calculated by evidence of sampled material. Geophysical methods for crust thickness determination based on electrical techniques are under development.

For crust sampling, three types of powerful remotely and TV-controlled hydraulic grabs were developed (Preussag AG, F.R. Germany, Fig. 7.15). Other standard tools for blind crust sampling are dredges.

At present, the following mining methods on an economic scale are discussed:

- Breaking-off of crusts from their substratum by means of sonar shock waves or by vibrations
- Mechanically operated scrapers similar to the Continuous Line Bucket System (CLB)
- Machinery like that in use for the removal of old road surfaces.

Collecting and uplift to the sea surface is regarded as a minor problem.

The metallurgical processing can rely on the methods developed for manganese nodule metallurgy. It would be possible to feed crust into the nodule processing as additional Co upgrading, for example, in the case of the large Peru Basin nodules usually low in cobalt.

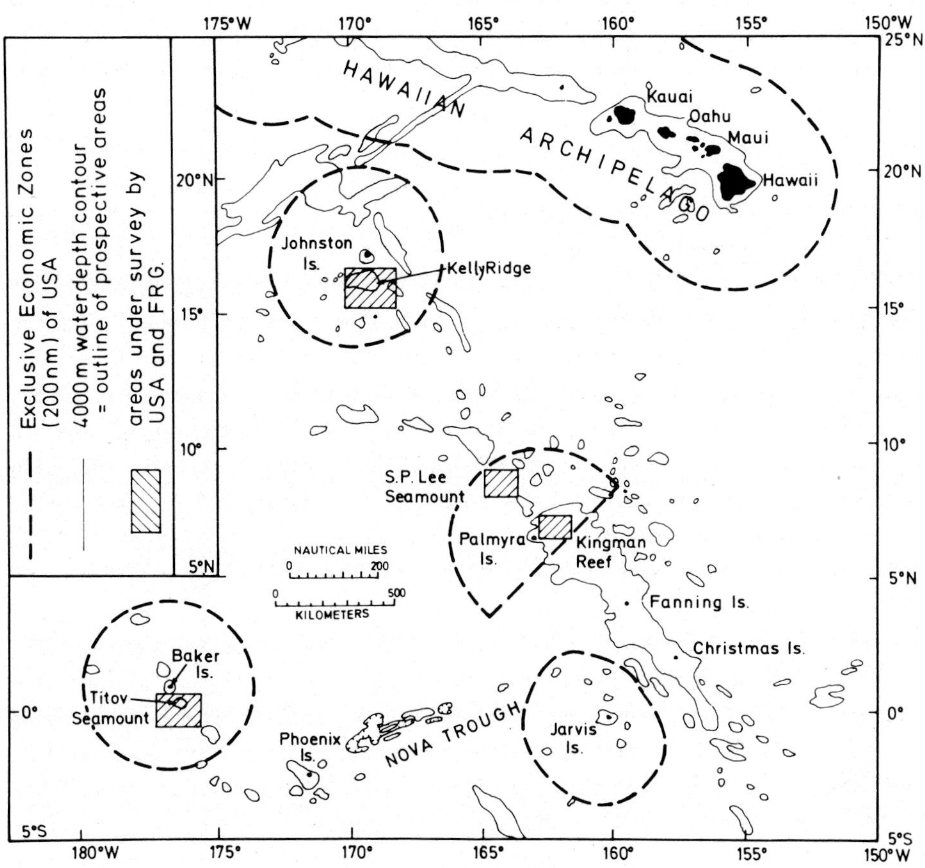

Fig. 7.25. Areas prospective for cobalt-crusts in the Central Pacific which are under survey by U.S.A. and F.R.G.

Economically interesting crust occurrences are found mainly within Exclusive Economic Zones. German activities were predominantly concentrated on the submarine ridges of the Mid-Pacific Island area, Baker Island area and Line Island area. The exploration survey on the Kelly Ridge south of Johnston Island, within the EEZ of the USA (see Fig 7.25)is at a very advanced state.

The best mineralisations so far were found on plateaus which have submerged into water depths of between 1000 to 2000 m. These plateaus often are topped by smaller volcanos elevating some hundreds of m and having gradients of ore-bearing slopes of 5 to 20 degrees. For some of the occurrences careful ratings arrive at reserves of 20 Mt of visible and probable ore. In 1985, three companies from the USA, Japan and the F.R. of Germany formed the Hard Mineral Consortium (HMC) for the development of these crusts to a potential metal resource.

PHOSPHORITE CONCRETIONS OFF NEW ZEALAND

History and occurrence

Phosphorite concretions are preferably forming in areas of upwelling of deep ocean water at depths of some hundreds of m (see Chapter 2). Recent occurrences have become known mainly from the continental shelf regions off Mexico, Peru, Chile, South Africa, and also from the submarine slopes of Pacific archipelagos (e.g. Johnston Island). They are very common and widespread and often occur in coexistence with manganese crusts replacing phosphorites with increasing water depths. The history of a phosphorite prospect is given in Table 7.4.

An occurrence of phosphorite concretions that is relatively well-known is situated southeast of New Zealand on the Chatham Rise, at a water depth bewteen 350 and 420 m (Fig.7.26). Since 1978, the deposit was explored in detail because these concretions turned out to be a very useful long-term fertiliser for the fields and meadows of New Zealand concretions being thought to be used directly after granulation and grinding without additional processing. Marine mining of the Chatham Rise deposit could meet New Zealand requirements for fertiliser and substitute biogenic phosphate occurrences which will be depleted within a few years on a number of Pacific islands. The development of the deposit has been managed as a New Zealand/German cooperation.

Phosphorite concretions of the Chatham Rise are found within the EEZ of New Zealand. An exploration licence is held by the domestic Fletcher Challenge Corp.

Chatham phosphorites have the shape of nodular, subangular pebbles varying in size between a few millimeters and 25 cm. They are of Miocene age (about 10 mill. years), originated by phosphatisation of chalk debris and they are embedded like gravel in a matrix of glauconite clayey fine sands and silts of Miocene to Pliocene age. This ore-bearing layer in some places may reach a thickness of 0.5 m. It is underlayn by more or less uniform light-coloured Lower Oligocene ooze and chalk (Meyer et al., 1985). The phosphorites average 21.5% P_2O_5 (9.4% P) and contain additionally 3% F and 150 ppm U.

TABLE 7.4
History and development of the New Zealand phosphorite project.

Year	Event
1952	First report on the phosphorites of the Chatham Rise, off New Zealand
1967/68	Reconnaissance survey by Global Marine Inc., USA, off California
1975-78	Research work by the New Zealand Oceanographic Institute (NZOI)
1978 and 1981	Detailed joint NZ and FRG surveys with German exploration vessels FS "Valdivia" and MS "Sonne" (NZOI, NZ Geological Survey, Fletcher Challenge Inc., BGR, and German companies Preussag and Deutsche Schachtbau)

The phosphorite occurrences are irregular and patchy on a large scale and on a small scale. Also, amounts of phosphorite (coverage) within such occurrences is highly variable, between 0 and 150 kg/m^2, even over short distances, depending on both the intensity of matrix mineralisation and the thickness of the ore-bearing sand. Within the exploitable occurrences which are very limited in size, ore density is on average 66 kg/m^2. Considerably extended areas show a mean coverage of only 11 kg/m^2 thus being very much below the present cutoff grade.

Exploration tools and methods

Contrary to the two-dimensional manganese nodule deposits being visually explorable, the phosphorite concretions are embedded in a matrix of thickness up to 0.5 m. Thus, the deposits are not laminae but a geological body which has to be penetrated by the exploration steps. Some disadvantages for systematic exploration are the result. For example, considerably smaller areas have to be explored and harvested because the required production rate of 0.5 Mt per year is essentially lower than in the case of manganese nodules (annual production of 4.3 Mt). However, average coverage is considerably higher than in the case of manganese nodules. Also, exploration takes place in water depths of 400 m only.

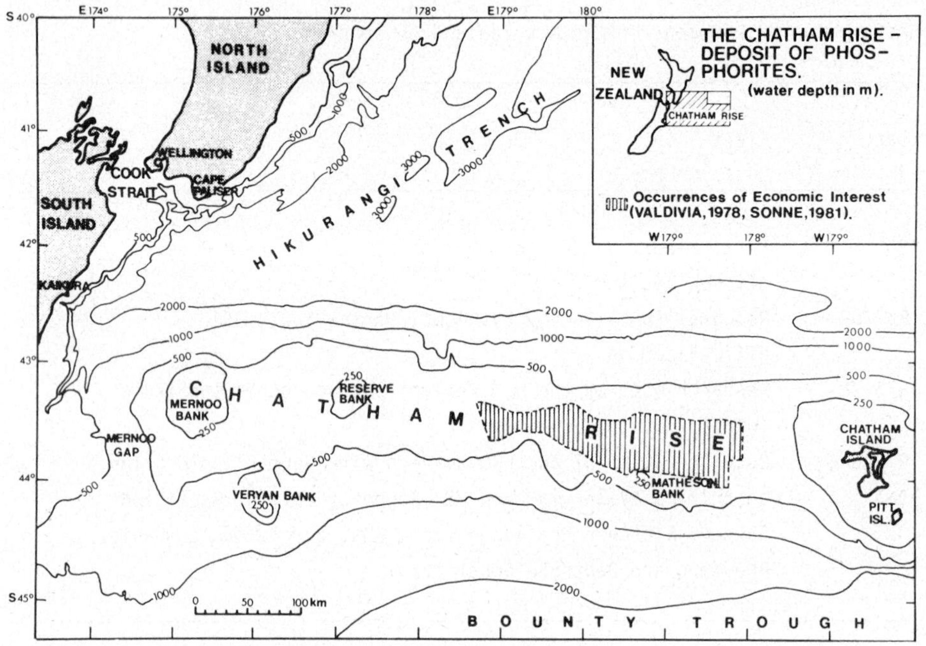

Fig. 7.26. Geographic and seafloor-morphological situation of the Chatham Rise phosphorite prospect.

Since only parts of the phosphorite concretions are visible on the sediment surface and because they generally are embedded in a matrix, the application of optical methods (photo cameras, TV) except for biologic and environmental investigations is efficient with restrictions, and applicable only in certain cases. Spot checks regarding presence or absence of phosphorite mineralisation would have to be carried out by means of time- and cost-consuming sampling if more suitable faster methods were not developed during recent times.

During the exploration activities it became evident that a deep-towed shallow seismic boomer with a frequency spectrum between 0.5 to 6 kHz produces sedimentary sequences of excellent detail. The boundary between the ore-bearing matrix on top and the underlying sterile chalk, i.e. the thickness of the prospective layer, is excellently resolved (Hansen, 1984; Falconer et al., 1984). This information was supported by a hull-mounted 30 kHz echosounder, a 3.5 kHz subbottom profiler and and deep-towed side-scan sonar yielding additional data on reflectivity, small-scale morphology of the seafloor and lensing-out of the ore layer (e.g. chalk exposure). Furthermore, the boomer gives valuable indications on the presence or absence of phosphorite concretions within this matrix by evidence of the reflection pattern (Meyer, 1985). Sterile areas show a smooth-rolling topography without refraction hyperbolae (Fig. 7.27) in their boomer reflection behaviour, whereas those with a high phosphorite yield exhibit a knobby relief as its seismic facies which is composed of continuous hyperbolae (Fig. 7.28).

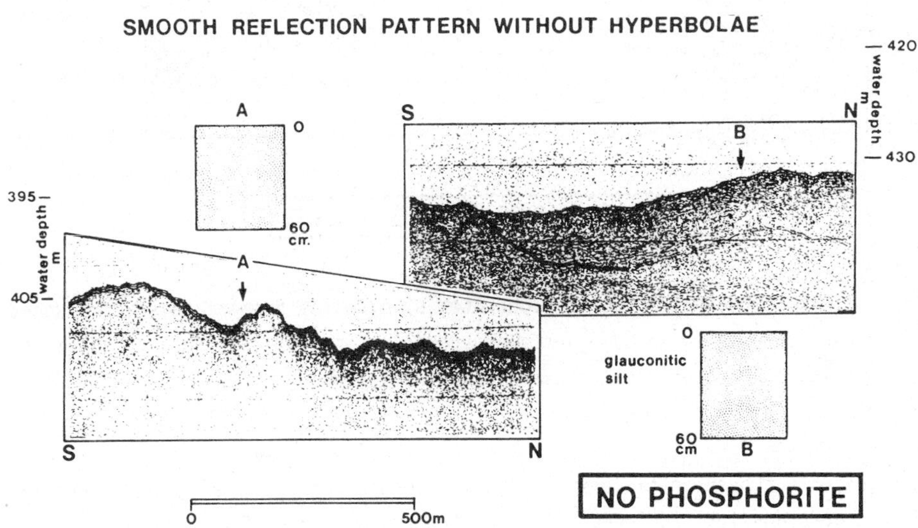

Fig. 7.27. Exploration for phosphorites by means of boomer seismic: smooth reflection facies without small-scale hyperbolae is typical for absence of phosphorites.

Sampling has to cope with the necessity of obtaining a quantitative sample with a volume as large as possible in order to calculate the total coverage per square unit which may vary considerably over short distances. For this task a heavy pneumatic grab sampler (Preussag) with penetration depth of 0.7 m (volume 0.8 m^3) and a sediment recovery of up to 1600 kg per

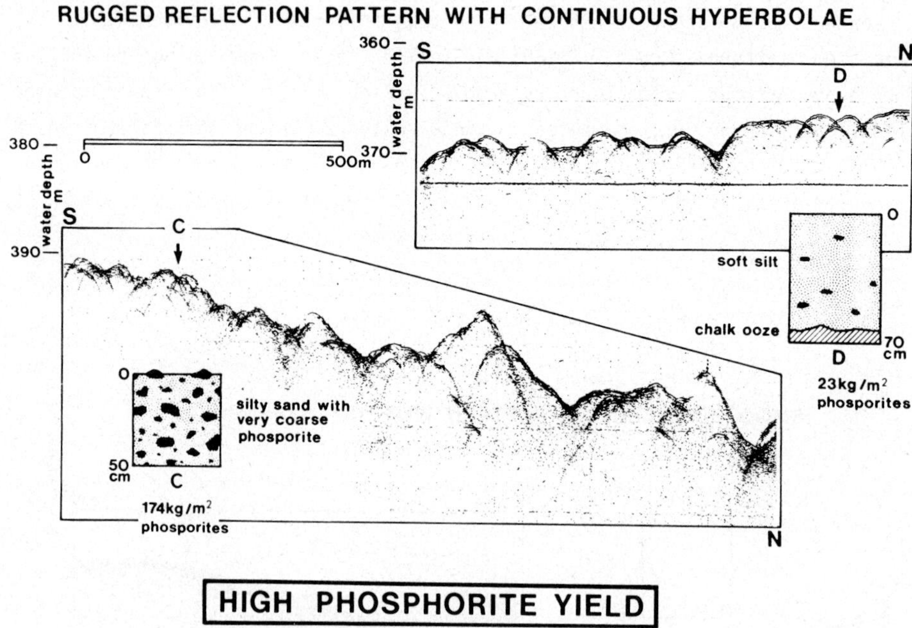

Fig. 7.28. Exploration for phosphorites by means of boomer seismic: rugged reflection facies with small-scale hyperbolae is typical for presence of phosphorite concretions in the upper sediment layer.

station has proved efficient. Advanced grabs of the second generation are additionally equipped with a TV system for seafloor surveying and remote control. The grabs of the third generation will have self-propulsion for exact positioning. Grab sampling as the main tool for phosphorite exploration was accompanied by heavy box-coring (25 by 25 cm, 1.5 m penetration depth). Such undisturbed cores were needed for soil-mechanical investigations necessary for the design of a mining unit. Greater penetration depths were obtained with a vibracorer but this procedure is much more costly in operation.

For the positioning of the survey tracks and the sampling locations, arrays of underwater acoustic transponders were employed which were calibrated by means of the ship satellite navigation system (including dead reckoning by Doppler sonar and radio navigation Omega/Loran C). By this method an absolute navigation accuracy of 200 m and a relative positioning accuracy of 5 to 10 m was possible.

At the present stage of exploratory work, 14 000 km^2 have been covered by a reconnaissance survey and 440 km^2 thereof by a very detailed survey. An area of 380 km^2 contains 25 Mt of proved reserves of phosphorite concretions with an average P content of 9.4%. Additional reserves of at least 25 Mt are probable in adjacent fields. They are of a lower coverage however.

REFERENCES

Metalliferous sediments

Bäcker, H., 1975. Exploration of the Red Sea and Gulf of Aden during the M.S. Valdivia cruises "Erzschlämme A" and "Erzschlämme B". Geol. Jb., D13: 3-78.

Bäcker, H., Lange, K. and Richter, H., 1975. Morphology of the Red Sea central graben between Subair Islands and Abul Kizaan. Geol. Jb., D13: 70-123.

Bignell, R.D., Tooms, J.S., Cronan, D.S. and Horowitz, A., 1974. An additional location of metalliferous sediments in the Red Sea. Nature, 248: 127-128.

Bruneau, L., Jerlow, N. G. and Koczy, F., 1953. Physical and chemical methods. Swedish Deep-Sea Expedit. Rept. III.

Degens, E.T. and Ross, D.A., 1969. Hot brines and recent heavy metal deposits in the Red Sea. Springer, Berlin, 600 pp.

Hartmann, M., 1972. Sound velocity data for the hot brines and corrected depth of interfaces in the Atlantis II Deep. Marine Geol., 12: M16-M20.

Meylan, M.A., Glasby, G.P., Knedler, K.E. and Johnston, J.H., 1981. Metalliferous deep-sea sediments. In. K.H. Wolf (Editor), Handbook of strata-bound and stratiform ore deposits. Elsevier, Amsterdam, Vol. 9, pp. 77-178.

Monin, A.S., Litvin, V.M., Podrazhansky, A.M., Sagalevich, A.M., Sorokhtin, O.G., Voitov, V.I., Yastrebov, V.S. and Zonensha, L.P., 1982. Red Sea submersible research expedition. Deep-Sea Res., 29: 361-373.

Natterer, K. and Luksch, J., 1901. Expedition S.M. Schiff "Pola" in das Rote Meer. Denkschriften der Kaiserlichen Akademie der Wissenschaften, Math.-Naturwiss. Kl., 69: 297-309.

Roeser, H.A., 1975. A detailed magnetic survey of the southern Red Sea. Geol. Jb., D13: 131-153.

Rona, P.A., 1983. Exploration of hydrothermal mineral deposits at seafloor spreading centers. Marine Mining, 4, 1: 7-38.

Massive sulfides

Anderson, R.N., 1975. Heat flow in the Marianna marginal basin. J. Geophys. Res., 80: 4043-4048.

Bäcker, H., Lange, J. and Marchig, V., 1985. Hydrothermal activity and sulphide formation in axial valleys of the East Pacific Rise crest bewteen 18 and 22° S. Earth Planet. Sci. Lett., 72: 9-22.

Ballard, R.D. and Francheteau, J., 1982. The relationship between active sulfide deposition and the axial processes of the Mid-Ocean Ridge. In: P. Halbach and P. Winter (Editors), Marine mineral deposits - New Research Results and Economic Prospects. Glückauf, Essen, pp. 137-176.

Ballard, R.D., Francheteau, J., Juteau, T., Rangan, C. and Normark, W., 1981. East Pacific Rise at 21° N: the volcanic, tectonic, and hydrothermal processes of the central axis. Earth Planet. Sci. Lett., 55: 1-10.

Baross, J.A. and Deming, J.W., 1983. Growth of "black smoker" bacteria at temperatures of at least 250° C. Nature, 303: 423-426.

Batiza, R., 1983. Abundances of hydrothermal mineral deposits on seamounts inferred from seamount petrologic characteristics and abundances. Proceed. Oceans'83. Marine Technology Society, pp. 797-800.

Batiza, R., Oestrike, R. and Futa, K., 1982. Chemical and isotopic diversity in basalts dredged from the East Pacific Rise at 10° S, the fossil Galapagos Rise and the Nazca Plate. Marine Geol., 49: 115-132.

Bertine, K.K. and Keene, J.B., 1975. Submarine barite-opal rocks of hydrothermal origin. Science, 188: 150-152.

Boström, K., 1973. The origin and fate of ferromanganoan active ridge sediments. Acta Univ. Stockholmiensis, 27: 149-243.

Boström, K., Peterson, M. N. A., Joenssen, O. and Fisher, D. E., 1969. The origin of aliminium - poor ferro - manganoan sediments in area of high heat flow on the East Pacific Rise, Mar. Geol., 7:427-447.

Boulegue, J., Perseil, E.A., Bernat, M., Dupre, B., Stouff, P. and Francheteau, J., 1984. A high-temperature hydrothermal deposit on the East Pacific Rise near 7° N. Earth Planet. Sci. Lett., 70: 249-259.

Bruneau, L., Jerlov, N.G. and Koczy, F.F.. Swedish Deep-Sea Expedition 1947-48, Vol. 3: Physics and chemistry.

Burkhardt, J., 1984. Entwicklung und Einsatz eines elektrohydraulischen Tiefseegreifers mit Fernsehkamera. Meerestechnik MT, 15: 135-138.

Crawford, A.M., Hollingshead, S.C. and Scott, S.D., 1984. Geotecnical engineering of deep-ocean polymetallic sulfides from 21° N, East Pacific Rise. Marine Mining, 4: 337-354.

Cyamex 1978. First named submersible dives on the Each Pacific Rise, 21° N: general results. Mar. Geophys. Res., 4: 345-379.

Desbruyeres, D. and Laubier, L., 1980. Alvinella pompejana gen. sp. nov., Ampharetidae aberrant des sources hydrothermales de la ride Est-Pacifique. Oceanologica Acta, 3: 267-274.

Dymond, J., 1981. Geochemistry of the Nazca plate surface sediments: an evaluation of hydrothermal, biogenic, detrital, and hydrogeneous sources. Geol. Soc. Am. Mem., 154: 133-173.

Ergunalp, D. and Weber, H., 1985. Massive sulfides on the East Pacific Rise and their benefication. Erzmetall, 36: 238-242.

Felbeck, H. and Somero, G.N., 1982. Primary production in deep-sea hydrothermal vent organisms: roles of sulfide-oxidizing bacteria. Trends Biochem. Sci., 7: 201-204.

Francheteau, J. and Ballard, R.D., 1983. The East Pacific Rise near 21° N, 13° N and 20° S: inferences for along-strike variability of axial processes of the Mid-ocean Ridge. Earth Planet. Sci. Lett., 64: 93-116.

Hekinian, R., Francheteau, J., Renard, V., Ballard, R.D., Choukroune, P., Cheminee, J.L., Albarede, F., Minster, J.F., Charlou, J.L., Marty, J.C. and Boulegue, J., 1983. Intense hydrothermal activity at the axis of the East Pacific Rise near 13° N: submersible witnesses the growth of sulfide chimney. Marine Geophys. Res., 6: 1-14.

Honnorez, J., 1969. La formation actuelle d'un gisement sousmarin de sulfures fumeroliens a Vulcano (mer tyrrhenienne). Mineral. Deposita, 4: 114-131.

Jones, M.L., 1981. Riftia pachyptila Jones: observations on on the vestimentiferan worm from the Galapagos Rift. Sience, 213: 333-336.

Kingston, M.J., Delaney, J.R. and Johnson, H.P., 1983. Sulfide deposits from the Juan de Fuca Ridge at 47°57' N, 129°067 W. Proceed. Oceans'83. Marine Technology Society, Vol.2, pp. 811-815.

Koski, R. A., Normark, W.R., Morton, J.L., Delaney, J.R., 1982. Metal sulfide deposits on the Juan de Fuca Ridge. Oceanus, 25: 42-48.

Kunzendorf, H., Walter, P., Stoffers, P. and Gwozdz, R., 1985. Metal Variations in divergent plate-boundary sediments from the Pacific Chem. Geol., 47: 113-133.

Leinen, M. and Anderson, R.N., 1981. Hydrothermal sediment from the Marianas Trough. EOS, 62: 914.

Lonsdale, P., 1979. A deep-sea hydrothermal site on a strike-slip fault. Nature, 281: 531-534.

Lonsdale, P., Batiza, R. and Simkin, T., 1982. Metallogenesis at seamounts on the East Pacific Rise. MTS Journal, 16: 54-61.

Lupton, J.E., 1983. Fluxes of helium-3 and heat from submarine hydrothermal systems: Guayamas Basin versus 21° N EPR. EOS, 64: 723.

Lupton, J. E. and Craig, H., 1981. A major helium-3 source at 15° S on the East Pacific Rise. Science, 214: 13-18.

Lupton, J.E., Johnson, H.P. and Delaney, J.H., 1981. Hydrothermal helium-3 on the Juan de Fuca Ridge. EOS, 62: 913-914.

Lupton, J.E., Klinkhammer, G.P., Normark, W.R., Haymon, R., Mac-Donald, K.C., Weiss, R.F. and Craig, H., 1980. Helium-3 and manganese at the 21° N East Pacific Rise hydrothermal site. Earth Planet. Sci. Lett., 50: 115-127.

Lupton, J.E., Weiss, R.F. and Craig, H., 1977. Mantle helium in hydrothermal plumes in the Galapagos Rift. Nature, 267: 603-604.

Malahoff, A., Embley, R.W., Cronan, D.S. and Skirrow, R., 1983. The geological setting of hydrothermal sulfides and associated deposits from the Galapagos Rift at 86° W. Marine Mining, 4: 123-137.

Marchig, V., Gundlach, H., Möller, P. and Schley, F., 1982. Some geochemical indicators for discrimination between diagenetic and hydrothermal metalliferous sediments. Marine Geol., 50: 214-256.

Minnitti, M. and Bonavia, F.F., 1984. Copper-ore grade hydrothermal mineralization discovered in a seamount in the Tyrrhenian Sea (Mediterranean): is the mineralization related to porphyry-coppers or to base metal lodes? Marine Geol., 59: 271-282.

Normark, W.R., Lupton, J.E., Murray, J.W., Delaney, J.R., Johnson, H.P., Koski, R.A., Clague, D.A. and Morton, J.L., 1982. Polymetallic sulfide deposits and water-column of active hydrothermal vents on the southern Juan de Fuca Ridge. MTS Journal, 16: 46-53.

Oudin, E., 1982. Mineralogie des depot sulfures lies a des zones d'accretion oceaniques actuelles (ride Est Pacifique) et fossile (Chypre). BRGM Principaux Result. Sci. Techn., Pt. 2, pp. 97-99.

Skornyakova, I.S., 1965. Dispersed iron and manganese in Pacific Ocean sediments. Internat. Geol. Rev., 7: 2161-2174.

Tufar, W., Gundlach, H., and Marchig, V., 1984. Zur Erzparagenese rezenter Sulfid-Vorkommen aus dem südlichen Pazifik. Mittl. Österr. Geol. Gesell., 77: 185-245.

Wright, J.A. and Fang, C.L., 1984. A microprocessor instrument for real-time marine heat flow measurement. EOS, 65: 1120.

Manganese nodules

Fellerer, R., 1975. Bibliographie der Manganknollen-Literatur 1878 bis 1975. BMFT Forschungsbericht No. M75-03, 121 pp.

Fellerer, R., 1980. Manganknollen. In: W. Schott (Editor), Die Fahrten des Forschungsschiffes Valdivia 1971 bis 1978. Geol. Jb., 38D: 35-76.

Glasby, G.P., 1977. Marine manganese deposits. Elsevier, Amsterdam, pp.

Meylan, M.A., Glasby, G.P. and Fortin, L., 1981. Bibliography and index to literature on manganese nodules (1861-1979). Dep. of Planning and Economic Development, State of Hawaii, 530 pp.

Murray, J. and Irvine, R., 1895. On the manganese oxides and manganese nodules in marine deposits. Trans. Roy. Soc. Edinburgh, 37: 721-742.

Cobalt-rich crusts

Aplin, A.C. and Cronan, D.S., 1985. Ferromanganese oxide deposits from the central Pacific Ocean. Geochim. Cosmochim. Acta, 49: 427-436.

Clark, A., Johnson, C. and Chinn, P., 1984. Assessment of cobalt-rich manganese crusts in the Hawaiian, Johnston and Palmyra Island's Exclusive Economic Zones. Natural Resources Forum, 8: 163-174.

Commeau, R.F., Clark, A., Johnson, C., Manheim, F.T., Aruscavage, P.J. and Lane, C.M., 1984. Ferromanganese crust resources in the Pacific and Atlantic Oceans. MTS Journal, 1/2: 421-430.

Craig, J.D., Andrews, J.E. and Meylan, M.A., 1982. Ferromanganese deposits in the Hawaiian archipelago. Marine Geol., 45: 127-157.

Glasby, G.P. and Andrews, J.E., 1977. Manganese crusts and nodules from the Hawaiian Ridge. Pacif. Science, 31: 363-379.

Halbach, P., 1982. Co-rich ferromanganese seamout deposits of the Central Pacific Basin. In: P. Halbach and P. Winter (Editors), Marine mineral deposits, new research results and economic prospects. Glückauf, Essen, pp. 60-85.

Halbach, P. and Manheim, F.T., 1984. Potential of cobalt and other metals in ferromanganese crusts on seamounts in the Central Pacific Basin. Marine Mining, 4: 319-336.

Halbach, P., Manheim, F.T. and Otten, P., 1982. Cobalt-rich ferromanganese deposits in the marginal seamount regions of the Central Pacific Basin. Erzmetall, 35: 447-453.

Halbach, P., Puteanus, D. and Manheim, F.T., 1984. Platinum concentrations in ferromanganese seamount crust from the central Pacific. Naturwissenschaften, 71: 577.

Hein, J.R., Manheim, F.T., Schwab, W.C. and Davis, A.S., 1985a. Ferromanganese crusts from Necker Ridge, Horizon Guyot, and S.P. Lee Guyot: geological consideraions. Marine Geol., in press.

Hein, J.R., Manheim, F.T., Schwab, W.C. and Davis, A.S., 1985b. Geological and geochemical data for seamounts and associated ferromanganese crusts in and near the Hawaiian, Johnston and Palmyra Island's Exclusive Economic Zones. U.S. Geological Survey Open-File Report 85-292.

Manheim, F.T., 1985. Marine cobalt resources. Unpubl. manuscript.

Mero, J., 1965. The mineral resoures of the sea. Elsevier, Amsterdam, 312 pp.

Phosphorite concretions

Falconer, R.K.H., Von Rad, U. and Wood, R., 1984. Regional structure and high-reolution seismic stratigraphy of the central Chatham Rise (New Zealand). Geol. Jb., D65: 29-56.

Hansen, R.D., 1984. High resolution seismic results of the detailed Sonne-17 survey areas (Chatham Rise, New Zealand). Geol. Jb., D65: 57-67.

Kudrass, H.R., 1984. The distribution and reserves of phosphorites on the central Chatham Rise (Sonne-17 cruise, 1981). Geol. Jb., D65: 183-198.

Meyer, K.W., Kudrass, H.R. and Von Rad, U., 1985. The Chatham Rise, New Zealand, marine phosphorite deposit geology, reserves and development aspects. Proceed. Intern. Fertilizer, Raw Materials Conf., London, in press.

Reed, J.J. and Hornibrook, N., 1952. Sediments from the Chatham Rise. NZ J. Sci. Technol., 34B: 173-188.

CHAPTER 8

LEGAL ASPECTS OF MARINE MINERAL EXPLORATION

E. D. BROWN

INTRODUCTION

Following this introduction, which places recent developments on the law governing marine mineral exploration in historial perspective, this Chapter provides a survey of the international legal framework within which such exploration must be conducted. Brief reference is made to the regime of the "areas within national jurisdiction", landward of the outer limit of the continental shelf, but the bulk of the Chapter is concerned with the legal regime of marine mineral exploration in the "Area beyond the limits of national jurisdiction". An account is given of both the United Nations Convention on the Law of the Sea, 1982 and of the alternative regime based on the national legislation of a number of major industrialised States.

It is only since the mid-1960's that the international community has concerned itself with the problem of identifying the existing rules of international law, or of developing new rules, to govern the exploration and exploitation of the mineral resources of what is now known as "the Area beyond the limits of national jurisdiction", that is, the area of the ocean bed lying seaward of the legal outward limit of the continental shelf. Prior to that time, in the two decades following the Second World War, interest was focussed primarily on the areas within the limits of national jurisdiction and in particular on the development of a legal regime for the exploration and exploitation of the mineral resources of the continental margin. Not only did the two regimes develop over separate historical periods; they were also constructed on the basis of quite different political and economic ideologies. It is for this reason that they are dealt with separately in the brief account given below of their principal features.

Before turning to this task, however, it may be useful to provide a sense of historical perspective by reviewing the main landmarks in the development of the law since 1945 (Table 8.1). This was the year in which the celebrated Truman Proclamation (Proclamation, 1945) was issued, whereby the United States claimed exclusive jurisdiction and control for the contiguous coastal State

TABLE 8.1
Historical landmarks in the development of a legal regime for marine mineral exploration and exploitation.

Year	Event	Remarks
1945	Truman Proclamation	US claims control over natural resources of continental shelf for the contiguous coastal State.
1958	Geneva Convention	'Sovereign rights' over continental shelf resources for coastal State.
1967	Pardo initiative in UN	Advocated that 'area beyond the limits of national jurisdiction' should be common heritage of mankind.
1982	UNCLOS III	Law of the Sea Convention signed by 159 countries but not yet in force as of 1985.

over the natural resources of the continental shelf, thus marking the beginning of a new epoch in both law and technology and giving birth to a new legal doctrine under which States became entitled to push their frontiers seawards and enjoy exclusive rights to explore and exploit the natural resources of a vast new area of submarine territory contiguous to their coasts. Although, prior to this American initiative, the area of the continental shelf had been recognised simply as part of the high seas, in which no State could lawfully acquire exclusive mineral rights, the Truman Proclamation was nonetheless widely welcomed as reflecting the need to develop new law to accommodate new technology and was gradually taken as a model for similar claims by other coastal states. Between 1945 and 1958 rules of international customary law crystallised on the basis of this State practice and in 1958 these rules were transformed into treaty law in the form of the Geneva Convention on the Continental Shelf adopted at the first United Nations Conference on the Law of the Sea (UNCLOS I). As will be seen below, the basic rules governing the exploration and exploitation of continental shelf minerals have remained unchanged since that time and later developments have been concerned mainly with the determination of the legal outer limit of this zone and with its delimitation between neighbouring (opposite or adjacent) States.

The next major landmark had its origins in 1967, when Dr. Pardo of Malta advocated in the United Nations that the Area of the seabed beyond the limits of national jurisdiction and its resources should be considered the common heritage of mankind, over which no State or other juridical person (such as a mineral company or a multinational consortium) should be permitted to acquire exclusive rights (Note Verbale, 1967). Like the Truman Proclamation before it, the Pardo initiative seemed to catch the mood of the time and led to UNCLOS III and its adoption in 1982 of the UN Convention on the Law of the Sea (UNCLOS, 1983). Part XI of this Convention embodies the 'common heritage' regime of seabed mining.

It may be useful at this point to explain the processes whereby this Convention may become binding upon States, and through them, upon their mineral companies. The convention was open for signature until 9 December 1984 and by that time had attracted 159 signatures. Signature does not bring the Convention into effect, though it does oblige signatories to refrain from acts which would defeat the object and purpose of the treaty (Vienna, 1969). The Convention will enter into force 12 months after ratification (formal confirmation by signatory of consent to be bound) or accession (similar confirmation by non-signatory State) by 60 States (UNCLOS, 1982, Article 308). As at 1 May 1985, only 18 such instruments had been recorded and it may therefore be some considerable time before the convention enters into force. Even then, it will of course be binding only on the parties to it. Following this event, States parties to the Convention would be under an obligation to give effect to it by adopting the national legislation necessary to apply its terms to mineral companies under their jurisdiction.

Although it is true that the overwhelming majority of States and particularly the Group of 77 developing States strongly favour the common heritage doctrine, there is also a considerable degree of opposition in many of the principal industrialised States, either to the general concept or to the particular form it takes in the new Convention. It is indeed this dissatisfaction with the Convention's seabed mining regime which has persuaded the Federal Republic of Germany, the United Kingdom and the United States not to sign the Convention. Moreover, these 3 countries, together with France, Italy, Japan and the Soviet Union, have also introduced unilateral legislation outside the UN Convention framework, which would permit them to grant national licences to explore for seabed minerals free of many of the conditions which would be imposed by the UN Convention. Much of this legislation is based upon the conviction of many of those States that seabed mining remains a high seas freedom until they have consented to a detailed treaty regime giving effect to the common heritage doctrine. This is, however, a view which is hotly disputed by the great majority of States which argue that the common heritage doctrine has now been incorporated in general (customary) international law.

Following this historical introduction, an account will next be given of the rules of international law governing the exploration of marine minerals in the area within the limits of national jurisdiction (Brown, 1984).

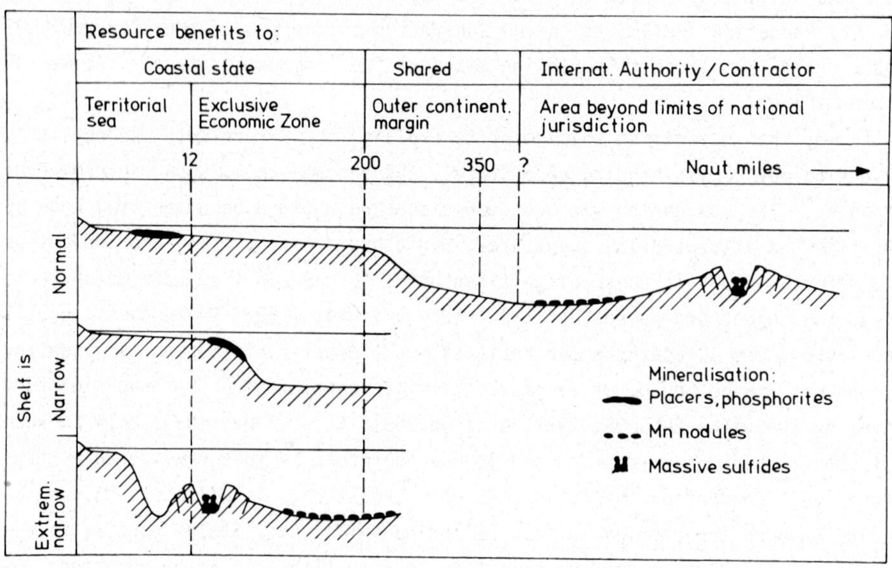

Fig. 8.1. Divisions of seafloor according to the UN Law of the Sea Convention.

AREAS WITHIN NATIONAL JURISDICTION

The areas within national jurisdiction include all maritime zones lying landward of the outer limit of the legal continental shelf (see Fig. 8.1.). In this context, it will suffice to distinguish two zones: (1) internal waters and the territorial sea, and (2) the continental shelf.

Internal waters and the territorial sea

In the zone landward of the outer limit of the territorial sea, the coastal State has sovereignty over all mineral resources, and foreign States or mineral companies would be entitled to carry out exploratory work only with the consent of the Government of the coastal State and subject to whatever terms and conditions it laid down. The precise breadth of the territorial sea has been a matter of controversy for many years but the trend in State practice is strongly towards recognition of a maximum breadth of 12 nautical miles and this breadth has been adopted in Article 3 of the UN Convention on Law of the Sea.

The continental shelf

Under the Geneva Convention on the Continental Shelf, 1958, the outer limit of the continental shelf has to be determined in accordance with a flexible and somewhat unsatisfactory formula laid down in Article 1, under which the term 'continental shelf' refers to submarine areas beyond the territorial sea out to a depth of 200 metres or, beyond that limit, to where the depth of the waters above admits of the exploitation of the natural resources of the said areas. In short, this 'elastic frontier' may be stretched further out, following advances in seabed mining technology. In practice, there has been a tendency for States to consider themselves entitled to claim as their legal continental shelf not only the geomorphic continental shelf but the whole of the geomorphic continental margin, thus adding also the continental slope and the continental rise. The new UN Convention reflects this State practice by adopting a definition of the outer limit of the 'continental shelf' which will entitle the coastal State to a legal continental shelf of a minimum breadth of 200 miles (370 km); where the geomorphic continental margin is of greater breadth than 200 miles, the legal continental shelf will extend to the seaward edge of the continental margin which is determined for this purpose by reference to a complex formula (UNCLOS, 1982, Article 76). It will be noted that there is now very little correspondence between the terminology of the lawyer and that of the geologist or the geomorphologist.

Subject to a few exceptions, none of which is important in this context, the regime for the exploration and exploitation of the mineral resources of the continental shelf established by the Geneva Convention on the Continental Shelf, 1958, has been re-incorporated in Part VI of the UN Convention on the Law of the Sea, 1982. Indeed, the most important Article in the present context, Article 2 of the Geneva Convention, has been taken over virtually unchanged as Article 77 of the UN Convention. It provides that the coastal State exercises over the continental shelf "sovereign rights" for the purpose of exploring it and exploitating its natural resources, including "the mineral and other non-living resources of the sea-bed and subsoil". It follows that, as in the case of the territorial sea, foreign governments and mining companies or mining consortia will not be able to conduct any exploratory work in the continental shelf except under licences granted in accordance with the national legislation of the coastal State concerned.

THE 'AREA' BEYOND THE LIMITS OF NATIONAL JURISDICTION

As was seen above, the law governing the exploration and exploitation of the mineral resources of the Area (see Fig. 1) is passing through a difficult formative phase at the present time. On the one hand, the vast majority of States believe that the Area and its resources are the common heritage of mankind and that their exploration and exploitation must be conducted in accordance with the new UN Convention on the Law of the Sea. On the other hand, however, a number of the most powerful and technologically most advanced States remain dissatisfied with the terms of the UN regime and maintain that they are entitled to adopt national legislation and grant seabed mining licences under it on the basis that seabed mining is a freedom of the high seas and will remain so until acceptable modifications are made to the UN regime.

In this section an account will be given of the principal features of, first, the UN regime, which is still under development, and, secondly, the unilateral legislation of a number of the leading industrialised States (Brown, in press).

The UN Regime

The United Nations regime for the Area comprises not only the provisions of Part XI of the UN Convention but also a scheme for Preparatory Investment Protection (PIP) which was grafted on to the Convention by way of Conference Resolutions during the final session of UNCLOS III. It had become clear in the Conference that the Convention would certainly not be acceptable to the United States and other leading industrialised States unless provision were made to safeguard the very heavy investments in seabed mining already made by a number of pioneer enterprises. The PIP Resolution may be regarded therefore as an attempt to achieve a compromise, whereby reasonable protection would be given to pioneer investors but, at the same time, the main principles of the Convention 'common heritage' regime, as already drafted, would be preserved.

The principal features of Part XI may be summarised as follows:

(i) The Area and its resources are the common heritage of mankind and no State or other entity may acquire rights in the minerals of the Area except under the system established in the Convention.

(ii) A parallel system is established under which all deep-sea mining would be under the control of an International Sea-Bed Authority ("the Authority"), which would be empowered both to undertake mining operations through an organ called the Enterprise and to enter into contracts enabling private and State ventures to acquire mining rights. The parallel system would work as follows: in applying for a contract, the applicant would identify two areas of equal estimated value, and in granting the contract, the Authority would allocate one of them to the applicant, while reserving the other for exploitation by either the Enterprise or by developing countries. This is the so-called 'site-banking' system.

(iii) The general policies of the Authority would be determined by the plenary Assembly, while their application in individual cases would be the responsibility of a 36-member Council. One of the Council's powers is that of approval of plans of work authorising miners to develop a mine-site.

(iv) The UN Convention contains a number of provisions designed to help the Enterprise to compete with private enterprise. They include a system of financing which places a considerable onus on States parties to the Convention, especially in the early years of operation, and a burdensome obligation upon contractors to transfer mining know-how to the Enterprise.

(v) In order to avoid causing hardship to producers of land-based minerals, an elaborate production control system is incorporated in the Convention, under which mineral production by sea-bed miners is restricted in accordance with a formula based on 60 per cent of the projected annual increase in the world's demand for nickel.

(vi) Provision is made for a review of the system at the end of a 15-year period.

A distinction is made in the Convention between 'prospecting' and 'exploration'. The Authority is to "encourage prospecting in the Area" and prospecting may be conducted simultaneously by more than one prospector in the same area. However, the proposed prospector would first have to provide the Authority with a written undertaking that it would comply with the Convention and be prepared to participate in training programmes designed to transfer technology and scientific knowledge to the Enterprise and developing States. He would at the same time have to notify the Authority of the approximate area in which prospecting was to be conducted. Even then, such prospecting would not confer on the prospector any rights with respect to the mineral resources except the right to recover a reasonable quantity of minerals to be used for testing (UNCLOS, 1982, Annex III, Art. 2).

Exploration is subject to quite different rules. Under the Convention, application has to be made to the Authority for a contract to carry out "activities in the Area" and such applications may be made only by "States Parties or state enterprises or national or juridical persons which possess the nationality of States Parties or are effectively controlled by them or their nationals, when sponsored by such States, or any group of the foregoing (UNCLOS, 1982, Art. 153(2)). Applicants also have to meet "qualification standards" relating to their financial and technical capabilities and their performance under any previous contracts with the Authority (UNCLOS, 1982, Annex III, Art. 4). The contract takes the form of a plan of work which may cover exploration only or exploration end exploitation. Exploration may proceed as soon as the plan of work has been approved.

The above brief description of the UN regime refers only to the original regime as laid down in the Convention itself and its Annexes. Attention must now be turned to the important modifications in the UN regime introduced by Conference Resolutions I and II, which respectively established a Preparatory Commission and made provision for the PIP scheme (Resolutions, 1983). The practical effect of these Resolutions is that, if seabed mining proceeds on the basis of the UN regime rather than under unilateral legislation, it is almost certain that the first generation mining ventures will be undertaken not under the Convention regime pure and simple but under the Convention regime as substantially modified by the PIP Resolution. Under the PIP scheme, it will be open to a consortium to secure a preferential status by registering as a "pioneer investor", provided it meets specified conditions relating to signature of the Convention by parent States and the level of pioneer investment. The registered pioneer investor may then have a "pioneer area" allocated to it, with the exclusive right to conduct "pioneer activities" therein. Pioneers may not proceed beyond the preliminary "pioneer activity" stage until the authority has granted them a contract approving their plan of work. This is practically guaranteed, however, provided the parent States of the consortia members have all ratified the Convention. The Authority is bound also to grant one further permit - a production authorisation - provided the provisions of the regime are followed.

Although applications have been made by four States for pioneer investor registration, the Preparatory Commission, which has the task of administering the PIP scheme, will not be in a position to grant any registration until agreement has been reached in the Preparatory Commission on the rules, regulations and procedures to be applied. At the time of writing (September 1985), this task has still not been completed.

The PIP scheme incorporated a 'site-banking' system similar to that in the Convention and referred to above. The Preparatory Commission designates the part of the total area submitted which is to be reserved for the Authority and the remaining part is allocated to the pioneer investor. Each pioneer is entitled to only one pioneer area which is not to exceed, in the beginning, 150,000 square kilometres.

The pioneer activities which the pioneer investor has the exclusive right to carry out include the elements of what is normally known as exploration, despite the fact that the pioneer investor has later to apply for approval of a plan of work for both exploration and exploitation. Pioneer activities include, for example, "the recovery from the Area of polymetallic nodules with a view to the designing, fabricating and testing of equipment which is intended to be used in the exploitation of polymetallic nodules" (Resolution II, Para. 1(b)(ii)).

Onerous obligation are imposed upon the pioneer investor. First, it has to pay a US$ 250,000 registration fee, plus a further US$ 250,000 when it applies to the Authority for a plan of work. An additional fee of US$ 1 million per year is payable from the time the pioneer area is allocated, payable once the Authority approves the plan of work. Nor are the pioneer investors' obligations only of a financial nature. They may be required to carry out exploratory work for the Enterprise on a cost-reimbursible basis plus interest at ten per cent; to provide training for Enterprise personnel; and they must undertake, before the entry into force of the Convention, to perform the obligations prescribed in the Convention relating to transfer of technology (Resolution II, Paras. 7(a), 7(b), 12(a)(i) to (iii)).

The 'Reciprocating States' regime

Unilateral legislation on sea-bed mining has already been adopted by the following 7 States: United States (1980), Federal Republic of Germany (1980, amended 1982), United Kingdom (1981), France (1981), the Soviet Union (1982), Japan (1982) and Italy (1985). However, in this Chapter, attention will be confined to the 6 States other than the Soviet Union. Though not identical, the statutes of these six countries are based on the same assumption and their key provisions are very similar (e.g. Brown, 1983). All of the Governments concerned insist that their legislation is interim in nature; that it does not involve any claim to sovereignty or sovereign rights over the deep seabed or its mineral resources; that they remain committed to the entry into force of a Convention embodying the principle that seabed mineral resources are the common heritage of mankind - if an acceptable text can be agreed upon; that they are not legally bound by the terms of UN General Assembly Resolutions on the subject; and that deep-sea mining conducted with due regard to the interests of other States in the freedom of the high seas is, under the present law, a legitimate exercise of a high seas freedom.

The scope of the interim legislation of the six States is very similar in a number of ways. First, as regards their temporal scope, provision is made in most of the acts for their repeal upon entry into force for the States concerned of the UN convention on the Law of the Sea. In most cases too, it is specified that no exploitation licence may authorise commercial recovery of seabed minerals prior to January 1988, though exploration may take place prior to that time. As regards the duration of licences, the statutes differ but within fairly narrow limits. For example, under British and American legislation, exploration licences may be issued for an initial period of 10 years and extended for further 5-year periods. The German statute, on the other hand, envisages an initial period of 20 years, with the possibility of 5-year extensions.

The personal scope of the statutes is again similar, in that States provide for the issue of licences to their own nationals, variously defined, and recognise their right to operate under licences granted by another reciprocating state. The statutes are reciprocal in the sense that they each make provision for the designation as a "reciprocating country" of any State having seabed mining legislation similar in its aims and effect to that of the designating State. The principal benefit of a licence issued by a reciprocating State is that it provides the licensee with an assurance that no conflicting licence to any part of the same area will be issued by any other reciprocating State.

Unlike the UN Convention regime, the legislation of the reciprocating States makes no provision for site-banking, nor do these States follow the UN pattern of placing limitations on the number of mining sites which may be held by a licensee, or of imposing production controls upon them. On the other hand, they do contain diligence provisions by, for example, requiring the licensee to make periodic and reasonable investments for exploration.

Five of the six statutes provide for the imposition of a levy on deep-sea mining operations and for payment of the proceeds into a fund intended to be the vehicle through which a share of the profits from the "common heritage" may be channelled to "mankind". The same rate of levy is adopted in the various national statutes - 3.75% of the value of the minerals recovered, a rate which is about half of that provided for under the UN regime.

It is of course beyond the scope of this Chapter to provide a detailed comparative analysis of the provision made for exploration of the Area and its mineral resources in these national statutes. It may, however, be useful for illustrative purposes to refer to some of the principal provisions of the British Act and Regulations under which an exploration licence has already been issued.

<u>Exploration licences under British Law</u>. The rules are laid down in the Deep Sea Mining (Temporary Provisions) Act 1981 (DSMA, 1981) and two statutory instruments issued under the Act, the Deep Sea Mining (Exploration Licences) (Applications) Regulations 1982 (S.I., 1982) and the Deep Sea Mining (Exploration Licences) Regulations 1984 (S.I., 1984).

Under the Act, the Secretary of State, in determining whether to grant an exploration licence, is empowered to have regard to any relevant factors and the licence may be granted for such period and contain such terms as he thinks fit.

Licences will not be granted in respect of any area already the subject of a licence or authorisation granted by another reciprocating State and it is made a criminal offence to interfere with operations carried on under an exploration licence or a reciprocal authorisation.

In determining whether or not to grant a licence and in determining the terms to which it is to be subject, the Secretary of State must have regard to the need to protect the marine environment.

When exercising his rights under an exploration licence, the licensee must act with reasonable regard for the interests of other persons in their exercise of the freedom of the high seas.

Provision is also made for the appointment of inspectors and for confidentiality of information received under the Act.

The 1982 Regulations specify the form in which applications must be made for exploration licences and fix the fee for grant of a licence at £ 10,000. Inter alia, a corporation applying for a licence would have to provide a description of the intended exploration area, the type of activity to be carried out, the proposed programme of activities and the environmental safeguards to be observed. Information would have to be provided too on the applicant's financial and technological capability and resources to undertake the proposed activities and justifying the size of the area applied for.

Further detail is provided in the 1984 Regulations and they also embody Model Clauses for incorporation in licences. Exploration licences are to be granted for an initial period of 10 years, with the possibility of extension for further periods of 5 years. The scale of fees payable, in addition to the initial £ 10,000 fee already mentioned, is laid down: £ 15,000 payable one year after the grant of the licence, a further £ 25,000 four years later and £ 25,000 for any extension of the licence.

The Model Clauses detail the obligations which may be imposed on licensees as terms of their licences. In exercising their rights to explore for the hard mineral resources of the 'exploration area' in accordance with an 'exploration plan', the licensee must conduct operations with due regard for safety, health and welfare and exercise all due skill, care and diligence. Elaborate regulations are laid down to protect the environment.

The Model Clauses also makes provision for inspection, confidentiality of data and the maintenance of detailed records of all operations.

Recent developments - conflict resolution

Whether a State with an interest in seabed mining is a party to the UN Convention or not, it clearly will be concerned to avoid or resolve conflicts concerning overlapping seabed claims. It is not surprising, therefore, that there have been several attempts since 1982 to design a mechanism for this purpose. The first successful attempt resulted in the adoption by four States - France, the Federal Republic of Germany, the United Kingdom and the United States - of an Agreement concerning Interim Arrangements relating to Polymetallic Nodules of the Deep Sea Bed (UK, 1982) which entered into force on 2 September 1982. Its object was to facilitate the identification and resolution of conflicts which might arise from applications for authorisations made by Pre-Enactment Explorers on or before 12 March 1983. International conflicts not otherwise resolved had to be settled, "if a party so elects", by an arbitration procedure specified in Appendix 1, on the basis of Principles for the Resolution of Conflicts laid down in Appendix 2. Although this Agreement embodied only a conflict-resolution procedure, further developments were foreshadowed in the agreement of the Parties to consult together to consider an arrangement to facilitate mutual recognition of national authorisations.

The 1982 Agreement was soon followed by the Provisional Understanding Regarding Deep Seabed Matters signed on 3 August 1984 (ILM, 1984) by 8 States, including - as at that date - 3 signatories of the UN Convention (France, Japan and the Netherlands) and 5 non-signatories (Belgium, Federal Republic of Germany, Italy, the United Kingdom and the United States). In essence, each of the 8 parties to the Provisional Understanding had encouraged the pioneer investors in which they or their companies participated to resolve possible overlapping conflicts with the other pioneer investors. The purpose of the Provisional Understanding was "to assure on governmental level the results" of the industry agreement reached by the 6 pioneer investors concerned. The parties to the Provisional Undertaking sought to refute allegations that, contrary to the UN Convention, the Understanding involved an obligation for the parties to it to recognise or support claims of pioneer investors made outside the framework of the UN regime. It was argued, somewhat disingenuously, that the obligation of the contracting governments was rather one of self-restraint, whereby they undertook to refrain from issuing an authorisation to a pioneer investor for an area which overlapped one of the mine sites allotted to the other pioneer investors in accordance with the above-mentioned agreement among the six pioneer investors.

The Provisional Understanding entered into force on 2 September 1984, though, in the case of Belgium, Italy and the Netherlands, only in respect of parts of the Understanding other than those relating to the issue of authorisations, pending the adoption by those States of seabed mining legislation. However, Italy adopted such legislation on 20 February 1985 and is therefore now in a position to give full effect to the Understanding.

As was seen above, the British Act forbids the issue of licences for areas for which an authorisation has already been granted by a reciprocating State. France, the Federal Republic of Germany and the United States were designated reciprocating countries by a series of orders made in 1982 (RO, 1982). The orders were to come into operation upon the entry into force of agreements between the United Kingdom and the three States concerned "with respect to the recognition of licences" issued by each country. However, presumably because such "recognition" would be incompatible with the self-restraint rationale of the Provisional Understanding, these orders were revoked and replaced by a new order in 1984 (RO, 1984), whereby these three countries and Japan were designated as reciprocating countries in a manner compatible with the Provisional Understanding.

A third instrument for conflict resolution deserves brief mention. The Understanding on Resolution of Conflicts among Applicants for Registration as Pioneer Investors was concluded among France, India, Japan and the Soviet Union on 31 August 1984 (UN, 1984). The purpose of this Understanding was to provide a timetable and procedure for the resolution of overlapping claims, this opening the way for the registration of a first group of pioneer investors by the Preparatory Commission. However, at the time of writing (September 1985), it remains uncertain whether this objective will be achieved in the face of opposition from other States in the Preparatory Commission.

CONCLUSION AND PROSPECT

Prediction in this field is always hazardous, especially in relation to the Area. However, given the fact that the boundaries of the territorial sea, the continental shelf and the EEZ can now be determined by reference to relatively more precise rules, and bearing in mind the greater accessibility of the seabed mineral resources of these more landward areas, it seems fair to predict that current exploration activity will continue to expand in the years ahead. So far as the Area is concerned, the outlook is less promising. As at September 1985, exploration licences have been granted by the United Kingdom and the United States, neither of which has signed the UN Convention. The sites covered by the licences are situated in the Clarion-Clipperton zone of the Pacific Ocean. On the other hand, pending further work by the Preparatory Commission, registration of pioneer investors may not proceed and, accordingly, no exploratory work may be conducted under the UN regime.

Given current prospects for the markets in the minerals concerned, it is hardly surprising that the sense of urgency which previously inspired the work of the Preparatory Commission is no longer evident. Given also the disincentive provided by the risks created by the existence of the complex legal situation reviewed above, the future of marine mineral exploration in the Area remains decidedly uncertain.

REFERENCES

BROWN, E. D., 1983. Deep Sea-Mining: The Consequences of Failure to Agree at UNCLOS III, Natural Resources Forum, 7: 55-70.

BROWN, E. D., 1984. Sea-Bed Energy and Mineral Resources and the Law of the Sea, Vol. 1: The Areas Within National Jurisdiction, Graham & Trotman, London.

BROWN, E. D., in press. Sea-Bed Energy and Mineral Resources and the Law of the Sea, Vol. 2: The Area Beyond the Limits of National Jurisdiction, Graham & Trotman, London.

DSMA, 1981. Deep Sea Mining Act 1981.

ILM, 1984. International Legal Materials, XXIII: 1354.

NOTE VERBALE, 1967. Maltese Note Verbale of 17 August 1967 to UN Secretary-General (UN Doc. A/6695, 18 August 1967) and Dr. Pardo's speech in the General Assembly's First Committee on 1 November 1967 (UN Doc. A/CI./PV.1515, 1 November 1967).

PROCLAMATION, 1945. US Presidential Proclamation No. 2667 of 28 September 1945.

RESOLUTIONS, 1983. Resolutions I and II. The text of the two Resolutions are in UNCLOS 1983, at pp. 175 and 177.

RO, 1982. S.I. 1982/176, S.I. 1982/177 and S.I. 1982/178.

RO, 1984. S.I. 1984/1170.

S.I., 1982. Deep Sea Mining (Exploration Licences) (Applications) Regulations 1982, S.I. 1982/58.

S.I., 1984. Deep Sea Mining (Exploration Licences) Regulations 1984, S.I. 1984/1230.

UK, 1982. UK Treaty Series No. 46 (1982), Cmnd. 8685; 1982, International Legal Materials, XXI: 950-962.

UN, 1984. UN Doc. LOS/PCN/L.8, 31 August 1984.

UNCLOS, 1982. UN Convention on the Law of the Sea, 1982.

UNCLOS, 1983. The Law of the Sea, United Nations, New York, 1983, Sales No. E. 83.V.5.

VIENNA, 1969. Vienna Convention on the Law of Treaties, 1969, Art. 18.

APPENDIX I
Selected additional references mainly from after 1970.

Adede, A.O., 1980. Developing countries expectations from and responses to the seabed mining regimes proposed by the Law of the Sea Conference. In: J.T. Kildow (Editor), Deep Sea Mining. Massachusetts Institute of Technology, Cambridge, MA, pp. 193-215.
Agiorgitis, G. and Gundlach, H., 1978. Platin-Gehalt in Tiefsee-Manganknollen. Naturwissenschaften, 65: 534.
Agterberg, P., 1974. Automatic contouring of geological maps to detect target areas for mineral exploration. Jour. Math. Geol., 6: 373-395.
Albers, J.P., 1973. Seabed mineral resources: a survey. Bull. Atomic Scient., 29: 34-38.
Aleva, G.J.J., 1973. Aspects of the historical importance on the physical geology of the Sunda shelf essential to the exploration of submarine tin placers. Geol. Mijnbouw, 52: 79-91.
Altschuler, Z.S., 1980. The geochemistry of trace elements in marine phosphorites - part I. In: Y. Bentor (Editor), Marine Phosphorites. SEPM Spec. Publ. 29: 19-30.
Amann, H., 1982. Technological trends in ocean mining. Phil. Trans. Royal Soc. London, A307: 377-403.
Amann, H., 1983. The Atlantis II Deep project in the Red Sea as a source of technology transfer for the development of marine polymetallic sulfides. Proceed. Oceans`83, Marine Technology Society, Vol. 2: 802-810.
Aplin, A.C. and Cronan, D.S., 1985. Ferromanganese oxide deposits from the Central Pacific Ocean, I. Encrustations from the Line Island Archipelago. Geochim. Cosmochim. Acta, 49: 427-436.
Archer, A.A., 1970. Sub-sea minerals and the environment. New Scientist, 48: 372-373.
Archer, A.A., 1973. Economics of off-shore exploration and production of solid minerals on the continental shelf. Ocean Management, 1: 5-40.
Archer, A.A., 1973. Sand and gravel demands on the North Sea- present and future. In: E.D. Goldberg (Editor), North Sea Science. Massachusetts Institute of Technology, pp. 437-449.
Archer, A.A., 1979. Resources and potential reserves of nickel and copper in manganese nodules. In: Manganese Nodules: Dimensions and Perspectives. Reidel, Dordrecht, pp. 71-87.
Ardus, D.A., Skinner, A., Owens, R. and Pheasant J., 1982. Improved coring techniques and offshore laboratory procedures in sampling and shallow drilling. Proceed. Oceanology International 1982.
Arthur, M.A. and Jenkyns, H.C., 1981. Phosphorites and paleoceanography. Proceed. 26th Intern. Geological Congress, Paris, pp. 83-96.
Arrhenius, G., 1963. Pelagic sediments. In: M.N. Hill (Editor), The Sea. Wiley, New York, pp. 655-727.
Bäcker, H. and Schoell, M., 1972. New deeps with brines and metalliferous sediments in the Red Sea. Nature, 240: 153-158.
Bäcker, H. and Müller, D., 1974. Suche nach Erzknollen im Indischen Ozean. Meerestechnik, 5: 50-51.
Bäcker H., Glasby, G.P. and Meylan, M.A., 1976. Manganese nodules from the Southwestern Pacific Basin. NZOI Oceanogr. Field Report 6, 88 pp.
Bäcker, H., 1976. Fazies und chemische Zusammensetzung rezenter Ausfällungen aus Mineralquellen im Roten Meer. Geol. Jb., D17: 151-172.
Bäcker, H., 1982. Lagerstättenbildung an divergenten Plattengrenzen. Erzmetall 35: 91-97.
Ballard, R.D., Holmcomb, R.T. and Van Andel, T.H., 1979. The Galapagos Rift at 86° W: 3. Sheet flows, collapse pits, and lava lakes of the rift valley. J. Geophys. Res. 84: 5407-5422.
Ballard, R.D. and Bischoff, J.L., 1984. Panel IB- Assessment and scientific understanding of hard mineral resources in the EEZ. Proceed. A national

program for the assessment and development of the mineral resources of the United States. U.S. Geological Survey Circular 929: 185-208.

Barnes, H.L., 1978. Geochemistry of hydrothermal deposits. Wiley, New York, 798 pp.

Bascom, W., 1964. Exploring the diamond coast. Geotimes, 9: 9-12.

Batchelor, B.C., 1979. Geological characteristics of certain coastal and offshore placers and essential guides for tin exploration in Sundaland, S.E. Asia. Bull. Geol. Soc. Malaysia II: 283-313.

Bateson, J.H. and Stephens, E.A., 1967. An appraisal of diamond finds in the peninsular Thailand. Transact. Instn. Min. Metall., B76: B125-B126.

Baturin, G.N., Merkulova, K.I. and Chalov, P.I., 1972. Radiometric evidence for recent formation of phosphatic nodules in marine shelf sediments. Marine Geol., 13: 37-47.

Baturin, G.N., 1982. Phosphorites on the sea floor. Elesevier, New York, 343 pp.

Beauchamp, G. and Cruickshank, M.J., 1983. Placer minerals on the U.S. continental shelves- opportunity for development. Proceed. Oceans`83, Marine Technology Society, Vol. II, pp. 698-702.

Beck, R.H. and Lehner, P., 1974. Oceans, new frontiers in exploration. Am. Ass. Petr. Geol. Bull., 58: 376-395.

Beiersdorf, H., 1972. Erkundung mariner Schwermineralvorkommen. Meerestechnik, MT, 3, 6: 217-223.

Beiersdorf, H., Kudrass, H.R. and Von Stackelberg, U., 1980. Placer deposits of ilmenite and zircon on the Zambesi shelf. Geol. Jb., D36: 5-85.

Belknap, D.F. and Kraft, J.C., 1981. Preservation potential of transgressive coastal lithostones on the U.S. Atlantic shelf. Marine Geol., 42: 429-442.

Bender, F., 1985. Angewandte Geowissenschaften. Enke, Stuttgart,Vol. 1+2, 1394 pp.

Berkovitch, I., 1975. How the difficulties of subsea treasure recovery are tackled. Engineer, 240: 43.

Bezrukov, P.L. and Andrushchenko, P.F., 1973. The geochemistry of manganese nodules from the Indian Ocean. Int. Geol. Rev., 16: 1044-1061.

Birch, G.F., 1979. Phosphorite pellets and rock from the western continental margin and adjacent coastal terrace of South Africa. Marine Geol., 33: 91-116.

Bischoff, J.L. and Dickson, F.W., 1975. Seawater-basalt interaction at 200° C and 500 bars: implications for origin of sea-floor heavy-metal deposits and regulation of seawater chemistry. Earth Planet. Sci. Lett., 25: 385-397.

Bischoff, J.L., Heath, G.R. and Leinen, M., 1979. Geochemistry of deep-sea sediments from the Pacific manganese nodule province: DOMES sites A, B and C. In: J.L. Bischoff and D.Z. Piper (Editors), Marine Geology and Oceanography of the Pacific Manganese Nodule Province. Plenum Press, pp. 397-436.

Blissenbach, E. and Fellerer, R., 1973. Continental drift and the origin of certain mineral deposits. Geologische Rundschau, 62: 812-840.

Blissenbach, E. and Nawab, Z., 1982. Metalliferous sediments of the seabed: the Atlantis-II-Deep deposit of the Red Sea. In: E. Mann Borgese and N. Ginsburg (Editors), Ocean Yearbook 3. The University of Chicago Press, pp. 77-104.

Bon, E.H., 1979. Exploration techniques employed in the Pula Tuyuh tin discovery. Trans. Instn. Min. Metall., Sec. A, 88: A13-A22.

Bonatti, E. and Nayudu, Y.R., 1965. The origin of manganese nodules on the ocean floor. Am. J. Sci., 263: 17-39.

Bonatti, E., 1975. Metallogenesis at oceanic spreading centers. Annu. Rev. Earth Planet. Sci., 3: 401-431.

Bonatti, E., Guerstein-Nonnorez, B.M. and Honnorez, J., 1976. Copper-iron sulfide mineralizations from the equatorial Mid-Atlantic Ridge. Econ. Geol., 71: 1515-1525.

Bonatti, E., 1978. The origin of metal deposits in the oceanic lithosphere. Scient. Am., 2: 54-61.

Bonatti, E., 1980. Metal deposits in the oceanic lithosphere. In: C. Emiliani (Editor), The sea. Wiley, New York, Vol. 7, pp. 639-686.
Bonatti, E. and Crane, K., 1984. Oceanic fracture zones. Scientif. Am., 250: 36-47.
Bonatti, E., Colantoni, P., Talwani, M. and Rossi, P., 1985. Massive sulphides on the Red Sea floor. Episodes, in press.
Boström, K. and Peterson, M.N.A., 1966. Precipitates from hydrothermal exhalations on the East Pacific Rise. Econ. Geol., 61: 1258-1265.
Bowen, R.E. and Hennessy, T.M., 1984. Adjacent State issues for the United States in establishing an Exclusive Economic Zone: the cases of Canada and Mexico. Proceed. Oceans`83. Marine Technology Society, pp. 444-448.
Boxer, B., 1984. Marine pollution research in the Exclusive Economic Zone planning. Ibid., pp. 445-458.
Breaux, J., 1979. Technolgy transfer: a case study of the inequity of the new international economic order. MTS Journal, 13: 19-23.
Bremner, J.M., 1980. Concretionary phosphorite from SW Africa. J. geol. Soc. London, 137: 773-786.
Brown, G.A., 1971. Offshore mineral exploration in Australia. Underwater Journal, 3: 166-177.
Brown, G.M. and Crutchfield, J.A., (Editors), 1981. Economics of ocean resources- a research agenda. Univ. of Washington Press, 254 pp.
Budreau, B.P. and Scott, M.R., 1978. A model for the diffusion controlled growth of deep-sea manganese nodules. Am. J. Sci., 278: 903-929.
Bunich, P., 1975. The economics of developing the resources of the world's oceans. Problems of Econ., 18: 79-81.
Burger, H., Ehrismann, W. and Stala, W., 1980. Aspects of the statistical analysis of marine ore deposits. Mineral. Deposita, 15: 335-350.
Burnett, W.C., 1977. Geochemistry and origin of phosphorite deposits from off Peru and Chile. Geol. Soc. Am. Bull., 88: 813-823.
Burnett, W.C. and Lee, A.I.N., 1980. The phosphate supply system in the Pacific. Geo Journal, 4: 423-435.
Burnett, W.C., Roe, K.K. and Piper, D.Z., 1983. Upwelling and phosphorite formation in the ocean. In: E. Suess and J. Thiede (Editors), Coastal upwelling: its sediment record. Plenum, New York, pp. 377-397.
Burns, R.G., 1976. Uptake of cobalt into ferromanganese nodules, soils, and synthetic manganese (IV) oxides. Geochim. Cosmochim. Acta, 40: 95-102.
Burns, V.M., 1979. Marine placers. In: R.G. Burns (Editor), Marine minerals. Mineral. Soc. Am., Vol. 6, pp. 347-380.
Calvert, S.E. and Price, N.B., 1977. Geochemical variation in ferromanganese nodules and associated sediments from the Pacific Ocean. Marine Chemistry, 5: 43-74.
Calvert, S.E., 1978. Geochemistry of oceanic ferromanganese deposits. Phil. Trans. Royal Soc., 290A: 43-73.
Cann, J.R., Winter, D.K. and Pritchard, R.G., 1977. A hydrothermal deposit from the floor of the Gulf of Aden. Min. Mag., 41: 193-199.
Carter, L., 1980. Ironsand in continental shelf sediments off western New Zealand- a synopsis. NZ Journ. Geol. Geophys., 23: 455-468.
CASM, 1983. Hydrothermal vents and sulfide deposits, axial seamount, Juan de Fuca Ridge. Proceed. Oceans`83. Marine Technology Society, pp. 801.
Cathless, L.M., 1981. Fluid flow and genesis of hydrothermal ore deposits. Econ. Geol., 75th Anniversary Volume, pp. 424-457.
Charlier, R.H., 1978. Other ocean resources. In: E. Mann Borgese and N. Ginsburg (Editors), Ocean Yearbook 1. The University of Chicago Press, pp. 160-292.
Charlier, R.H., 1983. Water, energy, and non-living ocean resources. In: E. Mann Borgese and N. Ginsburg (Editors), Ocean Yearbook 4. The University of Chicago Press, pp. 75-120.
Chase, R.L., Delaney, J.R., Johnson, H.P., Juniper, S.K., Karsten, J.L., Lupton, J.E., Scott, S.D. and Tunicliffe, V., 1983. North caldera hydrothermal vent field, axial seamount, Juan de Fuca Ridge. EOS, 64: 723.

Clague, D.A., Frey, F.A., Thomson, G. and Rindge, S., 1981. Minor and trace element geochemistry of volcanic rocks dredged from the Galapagos spreading center: role of crystal fractionation and mantle heterogeneity. J. Geophys. Res., 86: 9469-9482.

Cobb, E.H., 1973. Placer deposits of Alaska. U.S. Geological Survey Bull. 1374, 200 pp.

Coleman, R.G., 1977. Ophiolites. Springer, Berlin, 227 pp.

CONMAR, 1979. Continental margins- geological and geophysical research needs and problems. National Academy of Sciences, Washington, D.C., 302 pp.

Cook, P.J. and Marshall, J.F., 1981. Geochemistry of iron and phosphorus-rich nodules from the east Australian continental shelf. Marine Geol., 41: 205-221.

Corell, R.W., Blidberg, D.R. and Westneat, A.S., 1983. An assessment of robotic system technology for research in the ocean sciences. Proceed. Oceans`83. Marine Technology Society, pp. 8-10.

Corliss, J.B., 1971. The origin of metal-bearing submarine hydrothermal solutions. J. Geophys. Res., 76: 8128-8138.

Corliss, J.B. and Ballard, R.D., 1977. Oasis of life in the cold abyss. National Geographic, 152: 441-453.

Corliss, J.B., Lyle, M., Dymond, J. and Crane, K., 1978. The chemistry of hydrothermal mounds near the Galapagos Rift. Earth Planet. Sci. Lett.:, 40: 12-24.

Corliss, J.B., Dymond, J., Gordon, L.I., Edmond, J.M., Von Herzen, R.P., Ballard R.D., Green, K., Williams, D., Bainbridge, A., Crane, K. and Van Andel, T.H., 1979. Submarine thermal springs on the Galapagos Rift. Science, 203: 1073-1083.

Cox, A., 1969. Geomagnetic reversals. Science, 163: 237-245.

Clark, A., Johnson, C.J. and Chin, P.J., 1984. Assessment of cobalt-rich manganese crusts in the Hawaiian, Johnston and Palmyra Isalnds Exclusive Economic Zones. National Resources Forum, 8, 2: 163-174.

Craig, J.D., 1979. The relationship between bathymetry and ferromanganese deposits in the north equatorial Pacific. Marine Geol., 29: 165-186.

Cronan, D.S. and Tooms, J.S., 1969. The geochemistry of manganese nodules and associated pelagic deposits from the Pacific and Indian Oceans. Deep Sea Res., 16: 335-359.

Cronan, D.S., 1972. Regional geochemistry of ferromanganese nodules in the world ocean. In: D.R. Horn (Editor), Ferromanganese deposits on the ocean floor. National Science Foundation, pp. 19-29.

Cronan, D.S., 1976. Implications of metal dispersion from submarine hydrothermal systems for mineral exploration on mid-ocean ridges and island arcs. Nature, 262: 567-569.

Cronan, D.S., 1978. Recent advances in geochemical exploration for undersea mineral deposits. Trans. Instn. Min. Metall., 87A: 143-146.

Cronan, D.S. and Thomson, B., 1978. Regional geochemical reconnaissance survey for submarine metalliferous sediments in the southwestern Pacific Ocean. Trans. Instn. Min. Metall., 87B:

Cronan, D.S., 1979. Geochemical methods of prospecting for manganese nodules and metalliferous sediments. GERMINAL, pp. 393-401.

Cronan, D.S., 1979. Metallogenesis at oceanic spreading centres. J. Geol. Soc. London, 136: 621-626.

Cronan, D.S. and Moorby, S.A., 1981. Manganese nodules and other ferromanganese oxide deposits from the Indian Ocean. J. geol. Soc. London, 138: 527-539.

Cronan, D.S. and Moorby, S.A., 1982. Ore grade manganese nodules from the Central Indian Ocean. Oceanological International, paper OI 82 1.5.

Cronan, D.S., Glasby, G.P., Moorby, S.A. Thomson, J., Knedler, K.E. and McDougall, J.C., 1982. A submarine hydrothermal manganese deposit from the south-west Pacific island arc. Nature, 298: 456-458.

Cronan, D.S., 1984. Criteria for the recognition of potentially economic manganese nodules and encrustations in the CCOP/SOPAC region of the Central and Southwestern tropical Pacific. South Pacific Marine Geol.

Notes, 3, pp. 16.
Cronan, D.S., 1985. Manganese on the ocean floor. Trans. Instn. Min. Metall., 94A: 166-167.
Cruickshank, M.J. and Marsden, R.W., 1973. Marine mining. In: A.B. Cummins and I.A. Given (Editors), SME Mining Engineering Handbook. Section 20-1 to 20-200.
Cruickshank, M.J., 1974. Mineral resources potential of continental margins. In: C.A. Burke and C.L. Drake (Editors), The geology of the continental margins. Springer, Heidelberg, pp. 965-1000.
Cruickshank, M.J. and Hess, H.D., 1975. Marine sand and gravel. Oceanus, 19: 34-35.
Crutchfield, J.A., 1979. Resources from the sea. In: T.S. English (Editor), Ocean Resources and Public Policy. Univ. of Washington Press, pp. 105-133.
Cullen, D.J. and Singleton, R.J., 1977. The distribution of submarine phosphorite deposits on central Chatham Rise, east of New Zealand. NZOI field report, 24 pp.
Cullen, D.J., 1978. The uranium content of submarine phosphorite and glauconite deposits on Chatham Rise, east of New Zealand. Marine Geol., 28: 67-76.
Cullen, D.J., 1979. Mining minerals from the sea floor: Chatham Rise phosphorite. NZ agricult. Sci., 13: 85-91.
Cullen, D.J., 1980. Distribution, composition and age of submarine phosphorites on Chatham Rise, east of New Zealand. Soc. Econ. Paleont. Mineral., Spec. Public., 29: 139-148.
Cullen, D.J. and Burnett, W.C., 1985. Phosphorite associations on seamounts in the tropical southwest Pacific Ocean. Marine Geol., in press.
Daily, A.F., 1973. Placer mining. In: A.B. Cummins and I.A. Given (Editors), SME Mining Engineering Handbook. New York, Vol. 2, Sect. 17.6.
D`Anglejan, B.E., 1967. Origin of marine phosphorites off Baja California, Mexico. Marine Geol., 5: 15-44.
Dawson, E.W., 1984. The benthic fauna of the Chatham Rise: an assessment relative to possible effects of phosphorite mining. Geol. Jb., D65: 209-231.
DeCarlo, E.H., McMurthy, G.M. and Morgan, C.L., 1985. Geochemistry of ferromanganese crusts from the Hawaiian archipelago Exclusive Economic Zone. In preparation.
Degens, E.T. and Ross, D.A., 1969. Hot brines and recent heavy metal deposits in the Red Sea. Springer, New York, 1009 pp.
De Groot, S.J., 1979. An assessment of the potential environment impact of large-scale sand-dredging from the building of artificial islands in the North Sea. Ocean Management, 5: 211-232.
De Groot, S.J., 1979. The potential environmental impact of marine gravel extraction in the North Sea. Ocean Management, 5: 233-249.
Delaney, J.R., Johnson, H.P. and Karsten, J.L., 1981. The Juan de Fuca Ridge hot spot propagating rifting system: new tectonic, geochemical and magnetic data. J. Geophys. Res., 86: 11747-11750.
Delaney, J.R. and Cosens, B.A., 1982. Boiling and metal deposition in submarine hydrothermal systems. MTS Journal, 16: 62-66.
Denisov, S.V., 1971. Perspectives in reconnaissance for coastal marine placers on southern and western continental shores of Okkhotsk sea. Internat. Geol. Rev., 13: 301-304.
Derkmann, K.J., Fellerer, R. and Richter, H., 1981. Ten years of German exploration activities in the field of marine raw materials. Ocean Management, 7: 1-8.
Dickson, R. and Lee, A., 1973. Gravel extraction: effects on seabed topography. Offshore Services, 6 (6): 32-39 and 6(7): 56-61.
Dietz,R.S., 1961. Continent and ocean basin evolution by spreading of the sea floor. Nature, 190: 854-857.
Dobinson, A., Roberts, P.R. and Williamson, I.R., 1982. The development of a control system for the simultaneous operation of seismic profiling equipment. Oceanology International 1982.

Duane, D.B., Field, M.E., Meisburger, E.P, Swift, D.J.P. and Williams, S.J., 1972. Linear shoals on the Atlantic inner continental shelf, Florida to Long Island. In: D.J.P. Swift, D.B. Duane and O.H. Pilkey (Editors), Shelf sediment transport- process and pattern. Dowden, Hutchinson and Ross, Stroudsburg, PA, pp. 499-575.

Duane, D.B., 1976. Sedimentation and ocean engineering: placer mineral resources. In: D.J. Stanley and D.J.P. Swift (Editors), Marine sediment transport and environmental management. Wiley, New York, pp. 535-556.

Duane, D.B., Henrichs, D.F. and Offield, T.W., 1983. Federal program in marine polymetallic sulfide research. Proceed. Oceans`83. Marine Technology Society, pp. 825-827.

Duane, D.B. and Padan, J.W., 1984. NOOA activities in marine mining. Ibid., pp. 482-487.

Dunham, K.C. and Sheppard, J.S., 1970. Superficial and solid mineral deposits of the continental shelf around Britain. In: M.J. Jones (Editor), Mining and Petroleum Geology. Instn. Min. Metall., London, pp. 3-25.

Dunlop, A.C. and Mayer, W.T., 1973. Influence of Late Miocene- Pliocene submergence on regional distribution of tin in stream sediments, South West England. Trans. Instn. Min. Metall., B82: 1362-1364.

Duyverman, H.J., 1981. The occurrence of heavy mineral sands along the Tanzanian coast. J. geol. Soc. India, 22: 78-84.

Dymond, J., Corliss, J.B., Heath, G.R., Dasch, C.W. and Veeh, H.H., 1973. Origin of metalliferous sediments from the Pacific Ocean. Geol. Soc. Am. Bull., 84: 3355-3372.

Earney, F.C.F., 1980, Petroleum and hard minerals from the sea. Winston & Sons, London, 281 pp.

Eden, R.A., 1970. Marine gravel prospects in Scottish waters. Cement Lime and Gravel, 45: 237-240.

Edmond, J.M., Von Damm, K.L., McDuff, R.E. and Measures, C.I., 1982. Chemistry of hot springs on the East Pacific Rise and their effluent dispersal. Nature, 297: 187-191.

Emery, K.O. and Noakes, L.C., 1968. Economic placer deposits of the continental shelf. ECAFE, 1: 95-111.

Emery, K.O. and Skinner, B.J., 1977. Mineral deposits of the deep ocean floor. Marine Mining, 1: 1-71.

Ellis, A.J. and Mahon, W.A.J., 1977. Chemistry and geothermal systems. Academic Press, New York, 392 pp.

EPRSG, 1981. Crustal processes of the mid-ocean ridge. Science, 213: 31-40.

Exon, N.F., 1982. Manganese nodules in the Kiribati region, equatorial western Pacific. South Pacif. mar. geol. Notes, 2: 77-102.

Exon, N.F., 1983. Manganese nodule deposits in the Central Pacific Ocean and their variation with latitude. Marine Mining, 4: 79-107.

Falconer, R.K.H., Von Rad, U. and Wood, R., 1984. Regional structure and high-resolution seismic stratigraphy of the central Chatham Rise (new Zealand). Geol. Jb., D65: 29-56.

Filipek, L.H. and Owen, R.M., 1978. Analysis of heavy metal distributions among different mineralogical states in sediments. Can. Jour. Spectrosc., 23: 31-34.

Filipek, L.H. and Owen, R.M., 1979. Geochemical associations and Grain- size partitioning of heavy metals in lacustrine sediments. Chem. Geol., 26: 105-117.

Flipse, J.E., 1974. Ocean mining - its promises and its problems. Ocean Ind., 10: 133-136.

Force, E.K., 1976. Metamorphic source rocks of titanium placer deposits- a geochemical cycle. U.S. Geological Survey Prof. Paper 959B, 13 pp.

Fornari, D.J., Ryan, W.B.F. and Fox, P.J., 1984. The evolution of craters and calderas on young seamounts: insights from Sea MARC I and Sea Beam sonar surveys of a small seamount group near the axis of the East Pacific Rise at 10° N. J Geophys. Res., 89: 11069-11083.

Frakes, L.A. and Bolton, B.R., 1984. Origin of manganese giants: sea level change and anoxic-oxic history. Geology, 12: 83-86.

Frazer, J.Z. and Wilson, L.L., 1980. Manganese nodule resources in the Indian Ocean. Marine Mining, 2: 257-292.

Frazer, J.Z. and Fisk, M.B., 1981. Geological factors related to characteristics of manganese nodule deposits. Deep-sea Res., 28A: 1533-1551.

Friedrich, G., Kunzendorf, H. and Plüger, W.L., 1973. Geochemical investigation of deep sea manganese nodules from the Pacific on board R/V Valdivia- an application of the EDX technique. In: M. Morgenstein (Editor), The origin and distribution of manganese nodules in the Pacific and prospects for exploration. Univ. of Hawaii, pp. 31-43.

Friedrich, G., 1974. Schwermineralsandvorkommen im Küstenbereich Australiens. Erzmetall, 27: 350-353.

Friedrich, G., Plüger, W.L. and Kunzendorf, H., 1976. Geochemisch-lagerstättenkundliche Untersuchungen von Manganknollen-Vorkommen in einem Gebiet mit stark unterschiedlicher submariner Topographie (Zentral-Pazifik). Erzmetall, 29: 462-468.

Friedrich, G., Glasby, G.P., Plüger, W.L. and Thijssen, T., 1981. Results of the recent exploration for manganese nodules in the South Pacific by R.V. Sonne. Inter Ocean'81: 72-81.

Friedrich, G., Glasby, G.P., Thijssen, T. and Plüger, W.L., 1983. Morphological and geochemical characteristics of manganese nodules collected from three areas on an equatorial Pacific transect by R.V. Sonne. Marine Mining, 4: 167-253.

Gauss, G.A., Eade, J. and Lewis, K., 1983. Geophysical and sea bed sampling surveys for constructional sand in Nuku'alofa Lagoon, Tongatapu, Kingdom of Tonga. South Pacific Mar. Geol. Notes, 2, 10: 155-184.

GERMINAL, 1979. Offshore mineral resources. Documents BRGM, No. 7, 584 pp.

Geyer, R.A. and Moore, J.R., 1983. CRC Handbook of geophysical exploration at sea. CRC Press, Boca Raton, FL, 445 pp.

Gibbs, R.J., 1973. Mechanisms of trace metal transport in rivers. Science, 180: 71-73.

Gilmore, G.A., 1971. Indonesian tin prospects. Min. Mag., 125: 331-343.

Glasby, G.P., 1975. Marine mining in New Zealand: prospects for development. N.Z. Sci. Rev., 32: 53-56.

Glasby, G.P., 1979. Minerals from the sea. Endeavour, 3: 82-85.

Glasby, G.P., 1981. Manganese nodule studies in the Southwest Pacific. South Pacific. Mar. Geol. Notes, 2, 3: 37-46.

Glasby, G.P., Friedrich, G., Thijssen, T., Plüger, W.L., Kunzendorf, H., Ghosh, A.K. and Roonwal, G., 1982. Distribution, morphology and geochemistry of manganese nodules from the Valdivia 13/2 area, equatorial North Pacific. Pacific Science, 36: 241-263.

Glasby, G.P., Friedrich, G., Plüger, W.L., Thijssen, T. and Kunzendorf, H., 1983. Manganese nodule studies in the equatorial and southwest Pacific, the ICIME project. NZOI Oceanographic Summary No. 22, 10 pp.

Gloria, 1985. Gloria surveys the U.S. sea floor. Min. Mag., 153: 83-84.

Goldfarb, M.S., Converse, D.R., Holland, H.D. and Edmond, J.M., 1983. The genesis of hot spring deposits on the East Pacific Rise, 21o N. Econ. Geol. Monogr. 5: 184-197.

Greenslate, J., 1974. Microorganisms in the construction of manganese nodules. Nature, 249: 181-183.

Gundlach, H. and Marchig, V., 1982. Ocean floor "metalliferous sediments" - two possibilities for genesis. In: G.C. Amstutz (Editor), Ore genesis- the state of the art. Springer, Berlin, pp. 200-210.

Haggerty, J.A., Schlanger, S.O. and Silva, I.P., 1982. Late Cretaceous and Eocene volcanism in the Southern Line Islands and implications for hotspot theory. Geology, 10: 433-437.

Hails, J.R., 1976. Placer deposits. In: K.H. Wolf (Editor), Handbook of stratabound and stratiform ore deposits. Elsevier, Amsterdam, Vol. 3: 213- 244.

Halbach, P. and Fellerer, R., 1980. The metallic minerals of the Pacific

seafloor. Geojournal, 4: 407-422.
Halbach, P., Marchig, V. and Scherhag, C., 1980. Regional variatio in Mn, Ni, Cu and Co of ferromanganese nodules from a basin in the southeast Pacific. Marine Geol., 38: M1-M9.
Halbach, P., Hebisch, U. and Scherhag, C., 1981. Geochemical variations of ferromanganese nodules and crusts from different provinces of the Pacific Ocean and their genetic control. Chem. Geol., 34: 3-17.
Halbach, P., Scherhag, C., Hebisch, U. and Marchig, V., 1981. Geochemical and mineralogical control of different genetic types of deep-sea nodules from the Pacific Ocean. Mineralium Deposita, 16: 59-84.
Halbach, P. and Winter, P. (Editors), 1982. Marine mineral deposits- new research results and economic prospects. Glückauf, Essen, 243 pp.
Halbach, P., Segl, M., Puteanus, D. and Mangini, A., 1983. Co-fluxes and growth rates in ferromanganese deposits from Central Pacific seamount areas. Nature, 304: 716-719.
Hammon, A.L., 1974. Manganese nodules II: prospects for deep sea mining. Science, 183: 644-646.
Harris, D.P., 1984. Mineral resources appraisal - mineral endowment, resources and potential supply - concepts, methods and cases. Oxford University Press, 445 pp.
Hartmann, M., 1985. Atlantis II Deep geothermal brine system. Chemical processes between hydrothermal brines and Red Sea deep water. Marine Geol., 64: 157-177.
Haymon, R.M. and Kastner, M., 1981. Hot spring deposits on the East Pacific Rise at 21° N: preliminary description of mineralogy and genesis. Earth Planet. Sci. Lett., 53: 363-381.
Haymon, R., 1983. Growth history of hydrothermal black smoker chimneys. Nature, 301: 695-698.
Haymon, R.M., Koski, R.A. and Sinclair, C., 1984. Fossils of hydrothermal vent worms from Crateceous sulfide ores of the Samail ophiolite, Oman. Science. 23: 1407-1409.
Haynes, B.W., Law, S.L. and Barron, D.C., 1983. Mineralogical and elemental description of Pacific manganese nodules. BuMines IC 8906, 60 pp.
Heath, G.R. and Dymond, J., 1977. Genesis and transformation of metalliferous sediments from the East Pacific Rise, Bauer Deep, and Central Basis, northwest Nazca plate. Geol. Soc. Am. Bull., 88: 723-733.
Heath, G.R., 1981. Ferromanganese nodules of the deep sea. Econ. Geol., 75: 736-765.
Heezen, B.C. and Hollister, B., 1971. The face of the deep. Oxford University Press, 631 pp.
Hein, J.R., Manheim, F.T., Schwab, A.S., Daniel, C.L., Bouse, R.M., Morgenson, L.A., Sliney, R.E., Clague, D., Tate, G.B. and Cacchione, D.A., 1985. Geological and geochemical data for seamounts and associated ferromanganese crusts in and near the Hawaiian, Johnston Island, and Palmyra Island Exclusive Economic Zones. U.S. Geological Survey Open File Report 85-292, 192 pp.
Hein, J.R., Manheim, F.T., Schwab, W.C. and Davis, A.S., 1985. Ferromanganese crusts from the Necker Ridge, Horizon Guyot, and S.P. Lee Guyot: geological considerations. Marine Geol., in press.
Hekinian, R., Fevrier, M., Bischoff, J.L., Picot, P. and Shanks, W.C., 1980. Sulfide deposits from the East Pacific Rise near 21° N. Science, 207: 1433-1444.
Hekinian, R. and Fevrier, M., et al., 1983. East Pacific Rise near 13° N: geology of new hydrothermal fields. Science, 219: 1321-1324.
Hekinian, R., Renard, V. and Cheminee, J.L., 1983. Hydrothermal deposits on the East Pacific Rise near 13° N: geological setting and distribution of active sulfide chimneys. In: P.A. Rona, K. Boström, L. Laubier and K.L. Smith (Editors), Hydrothermal processes at seafloor spreading centers. Plenum, New York, pp. 571-594.
Hekinian, R., 1984. Undersea volcanoes. Scientif. Am., 251: 34-43.
Hekinian, R. and Fouquet, Y., 1985. Vocanism and metallogenesis of axial and

off-axial structures on the East Pacific Rise near 13° N. Econ. Geol., 80: 221-249.

Hering, N., 1973. New knowledge on prospecting and exploration of ore nodules deposits. Meerestechnik MT, 4: 1-11.

Holcomb R.T. and Clague, D.A., 1983. Volcanic eruption patterns along submarine rift zones. Proceed. Oceans`83. Marine Technology Society, Vol. 2, pp. 787-790.

Holden, W.M., 1975. Miners under the sea - right now. Oceans, 8: 55-57.

Holdren, C., 1981. Getting serious about strategic minerals. Science, 212: 305-307.

Hosking, K.F.G., 1970. The primary tin deposits of southeast Asia. Min. Sci. Eng., 2: 24-50.

HOW, 1975. How barite is recovered from an offshore consolidated deposit. Min. Eng., 27: 46.

Howarth, P.F., 1972. Exploration for phosphorite in Australia. Econ. Geol., 67: 1181.

Hubred, G., 1975. Deep-sea manganese nodules: a review of the literature. Min. Sci. Eng., 7: 71-85.

Hudson, A.G., Klinkhammer, G.P. and Craig, H., 1981. The distribution of hydrothermal manganese over the East Pacific Rise near 20° S. EOS, 62: 912.

Hudson, T. and DeYoung, J.H., 1978. Maps and tables describing areas of mineral resource potential, Seward Peninsula, Alaska. U.S. Geological Survey Open File Report 78-1-C, 62 pp.

Inderbitzen, A.L., 1983. Scientific goals of the advanced ocean drilling program. Proceed. Oceans`83. Marine Technology Society, pp. 353-357.

Jaeger, J.E. and Wernli, R.L., 1983. A conference whose time had come: a report on ROV`83. MTS Journal, 17: 90-99.

Janecky, D.R. and Seyfried, W.E., 1984. Formation of massive sulfide deposits on oceanic ridge crests: incremental reaction models for mixing between hydrothermal solutions and seawater. Geochim. Cosmochim. Acta, 48: 2723-2738.

Jaritz, W., Ruder,J. and Schlenker, B., 1977. Das Quartär im Küstengebiet von Mocambique und seine Schwermineralführung. Geol. Jb., B26: 3-93.

Jenkins, W.J., Edmond, J.M. and Corliss, J.B., 1978. Excess ^3He and ^4He in Galapagos submarine hydrothermal vents. Nature, 272: 156-158.

Jones, E.J.W. and Goddard, D.A., 1979. Deep-sea phosphorite of Tertiary age from Annan Seamount, eastern equatorial Atlantic. Deep-Sea Res., 26A: 1363-1379.

Jones, H.A. and Davies, P.J., 1979. Preliminary studies of offshore placer deposits, eastern Australia. Marine Geol., 30: 243-268.

Karns, A.W., 1976. Submarine phosphorite deposits of the Chatham Rise near New Zealand. Am. Assoc. Petrol. Geol. Mem., 25: 395-398.

Kash, D.E., White, I.L., Bergey, K.H., Chartrock, M.A., Devine, M.D., Leonard, R.L., Salomon, S.N. and Young, H.W., 1974. Energy under the ocean. Bailey Brothers and Swinfen, Richmond, Surrey, 380 pp.

Kaufman, A., 1970. The economics of ocean mining. MTS Journal, 4: 58-64.

Katz, H.R. and Glasby, G.P., 1979. Mineral resources of the New Zealand offshore region. South Pacific. Mar. Geol. Notes, 1: 95-110.

Keen, M.J., 1968. An introduction to marine geology. Pergamon, Oxford, 218 pp.

Kennett, J.P. and Watkins, N.D., 1975. Deep sea erosion and manganese development in the southeast Indian Ocean. Science, 198: 1011-1033.

Kesterke, D.G., 1984. Ocean minerals: the Bureau of Mines role in the mining and resources recovery. Proceed. Oceans`84. Marine Technology Society, pp. 476-479.

Kildow, J.T., Bever, M.B., Dar, V.K. and Capstaff, A.E., 1976. Assessment of economic and regulatory conditions affecting ocean minerals resource development. Massachusetts Instutute of Technology, Cambridge, MA, 1146 pp.

Klinkhammer, G.P. and Bender, M.L., 1980. The distribution of manganese in the Pacific Ocean. Earth Planet. Sci. Lett., 46: 361-384.

Knauer, G.A., Martin, J.H. and Gordon, R.M., 1982. Cobalt in northeast Pacific waters. Nature, 297: 49-51.

Knecht, R.W. and Cicin-Sain, B., 1983. Ocean policy making in a changing world. Proceed. Oceans`83. Marine Technology Society, Vol. 2, pp. 645-648.

Kogan, B.S., 1978. Method of geochemical prospecting for coastal marine placers. Intern. Geol. Rev., 20: 1309-1318.

Kolodny, Y., 1981. Phosphorites. In: C. Emiliani (Editor), The sea. Wiley, New York, Vol. 7, pp. 981-1023.

Kosalos, J.G. and Chayes, D., 1983. A portable system for ocean bottom imaging and charting. Proceed. Oceans`83. Marine Technology Society, pp. 649-656.

Koski, R.A., Clague, D.A. and Oudin, E., 1984. Mineralogy and chemistry of massive sulfide deposits from the Juan de Fuca Ridge. Geol. Soc. Am. Bull., 95: 930-945.

Kudrass, H.R. and Cullen, D.J., 1982. Submarine phosphorite nodules from the central Chatham Rise off New Zealand. Geol. Jb., D51: 3-41.

Kudrass, H.R., 1984. The distribution and reserves of phosphorite on the central Chatham Rise (Sonne-17 cruise, 1981). Geol. Jb., D65: 179-194.

Kudrass, H.R. and Von Rad, U., 1984. Underwater television and photographic observations, side-scan sonar and acoustic reflectivity measurements of phosphorite-rich areas on the Chatham Rise (New Zealand). Geol. Jb., D65: 69-89.

Kunzendorf, H., Friedrich, G. and Plüger, W., 1974. Anwendung eines Radioisotop-EDX-Systemes bei der Analyse von Manganknollen. Erzmetall, 27: 85-89.

Kunzendorf, H. and Friedrich, G.H., 1976. The distribution of U and Th in growth zones of manganese nodules. Geochim. Cosmochim. Acta, 40: 849- 852.

Kunzendorf, H. and Friedrich, G.H., 1976. Uranium and thorium in deep-sea manganese nodules from the Central Pacific. Trans. Instn. Min. Metall., 85B: 284-288.

Kunzendorf, H. and Friedrich, G., 1977. Die Verteilung von Uran in Manganknollen in Abängigkeit von der Knollenfazies und der Morphologie des Meeresbodens. Erzmetall, 30: 590-592.

Kunzendorf, H., Glasby, G.P., Plüger, W.L. and Friedrich, G.H., 1982. The distribution of uranium in some Pacific manganese nodules and crusts. Uranium, 1: 19-36.

Kunzendorf, H., Plüger, W.L. and Friedrich, G.H., 1983. Uranium in Pacific deep-sea sediments and manganese nodules. J. Geochem. Explor., 19: 147-162.

La Prairie, Y., 1976. The French programme for ocean development. MTS Journal, 10: 15-17.

Langeraar, W., 1976. Applied marine science and technology programme of the Netherlands. MTS Journal, 10: 26-35.

Leclaire, L., Clocchiatti,M., Giannesini, P.J. and Caulet, J.P., 1977. Depots metalliferes dans l`ocean Indien austral. Bulletin BRGM, Sect. II, 1: 13- 42.

Leipziger, D.M. and Mudge, J.L., 1976. Seabed mineral resources and the economic interests of developing countries, Cambridge, MA. pp.

Lenoble, J.P., 1979. AFERNOD prospecting methods for polymetallic nodules survey. GERMINAL, pp. 403-425.

Lenoble, J.P., 1980. Polymetallic nodule resources and reserves in the North Pacific from the data collected by AFERNOD. Oceanology International, 80: 11-18.

Lemes, D., Gieskes, J.M., Campbell, A.C., Stout, P.M. and Brumsack, H.J., 1984. Geochemistry of hydrothermally altered surface sediments in the Guayamas Basin- Gulf of California. EOS, 65: 974.

Lewis, K.B., Utanga, A.T., Hill, P.J. and Kingan, S.G., 1980. The origin of channel-fill sands and gravels on an algal-dominated reef terrace, Rarotonga, Cook Islands. South Pacific Mar. Geol. Notes, 2: 1-23.

Loncarevic, B.D., 1981. Ocean bottom seismometry. In: R.A. Geyer and J.R. Moore (Editors), CRC handbook of geophysical exploration at sea. CRC Press, Boca Raton, FL, pp. 219-255.
Lonsdale, P., 1977. Deep-tow observations at the mounds abyssal hydrothermal field, Galapagos Rift. Earth Planet. Sci. Lett., 36: 92-110.
Lonsdale, P.F., Bischoff, J.L., Burns, V.M., Kastner, M. and Sweeney, R.E., 1980. A high-temperature hydrothermal deposit on the seabed at a Gulf of California spreading center. Earth Planet. Sci. Lett., 49: 8-20.
Ludwig, G. and Figge, K., 1979. Schwermineralvorkommen und Sandverteilung in der Deutschen Bucht. Geol. Jb., D32: 23-68.
Lyle, M., Dymond, J. and Heath, G.R., 1977. Copper-nickel enriched ferromanganese nodules and associated crusts from the Bauer Basin, northwest Nazca plate. Earth Planet. Sci. Lett., 35: 55-64.
Lyle,, M., Heath, G.R. and Robbins, J.M., 1984. Transport and release of transition elements during early diagenesis: sequential leaching of sediments from MANOP sites M and H. Part I. Geochim. Cosmochim Acta, 48: 1705-1715.
Mackay, A.D., Gregg, P.E.H. and Syers, J.K., 1980. A preliminary evaluation of Chatham Rise phosphorite as a driect-use phosphate fertlizer. N.Z.J. Agricult. Res., 23: 441-449.
Macdonald, E.H., 1973. Manual of beach mining practice. Austr. Governm. Publish. Serv., 120 pp.
Maddox, J., 1984. Antarctic mining regime at risk. Nature, 307: 105-106.
Malahoff, A., McMurtry, G.M., Wiltshire, J.C. and Yeh, H.W., 1982. Geology and geochemistry of hydrothermal deposits from active submarine volcano Loihi, Hawaii. Nature, 298: 234-239.
Mangini, A. and Kühnel, U., 1985. Depositional history in the Clarion-Clipperton Zone during the last 250 000 y, in reltionship of Mn nodules and sediments in the equatorial North Pacific. In press.
Manheim, F.T., 1972. Mineral resources off the northeastern coast of the United States. U.S. Geological Survey Circular 669, 28 pp.
Manheim, F.T., Pratt, R.M. and McFarlin, P.F., 1980. Composition and origin of phosphorite deposits of the Blake Plateau. In: Y.K. Bentor (Editor), Marine phosphorites. SEPM Spec. Publ. 29: 117-138.
Manheim, F.T., Popenoe, P., Siapno, W. and Lane, C., 1982. Manganese-phosphorite deposits of the Blake Plateau. In: P. Halbach and P.Winter (Editors), Marine mineral deposits- new research results and economic prospects. Glückauf, Essen, pp. 9-44.
Manheim, F.T., Ling, T.H. and Lane, C.M., 1983. An extensive data base for cobalt-rich ferromanganese crusts from the world oceans. Proceed. Oceans`83. Marine Technology Society, pp. 828-831.
Mann Borgese, E., 1983. The law of the sea. Scientif. Am., 248: 28-35.
Marchig, V., 1981. Marine manganese nodules. In: Topics in current chemistry, 99: 99-126.
Martin, J. and Blissenbach, E., 1982. Law of the Sea development and ocean mining. Meerestechnik MT, 13:
McArthur, J.M. and Walsh, J.N., 1985. Rare-earth geochemistry of phosphorites. Chem. Geol., 47: 191-220.
McDonald, G.C.R. and W.K. Tong, 1978. Exploration and development of a shallow coastal tin deposit by suction dredging at Takua Pa, West Thailand. Trans. Instn. Min. Metall., A87: 29-38.
McDuff, R.E. and J.M. Edmond, 1982. On the fate of sulfate during hydrothermal circulation at mod-ocean ridges. Earth Planet. Sci. Lett., 57: 117-132.
McGuiness, W.T. and Hamilton, J.R., 1972. Recovery of offshore Cornish tin sands. Oceanology International 72: 417-419.
McKellar, J.B., 1975. The eastern Australian rutile province. Austr. Inst. Min. Metall., Monograph Series 5, 1, pp. 1055-1061.
McKelvey, V.E., 1969. Progress in the exploration and exploitation of subsea petroleum resources and its implication for the development beyond the limits of national jurisdiction. U.S. Geological Survey Circular 619, pp.

10-13.
McKelvey, V.E., 1969. Progress in the exploration and exploitation of hard minerals from the seabed. Ibid., pp. 13.
McKelvey, V.E., 1969. Potential ill effects of subsea mineral exploitation and measures to prevent them. Ibid., pp. 14-16.
McKelvey, V.E., 1969. Implications of geologic and economic factors to seabed resource allocation, development, and management. Ibid., pp. 22-26.
McKelvey, V.E., Stoertz, G.E. and Vedder, J.G., 1969. Subsea physiographic provinces and their mineral potential. Ibid. pp. 1-10.
McKelvey, V.E., Wright, N.A. and Bowen, R.W., 1983. Analysis of the world distribution of metal-rich subsea manganese nodules. U.S. Geological Survey Circular 886, 55 pp.
McRae, S.G., 1972. Glauconite. Earth-Sci. Reviews, 8: 397-440.
Meiser, H.J. and Muller, E., 1973. Manganese nodules- a further resource to cover the mineral requirements. Meerestechnik MT, 4: 145-150.
Menard, H.W., 1976. Time, chance, and the origin of manganese nodules. Amer. Scientist, 64: 529.
Menard, H.W. and Frazer, J.Z., 1978. Manganese nodules on the sea floor: inverse correlation between grade and abundance. Science, 199: 969-971.
Mero, J.L., 1978. Ocean mining: an hsitorical perspective. Marine Mining, 1: 244-254.
Meyer, K., 1973. Uran-Prospektion vor Südwestafrika. Erzmetall, 26: 313- 317.
Meyer, K., 1977. Wirtschaftlich interessante Schwermineral-Anreicherungen vor Mocambique. Erzmetall, 30: 453-456.
Meyer, K., 1983. Titanium and zirconium placer prospection off Pulmoddai, Sri Lanka. Marine Mining, 4: 139-166.
MIN, 1983. Radiometric techniques for seabed mineral surveys. Mining Journal, 300: pp. 208.
Mitchell, A.H.G. and Garson, M.S., 1976. Mineralizations at plate boundaries. Miner. Sci. Engng., 8: 126-169.
Morgenstein, M., 1973. The origin and distribution of manganese nodules in the Pacific and prospects for exploitation. The University of Hawaii, 175 pp.
Mooers, C.N.K., 1983. Satellite remote sensing and ocean physics. Proceed. Oceans`83. Marine Technology Society, pp. 4-5.
Moorby, S.A. and Cronan, D.S., 1983. The geochemistry of hydrothermal and pelagic sediments from the Galapagos hydrothermal mounds field, DSDP Leg 70. Mineral. Mag., 47: 291-300.
Moore, J.R., 1972. Exploration of ocean minerals resources- perspectives and predictions. Proc. Royal Soc. Edinburgh, 72: 193-206.
Moore, J.R., 1976. Metal bearing sediments of economic interest, coastal Bering Sea. Alaska geol. Soc., pp. K1-K17.
Moore, J.R., 1979. Marine placers: exploration problems and sites for new discoveries. GERMINAL, pp. 131-163.
Moore, J.R., 1982. Placer mineral exploration in high lattitudes. Oceanology International 82: 1.7.
Moore, W.S. and Vogt, P.R., 1976. Hydrothermal manganese crusts from two sites near Galapagos spreading axis. Earth Planet. Sci. Lett., 29: 349-356.
Moores, E.M., 1982. Origin and emplacement of ophiolites. Reviews of Geophysics and Space Physics, 20: 735-760.
Morris, M.A., 1982. Military aspects of the Exclusive Economic Zone. In: E. Mann Borgese and N. Ginsburg (Editors), Ocean Yearbook 3. The University of Chicago, pp.
Mottl, M.J. and Holland, H.D., 1978. Chemical exchange during hydrothermal alteration of basalt by seawater. Geochim. Cosmochim. Acta, 42: 1103-1115.
Müller, D., 1979. Sulphide inclusions in manganese nodules of the northern Pacific. Mineralium Deposita, 14: 375-380.
Murray, J.W. and Dillard, J.G., 1979. The oxidation of cobalt (II) adsorbed on manganese dioxide. Geochim. Cosmochim. Acta, 43: 781-787.

Mustafa, Z. Amann, H., 1980. the Red Sea pre-pilot mining test 1979. OTC 3874, 13 pp.

NAS, 1975. Mining in the outer continental shelf and in the deep ocean. National Academy of Sciences, Wahington, DC, 119 pp.

Nelson, C.H. and Hopkins, D.M., 1972. Sedimentary processes and distribution of particulate gold in the Northern Bering Sea. U.S. Geological Survey Prof. Paper 689, 27 pp.

Nesteroff, W.D., 1982. The origin of the ferromanganese coatings of deep- sea rocks in the Atlantic Ocean. In: R.A. Scrutton and M. Talwani (Editors), The ocean floor- Bruce Heezen commemorative volume. Wiley, New York, pp. 129-146.

Noakes, L.C. and Jones, H.A., 1975. Mineral resources offshore. Austral. Inst. Min. Metall., 11: 1093-1104.

Norris, R.M., 1964. Sediments of the Chatham Rise. Bull. New Zealand Dep. Scient. Industr. Res., 159: 1-39.

Oele, E., 1978. Sand and gravel from shallow seas. Geologie en Mijnbouw, 57: 45-54.

Ohmoto, H., 1978. Submarine calderas: a key to formation of volcanogenic massive sulfide deposits? Mining Geology, 28: 219-231.

Ong, P.M., 1966. Geochemical investigations in Mount's Bay, Cornwall. Ph.D. thesis, Univ. of London.

Owen, R.M., 1978. Geochemistry of platinum-enriched sediments: applications to mineral exploration. Marine Mining, 1: 259-282.

Owen, R.M., 1979. Geochemistry of platinum-enriched sediments of the coastal Bering Sea. In: J.R. Watterson and P.K. Theobald (editors), Proceed. 7th Internat. Geochemical Exploration Symposium. The Association of Exploration Geochemists, pp. 347-356.

Owen, R.M., Meyers, P.A. and Mackin, J.E., 1979. Influence of physical processes on the concentration of heavy metals and organic carbon in the surface microlayer. Geophys. Res. Lett., 6: 147-150.

Owen, R.M., 1980. Quantitative geochemical models of sediment dispersal patterns in mineralised nearshore areas. Marine Mining, 2: 231-249.

Owen, R.M. and Mackin, J.E., 1980. Authigenic associations between selected rare earth elements and trace metals in lacustrine sediments. Environ. Geol., 3: 131-137.

Owen, R,M., 1982. Geochemical exploration for marine minerals- placer deposits. Oceanology International 82: 1.4.

Pasho, D.W., 1976. Distribution and morphology of Chatham Rise phosphorites. Mem NZOI, 77: 1-27.

Pautot, G. and Melguen, M., 1979. Influence of deep water circulation and sea floor morphology on the abundance and grade of Central South Pacific manganese nodules. In: J.L. Bischoff and D.Z. Piper (Editors), Marine geology and oceanography of the Pacific manganese nodule province. Plenum, New York, pp. 621-649.

Peterson, G., 1979. Manganese nodules; future source of raw materials; supplies and profitability. In: F. Bender (Editor), The mineral resource potential of the earth. Schweizerbart'sche Verlagsbuchhandlung, Stuttgart, pp. 69-78.

Perseil, E.A. and Leclaire, L., 1979. Mineralogie et milieux de formation de depots polymetalliques dans l'Ocean Indien Occidental. In: Recherches oceanographiques dans l'Ocean Indien, pp. 151-168.

Pheasant, J., 1984. A microprocessor controlled seabed rockdrill/vibrocorer. Q.J. Undewater Technology, 10.1.

Piper, D.Z., Veeh, H.H., Bertrand, W.G. and Chase, R.L., 1975. An iron- rich deposit from the Northeast Pacific. Earth Planet. Sci. Lett., 26: 114-120.

Piper, D.Z., Burnett, W.C. and Basler, J.R., 1985. Elemental composition of seamount and shelf phosphorites from the Pacific Ocean. Marine Geol., in press.

Plüger, W.L. (Sc. cruise leader) and shipboard sc. party, 1981. Exploration von Manganknollen im Südwestpazifischen Becken. Fahrtbericht SO-14. Unpubl.

Rept., RWTH Aachen, 267 pp.
Plüger, W.L., Friedrich, G. and Stoffers, P., 1985. Environmental controls on the formation. Monograph Series Mineral Deposits, Gebrder Borntraeger, Berlin, 25: 31-52.
Price, N.B. and Calvert, S.E., 1978. The geochemistry of phosphorites from the Namibian shelf. Chem. Geol., 23: 151-170.
Puchelt, J. and Laschek, D. (1984). Marine Erzvorkommen im Roten Meer. Fridericiana, Univ. Karlsruhe.
Putzer, H. and Von Stackelberg, U., 1973. Exploration von Titanseifen vor Mocambique. Interocean 73, 1: 168-174.
Quasim, S.Z., 1983. Mining of phosphorite resources from the Indian continental shelf will help food production. J. Mines, Metals & Fuels, 5: 236-239.
Raab, W., 1972. Physical and chemical features of Pacific deep sea manganese nodules and their implications to the genesis of nodules. In: D.R. Horn (Editor), Ferromanganese deposits on the ocean floor. NSF, Washington, DC, pp. 31-49.
Reddy, B.J. and Clark, J.P., 1980. Effects of deep sea mining on international markets for copper, nickel, cobalt and manganese. In: J.T. Kildow (Editor), Deep sea mining. Massachusetts Institute of Technology, pp. 107-123.
Reimnitz, E. and Plafker, G., 1976. Marine gold placers along the Gulf of Alaska margin. U.S. Geological Survey Bull., 1415, 16 pp.
Renard, V., Hekinian, R., Francheteau, J. Ballard, R.D. and Bäcker, H., 1985. Submersible observations at the axis of the ultra-fast spreading East Pacific Rise (17°30' to 21°30'). Earth Planet. Sci. Lett., in press.
Riggs, S.R., 1979. Phosphorite sedimentation in Florida- a model phosphogenic system. Econ. Geol., 74: 285-314.
Riggs, S.R., Hine, A.C., Snyder, S.W., Lewis, D.W., Ellington, M.D. and Stewart, T.L., 1982. Phosphate exploration and resource potential on the North Carolina continental shelf. OTC 4395: 737-742.
RISE, 1980. East Pacific Rise: hot springs and geophysical experiments. Science, 207: 1421-1433.
Roe, K.K. and Burnett, W.C., 1985. Uranium geochemistry and dating of Pacific island apatite. Geochim. Cosmochim. Acta, 49: in press.
Rona, P., 1983. Hydrothermal mineralization at seafloor spreading centers: Atlantic Ocean and Indian Ocean. Proceed. Oceans`83. Marine Technology Society, pp. 817.
Ross, D.A., 1983. Effective use of the sea overcoming the Law of the Sea problems. Proceed. Oceans`83. Marine Technology Society, Vol. 1, pp. 1- 3.
Rowland, T.J. and Cruickshank, M.J., 1983. Mining for phosphorites on the United States outer continental shelf- opportunities for development. Ibid., Vol. 2, pp. 703-707.
Schanz, J.J., 1980. The United Nations' endeavour to standardize mineral resource classification. Natural Resources Forum, 4: 307-313.
Savit, C.H., 1981. Acquisition of seismic data at sea. In: R.A. Geyer and J.R. Moore (Editors), CRC handbook of geophysical exploration at sea. CRC Press, Boca Raton, FL, pp. 69-76.
Schneider, J., 1975. Manganknollen-Rohstoffquelle und Umweltproblem für die Zukunft. Umschau, 75: 724-726.
Scott, S.D., Lonsdale, P.F., Edmond, J.M. and Simoneit, B.R.T., 1983. Guayamas Basin, Gulf of California: examples of a ridge crest hydrothermal system in a sedimentary environment. Abstract. Geol. Assoc. Canada Annual Meeting: A61.
Scott, S.D., 1983. Basalt and sedimentary hosted seafloor polymetallic sulfide deposits and their ancient analogues. Proceed. Oceans`83. Marine Technology Society, pp. 818-824.
Scott, S.D., Barrett, T.J., Hannington, M., Chase, R.L., Fouquet, Y. and Juniper, K., 1984. Tectonic framework and sulfide deposits of southern Explorer Ridge, northeastern Pacific Ocean. EOS, 65: pp. 1111.

Scrutton, R.A. and Talwani, M., 1982. The ocean floor. Wiley, Chichester, 315 pp.
Seibold, E., 1973. Rezente submarine metallogenese. Geologische Rundschau, 62: 641-684.
Seibold, E., 1978. Deep sea manganese nodules, the challenge since "Challenger". Episodes, 4: 3-8.
Segl, M., Mangini, A., Bonani, G., Hofmann, H.J., Morenzoni, E., Nessi, M., Suter, M. and Wölfli, W., 1984. ^{10}Be dating of the inner structure of Mn encrustations applying the Zürich tandem accelerator. Nucl. Instr. Meth., 85: 359-364.
Sharp, W.E. and Bølviken, B., 1979. Brown algae; a sampling medium for prospecting Fjords. In: J.R. Watterson and P.K. Theobald (Editors), Proceed. 7th intern. geochem. exploration symposium. The Association of Exploration Geochemists, pp. 347-356.
Siddique, H.N., Rajamanickam, G.V. and Almeida, F., 1979. Offshore ilmenite placers of Ratnagiri, Konkan Coast, Mahararashtra, India. Marine Mining, 2: 91-110.
Siddique, H.N., Gujar, A.R, Hashimi, N.H. and Valsangkar, A.B., 1984. Superficial mineral resources of the Indian Ocean. Deep-sea Res., 31: 763-808.
Siegel, F.R. and Pierce, J.W., 1978. Geochemical exploration using marine mineral suspensates. Modern Geology, 6: 221-227.
Sillitoe, R.H., 1972. Formation of certain massive sulphide deposits at sites of sea-floor spreading. Trans. Instn. Min. Metall., B81: 141-148.
Simoneit, B.R.T. and Lonsdale, P.F., 1982. Hydrothermal petroleum in mineralized mounds at the seabed of Guayamas Basin. Nature, 295: 198- 202.
Singer, D.A. and Mosier, D.L. (Editors). Mineral deposit grade-tonnage models. U.S. Geological Survey Open-File Report 83-623, 100 pp.
Slatt, R.M., 1975. Dispersal and geochemistry of surface sediments: application to mineral exploration. Can. Jour. Earth Sci., 12: 1346- 1361.
Sleep, N.H. and Morton, J.L., 1983. Hydrothermal resources at mid-oceanic ridge axes. Proceed. Oceans`83. Marine Technology Society, Vol. II, pp. 782-786.
Solomon, M., 1976. Volcanic massive sulphide deposits and their host rocks-a review and explanation. In: K.H. Wolf (Editor), Handbook of stratabound and stratiform ore deposits. Elsevier, Amsterdam, Vol. 6, pp. 20-54.
Spiess, F.N. and Lonsdale, P.F., 1982. Deep tow rise crest exploration techniques. MTS Journal, 16, 3: 67-75.
Stalp, H.G., 1982. Ein Regelwerk zur Abschreckung von Investitionen im Meeresbergbau. Handelblatt, 21.9.1982.
Standish-White, D.W., 1972. Diamonds in the surf. Compressed Air Mag., 77: 8-11.
Stoffers, P., Glasby, G,P., Thijssen, T., Shrivastava, P.C. and Melguen, M., 1981. The geochemistry of coexisting manganese nodules, micronodules, sediments and pore waters from five areas in the equatorial and S.W. Pacific. Chemie der Erde, 40: 273-297.
Strens, M.R. and Cann, J.R., 1982. A model of hydrothermal circulation in fault zones at mid-ocean ridge crests. Geophys. J. R. Austr. Soc., 71: 225-240.
Suess, E., 1973. Interaction of organic compounds with calcium carbonate. Geochim. Cosmochim. Acta, 37: 2435-2447.
Swift, D.J.P., 1968. Coastal erosion and transgressive stratigraphy. J. Geol., 76: 444-456.
Sylwester, R.E., 1981. Single-channel, high-resolution, seismic-reflection profiling: a review of the fundamentals and instrumentation. R.A. Geyer and J.R. Moore (Editors), CRC handbook of geophysical exploration at sea. CRC Press, Boca Raton, FL, pp. 77-121.
Tagg, A.R. and Greene H.G., 1973. High-reolution seismic survey of an offshore area near Nome, Alaska. U.S. Geological Survey Prof. Paper 759-A, 22 pp.

Tharp, M., 1982. Mapping the ocean floor. In: R.A. Scrutton and M. Talwani (Editors), The ocean floor. Wiley, Chichester, pp. 19-31.

Thijssen, T., Glasby, G.P., Schmitz-Wiechowski, A., Friedrich, G., Kunzendorf, H., Müller, D. and Richter, H., 1981. Reconnaissance survey of manganese nodules from the northern sector of the Peru Basin. Marine Mining, 2: 385-428.

Tixeront, M., LeLann, F., Horn, R. and Scolari, G., 1978. Ilmenite prospection on the continental shelf of Senegal: methods and results. Marine Mining, 1: 171-188.

Tooms, J.S., Smith, D.T., Nichol, I., Ong, P. and Wheildon, J., 1965. Geochemical and geophysical mineral exploration experiments in Mounts Bay, Cornwall. In: W.F. Whittard and R. Bradshaw (Ediroes), Submarine geology and geophysics. Butterworths, London, pp.363-391.

Touret, D.G., 1983. The French ceep-sea mining legislation. Ocean Development and International Law Journal, 13:115-120.

Tracey, J.I., 1969. Topographic and geologic mapping of the sea bottom. U.S. Geological Survey Circular 619, pp. 16-19.

Tunnicliffe, V., Johnson, H.P. and Botros, M., 1984. Along-strike variation in hydrothermal activity on Explorer Ridge, N.E. Pacific. EOS, 65: 1124-1125.

Turner, S., 1981. Todorokite: a new family of naturally occurring manganese oxides. Science, 212: 1024-1027.

Turner, S.M., Siegel, M. and Buseck, P.R., 1982. Structural features of todorokite intergrowth in manganese nodules. Nature, 296: 841-842.

UN, 1984. Analysis of exploration and mining technology for manganese nodules. Graham & Trotman, London, 140 pp.

UP, 75. Update on offshore mining: the unheralded mineral producer. Min. Eng., 27: 42-44.

Urabe, T. and Sato, T., 1978. Kuroko deposits of the Kosaka mine, northeast Honshu, Japan - products of submarine hot springs and Miocene sea floor. Econ. Geol., 73: 161-179.

Uyeda, S. and Kanamori, H., 1979. Back-arc opening and the mode of subduction. J. Geophys. Res., 84: 1049-1061.

Uyeda, S., 1982. Subduction zones: an introduction to comparative subductology. Tectonophysics, 81: 133-150.

Van Andel, T.H. and Ballard, R.D., 1979. The Galapagos Rift at 86° W: 2. Volcanism, structure and evolution of the rift valley. J. Geophys. Res., 84: 5390-5406.

Van Overeem, A.J.A., 1970. Offshore tin exploration in Indonesia. Trans. Instn. Min. Metall., A79: 81-85.

Veeh, H.H., Burmett, W.C. and Sontar, A., 1973. Contemporary phosphorites on the continental margin of Peru. Science, 181: 844-845.

Veenstra, H.J., 1969. Gravels of the southern North Sea. Marine Geol., 7: 449-469.

Von Damm, K.L., 1983. Chemistry of submarine hydrothermal solutions at 21° North, East Pacific Rise and Guayamas Basin, Gulf of California. Ph.D. diss., Woods Hole Ocenaographic Institution.

Von der Borch, C.C., 1970. Phosphatic concretions and nodules from the upper continental slope, northern New South Wales. J. geol. soc. Austr., 16: 755-759.

von Rad, U. and Kudrass, H.R. (Editors), 1984. Geology of the Chatham Rise phosphorite deposits east of New Zealand: results of a prospection cruise with R/V Sonne (1981). Geol. Jb., D65.

von Rad, U., 1984. Outline of Sonne cruise SO-17 on the Chatham Rise phosphorite deposits east of New Zealand. Ibid., pp. 5-23.

von Rad, U. and Rosch, H., 1984. Geochemistry, texture, and petrography of phosphorite deposits east of New Zealand. Ibid., pp. 129-178.

von Stackelberg, U., 1979. Sedimentation, hiatuses, and development of manganese nodules: Valdivia site VA 13/2, northern Central Pacific. In: J.L. Bischoff and D.Z. Piper (Editors), Marine geology and oceanography of the Pacific manganese nodules province. Plenum, New York, pp. 559-586.

von Stackelberg, U. and Riech, V., 1981. Seifen-Exploration vor Ost-Australien. Proceed. Interocean 81, pp. 41-53.
von Stackelberg, U., 1982. Influences of hiatuses and volcanic ash rains on the origin of manganese nodules of the equatorial North Pacific (Valdivia cruises VA 13/2 and VA 18). Marine Mining, 3: 297-314.
von Stackelberg, U. (Editor), 1982. Heavy mineral exploration of the East Australian shelf, Sonne cruise SO-15, 1980. Geol. Jb., D56, 215 pp.
von Stackelberg, U., Kunzendorf, H., Marchig, V. and Gwozdz, R., 1984. Growth history of a large ferromanganese crust from the equatorial North Pacific nodule belt. Geol. Jb., A75: 213-235.
von Stackelberg, U. (Editor), 1985. Hydrothermal sulfide deposits in back-arc spreading centers in the Southwest Pacific. BGR Circular 2: 3-14.
Webb, P., 1979. Legal and economic aspects of dredging marine aggregates in the U.K., GERMINAL, pp. 49-60.
Wedepohl, K.H., 1980. Geochemical behaviour of manganese. In: I.M. Varentsov and G. Grassely (Editors), Geology and geochemistry of manganese. Hungarian Academy of Science, Budapest, pp. 335-351.
Wedepohl, K.H., 1980. Potential sources for manganese oxide prcipitation in the oceans. Ibid., pp. 13-22.
White, D.E., 1982. Active geothermal systems and hydrothermal ore deposits. Econ. Geol., 75th Anniv. Vol., pp. 392-423.
Whitehead, J.A., Dick, H.J.B. and Schouten, H., 1984. A mechanism for magmatic accretion under spreading centers. Nature, 312: 146-148.
Woolery, T.J. and Sleep, N.H., 1976. Hydrothermal circulation and geochemical flux at mid ocean ridges. J. Geol., 84: 249-275.
Wynn, J.C., Grosz, A.E. and Foscz, V.M., 1985. Induced polarization and magnetic response of titanium-bearing placer depsoits in the southeastern United States. Geophysics, in press.
Yin, W.W.S., 1975. Geochemical determination of tin in sediments off north Cornwall. Trans. Instn. Min. Metall., B84: 64-65.
Yin, W.W.S., 1979. Geochemical exploration for tin placers in St. Ives Bay, Cornwall. Marine Mining, 2: 59-78.
Zumberge, J.H., 1979. Mineral resources and geopolitics in Antarctica. Amer. Scient., 67: 73-74.

APPENDIX II
Hypothetical marine jurisdictional claims (OY, 1983) and countries sovereignty gain according to UNCLOS III.

Country/territory	Area (km^2)	Coastline (km)	Hypothetical 200-nm (km^2)	Gain (%)
Albania	28749	418	3600	12.5
Algeria	2460500	1183		
Angola	1245790	1600	147000	11.8
Antigua	280	153		
Argentina	2771300	4989	339500	12.3
Australia	7692300	25760	1854000	24.1
Bahamas	11396	3542	221400	1942
Bahrain	596	161	1500	252
Bangladesh	142000	580	22400	15.8
Barbados	430	97	48800	1135
Belgium	30562	64	800	2.6
Belise	22973	386	9000	39.2
Benin	115773	121	7900	6.8
Bermuda	54	103		
Brazil	8521100	7491	924000	10.8
Brunei	5776	161	7100	122.9
Bulgaria	11852	354	9600	8.6
Burma	678600	3060	148600	21.9
Cameroon	47400	402	4500	9.5
Canada	9971500	90908	1370000	13.7
Cape Verde	4040	965		
Chile	740740	6425	667300	90.0
China	9600000	14500	28100	0.3
Columbia	1139600	2414	175000	15.4
Comoros	2170	340		
Congo	349650	169	7200	2.1
Cook Islands	240	120	556100	231708
Costa Rica	51000	1290	75500	148
Cuba	114478	3735	105800	92.4
Cyprus	9251	537	29000	313
Denmark	42994	3379	20000	46.5
Djibouti	23310	314		
Dominica	790	148		
Dominican Republic	48692	1288	78400	161
Ecuador	274540	2237	338000	123
Egypt	1000258	2450	5060	5.1
El Salvador	21400	307	26800	125
Equat. Guinea	27972	296	82600	295
Ethiopia	1178450	1994	21600	1.8
Falkland Islands	12168	1288		
Faroe Islands	1340	764		
Fiji	18272	1129	368900	2019
Finland	336700	1126	28600	8.5
French Guiana	90909	378		
French Polynesia	4000	2525		
France	551670	3427	99500	18.0
Gabon	264180	385	62300	23.6

Country/territory	Area (km^2)	Coastline (km)	Hypothetical 200-nm (km^2)	Gain (%)
Gambia	10360	80	5700	55.0
GDR	108262	901	2800	2.6
Germany	248640	1488	11900	4.8
Ghana	238280	539	63600	26.7
Gibraltar	6	12		
Greece	132608	13676	147300	111
Greenland	2175600	44087		
Grenada	344	122		
Guadeloupe	1779	306		
Guatemala	108880	400	28900	26.5
Guinea	246	346	2070	8.4
Guinea-Bissau	36260	274	43900	121
Guyana	214970	459	38000	17.7
Haiti	27713	1771	46800	168.9
Honduras	112150	820	58600	52.3
Hong Kong	1036	733		
Iceland	102952	4988	252800	245.6
India	3136500	7003	587600	18.7
Indonesia	1906240	54716	1577000	82.7
Iran	1647240	3180	45400	2.8
Iraq	445480	58	200	0.04
Ireland	68894	1448	110900	161
Israel	20720	273	6800	32.8
Italy	301217	4996	161000	53.4
Ivory Coast	323750	515	30500	9.4
Jamaica	11422	1022	86800	759.9
Japan	370370	12075	1126000	304
Jordan	96089	26		
Kampuchea	181300	443	16200	8.9
Kenya	582750	536	34400	5.9
Kiribati	684	1143		
Korea, D.P.R.	121730	2495	37800	31.1
Korea, Rep.	98400	2413	101600	103
Kuwait	16058	499	4100	25.5
Lebanon	10360	225	6600	63.7
Liberia	111370	579	67000	60.2
Libya	1758610	1770	98600	5.6
Macao	16	40		
Madagascar	595700	4818	376800	63.3
Malaysia	332556	4675	138700	41.7
Maldives	298	644	279700	93859
Malta	313	140	19300	6166
Martinique	1100	290		
Mauritania	1085210	754	45000	4.1
Mauritius	1856	177	345000	18588
Mexico	1978800	9330	831500	42.0
Monaco	2	4		
Mozambique	786762	2470	163900	20.8
Namibia	823620	1489	145900	17.7
Nauru	21	24	92500	440476
Netherlands	33929	451	24700	72.8
Neth. Antilles	1020	364		
New Caledonia	22015	2254	382400	1737
New Zealand	268276	15134	1058000	394.4
Nicaragua	147900	910	46600	31.5

Country/territory	Area (km^2)	Coastline (km)	Hypothetical 200-nm (km^2)	Gain (%)
Nigeria	924630	853	61500	6.7
Norway	323750	5832	590500	182.4
Oman	212380	2092	163800	77.1
Pakistan	803000	1046	92900	11.6
Panama	75650	2490	89400	118.2
Papua New Guinea	475369	5152	684200	143.9
Peru	1284640	2414	229400	17.9
Phillipines	300440	22540	551400	183.5
Poland	312354	491	8300	2.7
Portugal	94276	1793	517400	548.8
Qatar	10360	563	7000	67.6
Reunion	2512	201		
Romania	273503	225	9300	3.4
St. Lucia	616	158		
St. Vincent	389	81		
Sao Tome and Prin.	964	200		
Saudi Arabia	2331000	2510	51900	2.2
Senegal	196840	531	60000	30.5
Seychelles	404	491		
Sierra Leone	72261	402	45400	62.8
Singapore	583	193	100	17.2
Solomon Islands	29785	5313	458400	1539
Somalia	637140	3025	288300	45.2
South Africa	1222480	2881	296500	24.3
Spain	505050	4964	355600	70.4
Sri Lanka	65500	1340	150900	230.4
Sudan	2504530	853	26700	1.1
Suriname	142709	386		
Sweden	448070	3218	45300	10.1
Syria	186480	196		
Taiwan	32260	990	114400	354.6
Tanzania	939652	1424	65100	6.9
Thailand	512820	3219	94700	18.5
Togo	56980	56	300	0.5
Tonga	997	419	158400	15888
Trinidad/Tobago	5128	362	22400	436.8
Tunisia	164206	1408	25000	15.2
Turkey	766640	2574	69000	9.0
Tuvalu	26	24	211500	813462
USSR	22274000	46670	1309000	5.9
United Arab Emir.	83880	1448	17300	20.9
United Kingdom	243978	12429	274800	112.6
USA	9363396	19924	2220000	23.7
Uruguay	186998	660	34800	18.6
Vanatu	14763	2528	179900	1219
Venezuela	911680	2800	106600	11.7
Vietnam	329707	3444	210600	63.9
Wallis/Futuna	207	129	71900	34734
Western Sahara	266770	1110		
Western Samoa	2849	403	38100	1337
Yemen, Arab Rep.	194250	1528	9900	5.1
Yemen, Democ. Rep.	287490	1383	160500	55.8
Yugoslavia	255892	3935	15300	6.0
Zaire	2343950	37	300	0.01

Compilation of the 10 largest hypothetical 200-nautical-miles areas.

Country/territory	Hypothetical 200-nm (km^2)	Gain in area of sovereignty (%)
USA	2220000	23.7
Australia	1854000	24.1
Indonesia	1577000	82.7
Canada	1370000	13.7
USSR	1309000	5.9
Japan	1126000	304
New Zealand	1058000	394
Brazil	924000	10.8
Mexico	831500	42.0
Papua New Guinea	684200	144

Compilation of the 10 largest sovereignty-gain countries/territories.

Country/territory	Area (km^2)	200-nm sovereignty gain (%)
Tuvalu	26	813462
Nauru	21	440476
Cook Islands	240	231608
Maldives	298	93859
Wallis/Futuna	207	34734
Mauritius	1856	18588
Tonga	997	15888
Malta	313	6166
Fiji	18272	2019
New Caledonia	22015	1737

OY, 1983. Appendix G: Tables, general information. In: E. Mann Borgese and N. Ginsburg (Editors), Ocean Yearbook 3. The University of Chicago Press, pp. 563-568.

APPENDIX III

Proposed strategies for the exploration of metalliferous sediments.

Marine exploration methods and equipment have made significant progress in recent years, which also may influence the future search for metalliferous sediments. Some of these methods are also applicable for massive sulfides since both resources are linked to submarine hydrothermal activity.

Usually a regional reconnaissance survey is followed by site-specific operations which in turn may be accompanied by environmental surveys and a final evaluation of the deposit. The proposed steps for the exploration of metalliferous sediments are summarised in Table III.1.

TABLE III.1

Proposed exploration steps for metalliferous sediments.

Progressive exploration steps ------------			
I Reconnaisance	II Detailed expl.	III Environm. survey	IV Depos. eval.
Geotect. analys.		Survey	Process. tests
	Operational plann. & contract negotiations	Analysis & assessm.	Logist. f. mining steps
	Geophysical survey		Final evaluation
	Sampling and analysis		
	Evaluation		

Geotectonic analysis. The choice of the exploration is made on the basis of present geotectonic knowledge mainly: plate boundaries, fault systems, basin fillings, volcanic belts, etc. However, economic and political considerations are to be taken into account as well. Today, raw-material related research is not possible without first obtaining authorisations and licences, even when the necessary funding and technical means are available.

Geophysical survey. During a first geophysical survey all those methods should be employed that can be run at normal cruising speeds, i.e. at 5 to 12 knots.

The geophysical survey includes a regional bathymetric survey with both both narrow- and multiple-beam echosounders applied at exploration step I. Large-scale side-scanning is also proposed and a subbottom profiling should be conducted. Magnetic and gravimetric methods are applied in some selected areas. At the detailed exploration (step II), bathymetry is extended to cover the whole target area including additional deep-tow geophysical measurements. Although not all of these methods are necessary to provide a sufficient database for the exploration steps described below, the interpretation of bottom features becomes much easier if several independent methods are used.

The geophysical survey terminates in morphological and structural maps, free brine surfaces, sediment ponds and magnetic anomalies. At the detailed exploration stage, the shape and size of a deposit, sediment thickness, slope angles and obstacles are defined. The geophysical survey should provide an accurate, well-positioned bathymetric map at a scale 1:10000 with 5 to 10 m contour intervals. It is important to recognise areas which can be ruled out for further exploration (e.g. unfavorable morphology). A determination of sediment thickness and distribution of obstacles can best be performed by deep-towed instrument packages including subbottom profiler and side-scan sonar.

Sampling and analysis. Based upon maps produced in the course of the geophysical survey a first station work is planned.

The use of geochemical indicators in seawater to detect seafloor hydrothermalism has developed considerably during recent years. Mantle-derived gasses, such as ^{3}He, ^{220}Rn and methane can be traced over tens and hundreds of kilometers away from the source. Another indicator is dissolved and suspended Mn having a range of some tens of kilometers from the source. Both methane and Mn have the advantage of being analysable onboard ship.

Water sampling at present can best be facilitated by a multiple water sampling array including a CTD sonde. Temperature, salinity and light attenuation anomalies may register the proximity of a hydrothermal source.

Sediment sampling during the reconnaissance survey is only for establishing geochemical trends in surface and surface-near sediments by metals that usually derive from hydrothermal sources (e.g. Fe, Mn, Zn, Au, Ag, As, Ba). To compare time-corresponding sediments the preservation of the sediment surface and the knowledge of sedimentation rates is important. The most suitable sampling gear is the box corer (e.g. Fig. 7.8).

Heat flow values that are largely scattered indicate convective flow, and geothermal measurements are therefore useful to locate hydrothermal activity. Heat flow measurements may be carried out in combination with long box corers or piston coring operations, or they may be conducted separately along profiles.

Evaluation. Station work should lead to the elimination of areas with metalliferous sediments. The subsequent recovery of a number of long sediment cores should determine the approximate horizontal and vertical extension of a hydrothermal deposit. If the preliminary results indicate economically interesting quantities and qualities care should be taken to provide sufficient sediment material for complete analyses, which is carried out usually in the home laboratories. Some material is needed for bench-scale processing tests.

The coring program, which should provide long and undisturbed cores, is based on grid-sampling, the size of the grid being dependent on factors which influence the value of the deposit (e.g. metal grade, solid contents, thickness). It depends also on performance parameters for exploitation (e.g. slope angle, geotechnical properties). The coring program should provide sufficient representative material for processing tests and undisturbed samples for geotechnical measurements.

The following analysis and evaluation should lead to the establishment of the analytical value of the deposit and provide information for the design of processing and mining methods.

Steps III and IV. These steps clearly do not belong to marine mineral exploration but they are often performed during exploration cruises.

In general, exploitation of metalliferous sediments will be impossible without ore preconcentration steps at sea, to a degree that is acceptable for later metallurgical recovery. Consequently, the disposal of the resulting tailings from preconcentration operations may cause environmental problems. A thorough environmental survey establishing the pre-mining conditions is therefore needed in areas that are insufficiently known. Furthermore, tailing characteristics and their behaviour in seawater should be investigated. The results of these investigations should then lead to a preliminary environmental impact report which at a later time has to be supplemented by measurements carried out during pilot plant tests.

All data collected during the exploration phases, supplemented by further economic information such as metal prices, logistics and scales will lead to a deposit evaluation, feasibility calculations and eventually a decision on pilot mining and processing operations.

INDEX